ライフサイエンス英語

類語使い分け辞典

編集／河本　健
監修／ライフサイエンス辞書プロジェクト

羊土社

序

本書は，医学・生物学（生命科学）分野の英語論文あるいは学会抄録を書くときに参照する辞書／参考書として制作された．制作にあたっては，ライフサイエンス辞書プロジェクト（http://lsd-project.jp）によって収集・制作された英語論文抄録データベースを利用し，そこから英語論文で高頻度に使われ，論文執筆に必要な重要単語を抽出して品詞と意味による分類を行った．本書のもっとも大きな特徴は，論文抄録中での使用頻度に基づいて用語の収集を行い，収録したすべての単語・連語にその頻度を記したことである．これによって類似表現の中で，どれがもっともよく使われるかをひと目で知ることができる．

英語論文を作成するときのコツの1つは，類語や類似表現を上手に使い分けることである．科学論文は，1つのことをいろいろな角度から検討して結論を導く場合が多いので，同じようなことを何度も書かなければならないことが少なくない．例えば，ある処置を行ったときにある遺伝子の発現が上昇したとしよう．これを，“処置をこう変えても上昇した．さらにああしても上昇した．でも，こうしたら上昇しなかった．また，メッセンジャーRNAレベルだけでなく，タンパク質レベルも上昇した．プロモータ活性も上昇した．……”など．このような場合，「上昇した」を英語で表現するために，increaseという単語が使えるであろうが，“… increased …, … increased …, … increased …, … did not increase …, … increased …, … increased …”とすべてを1つの単語だけで表現しようとしたら，どうだろう？ 滑稽であることは，誰の目にも明らかだ．しかし，本書を調べれば，簡単に“… increased …, … elevated …, … enhanced …, … did not increase …, … up-regulated …, … induced …”などの類義語に置き換えられることがわかる．表現にバリエーションをつけるためには，このような類語の使い分けがもっとも効果的な方法であろう．しかし，このような類似表現を探す際に，一般の類語辞典や和英辞典を使ったのではなかなかうまくいかない．たくさんの単語が記載されている割には，使えるものが少ないからだ．それにだいたい，どの単語が論文にふさわしいかを判断することも難しい．しかし，本書を使えば論文に使える表現を簡単に見つけることができるだけでなく，頻度情報や類義語との比較によって，どの単語がより的確であるかを判断することもできる．

論文執筆の際に気を付けなければならないもう1つの点は，それぞれの単語の用法である．本書では，ある単語の前後にどのような単語が用いられるかという共起表現について代表的なものを示し，表にしてまとめてある．これによって，他の単語との相性やそれぞれの単語の用法を知ることができる．さらに，大切な共起表現すべてに例文をつけた．採用した例文は，すべてPubMed論文抄録から引用した典型的な生の英文なので，本書にはそのまま論文に使える英文が満載されている．論文執筆の第一歩は，過去の類似の論文に学ぶことである．例文をよく読んで，それをまねながら自分の論文に取り入れていけば，文法や用法を間違えることなく，よく通じる英文が書けるであろう．

このように本書は，論文抄録データベースをフルに活用して編集されたものであり，英語論文執筆時には手元に置いて，収録されているたくさんのデータを有効に活用していただければ幸いである．

2006年5月

編著者を代表して
河本　健

編集 / 河本　健

広島大学大学院医歯薬学総合研究科講師

監修 / ライフサイエンス辞書プロジェクト

金子周司

京都大学大学院薬学研究科教授

鵜川義弘

宮城教育大学環境教育実践研究センター教授

大武　博

京都府立医科大学第一外国語教室教授

河本　健

広島大学大学院医歯薬学総合研究科講師

竹内浩昭

静岡大学理学部助教授

竹腰正隆

東海大学医学部基礎医学系分子生命科学講師

藤田信之

製品評価技術基盤機構バイオテクノロジー本部
ゲノム解析部門部門長

本書の特徴および使い方

われわれは，生命科学分野の論文で使われる用語を調査する目的で，この分野の主要学術雑誌89誌に2000年～2004年までの間に掲載された英語論文のうち，アメリカとイギリスから発表された論文の抄録を収集してコーパス解析を行った．構築された抄録データベース（約15万抄録：約3,000万語）に含まれる単語の種類は総計20万語にもなったが，3,000万語のうちの70％（約2,100万語）は3,000回以上登場する高頻出単語（語形変化による重複を含む約1,300語）によって占められていた．本書では，このような論文で高頻度に使われる動詞・名詞・形容詞・副詞・接続詞など，論文を書くために重要な単語や使い分けに注意すべき単語を約1,000語選び，それらの意味による分類と用法の解説を行った．これら1,000語の基本単語に自分の専門分野の用語（主に名詞）を加えるだけで，論文のほとんどの部分を執筆できるであろう．

❖本書の特徴

本書には次のような特徴がある．

- ●英語論文抄録で高頻度に用いられる単語を，品詞と意味や使われる状況によって分類した
- ●収集したすべての単語にその用例数（3,000万語のデータベース中での出現回数）を示した
- ●ある単語の前後にどのような単語が高頻度で用いられるかという共起表現について，その用例数とともに表示した
- ●PubMed論文抄録から典型的な例文を引用し，それを日本語訳とともに示した

本書は，動詞編，名詞編，形容詞編，副詞編，接続詞・接続語編の5つのパートから構成されており，それぞれ論文を書くために重要な単語〔動詞405語，名詞272語，形容詞153語，副詞151語，接続語（熟語）79語〕が，使われる状況と意味によって分類されている．意味による分類は，動詞86分類，名詞66分類，形容詞40分類，副詞34分類，接続語14分類に分けられている．詳しくは，この後の単語分類リストを参照していただきたい．各分類項目は，**図**に示すような構成になっている．

❖本書の構成

1 分類項目

品詞や意味による分類．検索に便利なよう，【 】内に代表的な単

図 ● 本書の構成

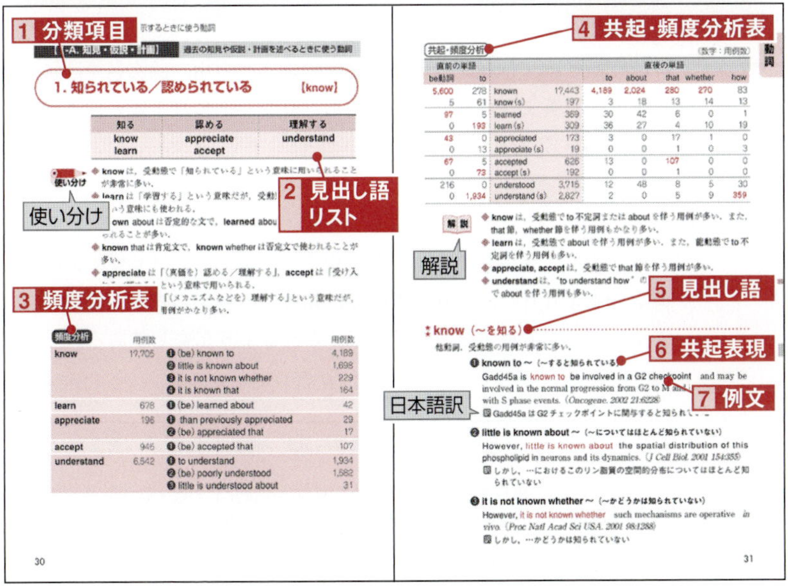

1 分類項目
2 見出し語リスト
3 頻度分析表
使い分け
4 共起・頻度分析表
解説
5 見出し語
6 共起表現
日本語訳
7 例文

語も掲載．

2 **見出し語リスト**

取り上げた類語とその日本語訳を一覧表にした．必要に応じて類語の**使い分け**の説明も加えてある．

3 **頻度分析表**

見出し語の用例数※1を表にして示した．また，どのような単語が使われるかという共起表現の用例数※2も合わせて明記した．

4 **共起・頻度分析表**

見出し語の直前／直後にくる主な単語と，その用例数をまとめた．必要に応じ，どのような単語とセットで使われやすいかという**解説**も加えた．

5 **見出し語**

重要度を★の数（0～2つ）で示した．

6 **共起表現**

代表的な共起表現をあげた．

7 **例文**

PubMed 論文抄録から典型的な生の論文を引用．論文執筆時に参考となる英文が満載である．なお，ゴシック体は下記の**日本語訳**（部分訳）に対応．

❖本書の使い方

本書では，見出し語の順番をアルファベット順にはせず，意味による分類を行った．これによって，類義語の中から書きたい内容によくマッチする単語を簡単に見つけられるよう工夫してある．

※1：3,000 万語のデータベース中での出現回数を示す．本書の大きな特徴の1つである．

※2：見出し語の出現頻度には，動詞や名詞の語形変化形の数も含まれる．一方，共起表現の出現頻度には，語形変化形の数は含まれない．つまり，know の数には known, know, knowing, knew, knows が含まれるが，known to の数には know to の数は含まれない．

① 収録したすべての見出し語の単語分類リストが13ページ以降にあるので，まずはここを探して使いたい単語のある項目を見つけよう．

② 次に，7ページの**図**に示すような該当する分類項目を調べて単語の使い方を理解し，適した単語を選択する．

③ 本文中のそれぞれの分類項目を参照する際には，その分類の前後にも近い意味の単語がある場合が多いので，近くの項目も合わせて調べてみるとよい．

④ 巻末にはアルファベット順および50音順の索引があるので，そこからも見出し語および類語を見つけることができる．

本書には，論文でよく使われる高頻出単語の多くを集めてあるが，それでも自分のイメージどおりの言葉が見つからないこともあるかもしれない．そのような場合であっても，本書に示した単語を使って何とか自分の意図を表現できないか工夫してみることが大切である．

本書へのご意見ご感想は，ruigo@lsd.pharm.kyoto-u.ac.jp までお寄せください．

（河本　健）

使用したデータベースとコーパス解析の方法について

❖コーパスとは

本書のもとになっているのは，われわれが独自に構築したライフサイエンス分野の専門英語のコーパスである．コーパスとは一定の基準に従って収集された言語データのことを言う．コンピュータによって大量のテキストが扱えるようになった今日では，コーパスは主に「コンピュータで扱えるように体系化された大量の言語データ」すなわちコンピュータ・コーパスの意味で使われている．規模の大きな汎用の英語コーパスとしては，British National Corpus や Bank of English（Cobuild corpus）があり，新聞，雑誌，書籍などの書き言葉から，テレビニュースや日常会話などの話し言葉まで，およそありとあらゆるタイプの英語の文章がテキストとして収録されている．これら大規模コーパスのコンピュータ分析をもとにした数量的な視点を取り入れることによって，辞書の編纂（へんさん）方法（見出し語の選択，意味の記載順，例文の選択など）が大きく様変わりしたと言われている．

❖ライフサイエンス英語コーパスについて

幸いライフサイエンス分野では PubMed と呼ばれる無料の文献データベースが利用できることから，われわれは PubMed に収録されている学術論文の抄録を主な言語資料とした．生化学，分子生物学などの基礎的な分野から臨床医学などの応用分野に至るまで，ライフサイエンスのさまざまな分野を網羅する主要な 89 の学術雑誌を選び，2000 年から 2004 年までの 5 年間にアメリカまたはイギリスの研究機関から出された論文約 14 万 6,000 報の抄録を収集した．そこから抽出した約 123 万件の文章を言語資料とした．専門英語に特化したコーパスとしては十分な情報量と考えられる．すべての文章には，付帯情報としてPubMedの登録番号をもたせており，分析結果から容易に元の論文抄録を参照できるように工夫されている．

❖本コーパスの有用性と活用法

このコーパスには総語数にして約3,000万語の情報が含まれている．名詞や動詞の語尾活用を考慮するとユニークな単語の数は約20万語と見積もられるが，そのうち出現頻度の高い上位5万語でコーパス全体の実に98.8％をカバーしている．5万語という数は，現在のライフサイエンス辞書のサイズ（英和見出し語として4万9,000語）にほぼ匹敵する．こうして構築されたライフサイエンス英語コーパスは，ライフサイエンス辞書プロジェクトのオンライン検索サービスWebLSDにおいて共起検索の形で実装されている（付録1参照）．任意の検索語に対してその場でライフサイエンスコーパスを検索し，語句の出現頻度を調べたり，前後の隣接語を数量的に捉えることができるようになっている．それだけにとどまらず，英和辞書における見出し語の選択や，複合語の抽出，例文の抽出などにもコーパスが活用されており，ライフサイエンス辞書のすべてがこのコーパスをベースにしていると言っても過言ではない．

❖コーパスをもとに活きた英語を提示

科学論文の執筆を目的とした例文集や活用辞典はこれまでにも多数出版されているが，本書がそれらと決定的に異なるのは，コーパスのコンピュータ分析によって得られた数量的なデータを基礎に置いている点である．これによって，実際の学術論文で好んで使用される「活きた英語」を提示できているものと思う．本書の編纂において使用したコーパス分析の方法は，基本的にはWebLSDで提供している共起検索と同じものである．書籍としてまとめる関係で割愛せざるを得なかった項目も多い．必要に応じてWebLSDも併用していただけると幸いである．

（藤田信之）

ライフサイエンス英語

類語使い分け辞典

目　次

単語分類リスト

動詞編

Ⅰ）研究内容・仮説・証明を提示するときに使う動詞

【Ⅰ-A. 知見・仮説・計画】 過去の知見や仮説・計画を述べるときに使う動詞

1. 知られている／認められている ……【know】 30
know, learn, appreciate, accept, understand

2. 考えられる／思われる ……【think】 33
consider, regard, believe, think, suspect, appear, seem, take into account

3. 予想する／推定する ……【predict】 36
predict, expect, assume, presume, estimate

4. 仮定する／提唱する ……【hypothesize】 39
hypothesize, postulate, propose

5. 計画する／予定する ……【design】 40
design, plan, schedule

6. 〜しようとする／試みる／取り組む ……【attempt】 41
attempt, try, seek, address, undertake, aim

【Ⅰ-B. 提示・報告・証明】 研究内容の提示や証明を述べるときに使う動詞

7. 示す／意味する ……【show】 44
show, indicate, exhibit, display, represent, suggest, imply, mean

8. 提示する／提供する ……【present】 47
present, provide, raise

9. 述べる／報告する／結論する ……【describe】 48
describe, state, mention, note, report, conclude

10. 説明する ……【explain】 50
explain, account for, illustrate

11. 証明する／確立する ……【demonstrate】 52
demonstrate, evidence, document, prove, establish

12. 明らかにする ……【reveal】 53
reveal, disclose, elucidate, clarify, uncover, manifest, solve

13. 確認する／検証する ……【confirm】 56
confirm, verify, validate, ascertain

14. 支持する／強化する ……【support】 58
support, favor, reinforce, strengthen

【Ⅰ-C. 研究・発見・評価・達成】 研究・発見・評価・達成に関する動詞

15. 研究する／調査する／調べる ……【study】 60
study, investigate, survey, explore, search, examine, test, analyze, dissect, look at, ask

16. 見つける／観察する ……【find】 65
find, identify, detect, discover, observe, see, note

17. 評価する／比較する ……【assess】 68
assess, evaluate, estimate, compare

18. 決定する／特徴づける ……【determine】 69
determine, define, characterize, mark, feature

19. 促す／可能にする／許す ……【allow】 71
lead, prompt, enable, allow, permit, make it possible, approve

20. 役立つ／助ける ……【help】 74
help, serve, assist, aid

21. 達成する／成す ……【achieve】 76
achieve, accomplish, make

Ⅱ）主に結果や現象を説明するために使う動詞

【Ⅱ-A. 発生・由来】 事象の発生・由来・獲得に関係する動詞

1. 現れる ……【emerge】 78
emerge, appear, express

2. 起こる／～が生じる／発生する ……【occur】 79
occur, take place, arise, develop

3. ～に由来する／～から生じる ……【derive】 81
derive, originate, stem, arise, come from, draw, emerge

4. ～に起因する／帰する ……【result from】 83
result from, attribute, ascribe

5. 得る／得られる ……【obtain】 85
obtain, acquire, gain, get

【Ⅱ-B. 誘発・産生】 事象の誘発・産生に関係する動詞

6. 引き起こす／～を生じる ……【cause】 87
cause, result in, lead to, give rise to, produce, yield, bring about, contribute

7. 誘導する／誘発する ……【induce】 90
induce, elicit, evoke, provoke

8. 産生する ……【produce】 91
produce, generate, raise

【Ⅱ-C. 増加・促進】増加・増強・促進を表す動詞

9. 上昇する／上昇させる／増加させる ……【increase】 93
increase, elevate, up-regulate, raise, induce

10. 増強する／強化する ……【enhance】 95
enhance, augment, potentiate, activate

11. 促進する ……【promote】 97
promote, accelerate, facilitate

12. 進行する／進行させる／進歩させる ……【proceed】 98
proceed, progress, advance

13. 増殖する／増幅する ……【grow】 99
grow, proliferate, replicate, propagate, amplify

14. 拡大する／延長させる ……【extend】 101
extend, expand, enlarge, spread, prolong

【Ⅱ-D. 低下・抑制・破壊】低下・抑制・抑止・破壊を表す動詞

15. 低下させる／低下する／減少させる ……【decrease】 104
decrease, reduce, diminish, down-regulate, lower, depress, decline, shorten, minimize

16. 抑制する／弱める／干渉する ……【inhibit】 107
inhibit, suppress, repress, attenuate, weaken, relieve, alleviate, mitigate, interfere

17. 抑止する／阻止する／妨げる ……【block】 111
block, abrogate, silence, prevent, hinder, occlude, hamper

18. 破壊する／損傷する ……【disrupt】 113
disrupt, destroy, lesion, impair, damage

19. 消滅させる／消失する／欠失している ……【abolish】 115
abolish, ablate, eliminate, remove, disappear, lose, delete

【Ⅱ-E. 変化・移動・影響】変化・移動・影響を表す動詞

20. 変化する／変化させる ……【change】 118
change, alter, shift, convert, transform, modify, vary

21. 移動する／移す ……【transfer】 121
transfer, shift, translocate, transport, migrate, move, recruit, mobilize, incorporate

22. 影響する ……【affect】 124
affect, influence, impact

【Ⅱ-F．疾患】病気や治療に関係する動詞

23. 病気にかかる ……………………………【develop】 126
develop, suffer, affect, predispose, infect

24. 治療する／治療を受ける ……………………………【treat】 128
treat, cure, administer, give, undergo, receive, take, vaccinate, admit, ameliorate

Ⅲ）主に研究対象の関係・性質・機能について述べるときに使う動詞

【Ⅲ-A．関連・異同・識別】関連・異同・識別に関する動詞

1. 関与する ……………………………【involve】 132
involve, implicate, engage, participate, take part in, play ～ role in, play ～ part in

2. 関連する／関連づける ……………………………【relate】 135
relate, associate, link, connect, couple, concern, correlate

3. 伴う／付随する ……………………………【accompany】 138
accompany, follow, associate

4. 一致する／似ている ……………………………【correspond】 140
correspond, coincide, agree, fit, match, resemble, represent

5. 異なる ……………………………【differ】 143
differ, vary

6. 区別する ……………………………【distinguish】 144
distinguish, differentiate, discriminate

【Ⅲ-B．性質】性質について述べるときに使う動詞

7. 持つ ……………………………【have】 146
have, possess

8. 含む ……………………………【contain】 147
contain, include, involve

9. 維持する／保持する ……………………………【maintain】 148
maintain, sustain, keep, hold, retain

10. 続ける／持続する ……………………………【continue】 150
continue, persist, last

11. 保存する／貯蔵する ……………………………【conserve】 151
conserve, preserve, store

12. 蓄積する／沈着する ……………………………【accumulate】 153
accumulate, deposit

13. 構成する／～から成る ……………………………【compose】 154
compose, comprise, constitute, consist

14. 必要とする ……【require】 155
require, need, necessitate
15. 分ける／隔てる／分離する ……【separate】 157
separate, divide, interrupt, segregate, dissociate
16. 制限する／限られる ……【limit】 160
limit, restrict, confine
17. 避ける ……【avoid】 161
avoid, circumvent, escape
18. ～のままである／存在する ……【remain】 162
remain, exist, survive
19. 局在する／位置する ……【localize】 163
localize, locate, position, map

【Ⅲ-C. 機能】機能に関係する動詞

20. 応答する／反応する ……【respond】 166
respond, react
21. 認識する／知覚する ……【recognize】 167
recognize, perceive, communicate
22. 結合する／接着する ……【bind】 168
bind, associate, engage, couple, bond, attach, adhere, connect, join, interact
23. 働く／機能する ……【function】 173
function, act, serve, behave, operate, work
24. 調節する ……【regulate】 175
regulate, control, modulate, drive, mediate, adjust
25. 形成する／集合する ……【form】 177
form, assemble, aggregate
26. 与える ……【confer】 179
confer, give, render, supply

Ⅳ）研究の方法や実施について述べるときに使う動詞

【Ⅳ-A. 方法・実施】研究の方法・実施に関係する動詞

1. 選択する ……【select】 181
select, choose
2. 使う／利用する ……【use】 182
use, employ, utilize, take advantage of, exploit, apply
3. 行う ……【perform】 184
perform, carry out, conduct, do, complete

4. 切断する ……【cleave】 186
truncate, cleave, break, digest
5. 置換する ……【substitute】 188
substitute, replace
6. 作製する／構築する ……【create】 189
construct, create, generate, synthesize, assemble, build, organize, develop
7. 導入する ……【transfect】 192
transfect, introduce, transform
8. 測定する／定量する ……【measure】 194
measure, assay, quantify, quantitate
9. 集める／回収する ……【collect】 196
collect, gather, harvest, recover, retrieve
10. 濃縮する ……【enrich】 198
enrich, concentrate
11. 単離する／精製する ……【isolate】 199
isolate, separate, dissociate, purify
12. 調製する ……【prepare】 200
prepare, process
13. 刺激する／処理する ……【stimulate】 201
stimulate, prime, treat, add
14. 暴露する ……【expose】 203
expose, challenge
15. 培養する／インキュベートする ……【culture】 204
culture, cultivate, incubate

名詞編

I）研究内容・仮説・証明を提示するときに使う名詞

【I-A. 知識・仮説・目的】 知識・仮説・目的などを表す名詞

1. 知識／理解 ……【knowledge】 208
knowledge, understanding, awareness
2. 予想 ……【prediction】 209
prediction, expectation
3. 可能性／確率 ……【possibility】 210
possibility, potential, probability, likelihood, chance
4. 仮説／概念 ……【hypothesis】 212
hypothesis, assumption, concept, notion, idea, view

5. 計画 ……【design】 214
design, program, schedule

6. 目的／試み ……【purpose】 215
purpose, aim, goal, objective, end, attempt, effort

【Ⅰ-B. 研究・発見・報告】 研究・発見・評価・証明・報告に関係する名詞

7. 研究／検査 ……【study】 218
study, investigation, research, work, survey, search, dissection, examination, test, analysis

8. 同定／発見／事実 ……【identification】 221
identification, detection, discovery, finding, observation, fact, characterization

9. 評価／比較 ……【assessment】 223
assessment, evaluation, estimation, comparison

10. 決定 ……【determination】 225
determination, decision

11. 証明／証拠 ……【evidence】 226
demonstration, proof, evidence

12. 報告／結論／説明／示唆 ……【report】 227
report, description, conclusion, explanation, interpretaion, suggestion

Ⅱ）主に結果や現象を説明するときに使う名詞

【Ⅱ-A. 発生・増加】 発生・増加などを表す名詞

1. 存在 ……【presence】 230
presence, existence

2. 発生／発現 ……【appearance】 231
appearance, occurrence, emergence, outbreak, expression

3. 由来／原因 ……【origin】 233
origin, source, cause

4. 産生 ……【production】 234
production, generation, yield

5. 増加／上昇 ……【increase】 235
increase, elevation, augmentation, up-regulation

6. 増強 ……【enhancement】 237
enhancement, potentiation, activation

7. 促進／誘導 ……【promotion】 238
promotion, acceleration, facilitation, induction

8. 進行／進歩 ……【progression】 239
progression, progress, advance

9. 増殖 ……【growth】 240
growth, proliferation, replication

10. 拡大／伸展 ……【expansion】 242
expansion, extension

【Ⅱ-B. 低下・消滅】低下・消滅などを表す名詞

11. 低下／減少 ……【decrease】 243
decrease, reduction, fall, loss, decline, attenuation, down-regulation

12. 抑制 ……【inhibition】 245
inhibition, suppression, repression, depression, interference

13. 阻止／遮断 ……【block】 247
block, blockade, absence

14. 破壊／喪失／切断 ……【disruption】 248
disruption, destruction, ablation, loss, deletion, defect, deficiency, cleavage, truncation, elimination, removal

【Ⅱ-C. 変化・移動・影響】変化・移動・影響などを表す名詞

15. 変化／置換 ……【change】 252
change, alteration, modification, conversion, substitution, replacement

16. 移動／移行／伝達 ……【transfer】 254
transfer, movement, migration, shift, transition, translocation, entry, transmission, transduction

17. 取り込み ……【uptake】 257
uptake, incorporation

18. 影響 ……【influence】 258
influence, impact, effect

【Ⅱ-D. 障害・疾患】障害・疾患などを表す名詞

19. 障害／疾患 ……【disorder】 260
disease, illness, disorder, impairment, deficit, disturbance, barrier, dysfunction, damage, occlusion

20. 発症／感染 ……【development】 263
development, onset, morbidity, infection

21. 症状／徴候 ……【symptom】 264
symptom, manifestation, sign, indication

Ⅲ）主に研究対象の関連・性質・機能や研究の方法を述べるときに使う名詞

【Ⅲ-A．関連・異同・識別】関連・異同・識別を表す名詞

1．関与 ……【involvement】 267
involvement, participation

2．関連 ……【relation】 268
relation, relationship, association, connection, link, correlation

3．一致／類似 ……【agreement】 270
agreement, concordance, similarity, homology

4．違い／識別 ……【difference】 272
difference, variation, discrimination, distinction

【Ⅲ-B．性質】性質などを表す名詞

5．維持／保持 ……【maintenance】 274
maintenance, retention

6．耐性／寛容 ……【resistance】 275
resistance, tolerance

7．保存 ……【conservation】 276
conservation, preservation

8．貯蔵 ……【pool】 277
pool, storage, store

9．蓄積／沈着 ……【accumulation】 278
accumulation, deposition, deposit

10．組成 ……【composition】 279
composition, component, constituent

11．構造 ……【structure】 280
structure, conformation, architecture

12．要求／必要 ……【requirement】 281
requirement, demand, need, necessity

13．制限／限界 ……【restriction】 283
restriction, limitation

14．領域／部位 ……【region】 284
region, area, site, locus, location, localization, position

15．状態 ……【state】 287
state, status, condition, situation

16．重要性 ……【importance】 288
importance, significance, implication

17．特徴 ……【feature】 289
feature, characteristic, character, profile

18. 効果／効力／効率 ……【effect】 291
effect，efficacy，potency，efficiency

19. 能力／潜在力 ……【ability】 293
ability，capacity，capability，competence，power，potential

【Ⅲ-C. 機能】機能に関係する名詞

20. 応答／反応 ……【response】 295
response，reaction

21. 認識／認知 ……【recognition】 296
recognition，perception，cognition

22. 結合／接着 ……【binding】 297
binding，bond，bonding，association，connection，interaction，adhesion，attachment

23. 機能／作用 ……【function】 300
function，action，operation

24. 調節 ……【regulation】 301
regulation，control，modulation，adjustment

25. 形成／集合 ……【formation】 302
formation，assembly，aggregation

26. 機構／役割 ……【mechanism】 304
mechanism，machinery，mode，basis，role

【Ⅲ-D. 方法】研究の方法などを表す名詞

27. 選択 ……【selection】 306
selection，choice，option

28. 使用／利用 ……【use】 307
use，usage，utilization

29. 置換／交換 ……【substitution】 308
substitution，replacement，exchange

30. 測定／定量 ……【measurement】 309
measurement，assay，quantification，quantitation

31. 単離／精製 ……【isolation】 311
isolation，separation，segregation，purification

32. 処理 ……【treatment】 312
treatment，addition，stimulation

33. 培養／インキュベーション ……【culture】 313
culture，incubation

形容詞編

I）使い分けに注意したい形容詞

【I-A. 程度】程度を表す形容詞

1. 著しい ……【marked】 316
significant, marked, striking, prominent, intense, dramatic, drastic, remarkable, notable, noteworthy

2. 強力な ……【strong】 319
strong, powerful, robust, potent

3. 大きな ……【large】 320
large, huge, great, massive

4. 主な ……【major】 322
major, main, dominant, predominant

5. 完全な ……【complete】 323
complete, full, perfect

6. 十分な／適当な ……【sufficient】 325
sufficient, adequate, appropriate, proper, reasonable

7. 明らかな ……【clear】 327
clear, apparent, evident, pronounced, obvious, overt

8. 広い ……【wide】 329
wide, broad, extensive, widespread

9. 全体の ……【whole】 330
whole, overall, entire, total

10. 多数の ……【many】 331
many, numerous, large number of, multitude of

11. いくつかの ……【several】 332
several, a number of

12. かなりの ……【substantial】 333
substantial, considerable

13. 中程度の ……【moderate】 334
moderate, modest

14. ほとんどない／わずかな／ない ……【little】 335
little, few, a few, subtle, slight, insignificant, less, small, weak, deficient, free

15. 普通の／一般的な ……【common】 338
common, general, conventional, normal

16. 特別の／特異的な ……【particular】 340
particular, special, specific

【Ⅰ-B. 性質・状態】性質・状態を表す形容詞

17. 可能な／ありそうな ……【possible】 342
probable, likely, possible, feasible, potential

18. 重要な／決定的な ……【important】 344
important, key, vital, critical, crucial, definitive, conclusive, serious

19. 中心的な ……【central】 348
central, pivotal

20. 必要な ……【necessary】 349
necessary, required for, fundamental, essential

21. 有用な ……【useful】 350
useful, available

22. 効果的な／活発な ……【effective】 352
effective, efficacious, efficient, active

23. 調節性の／誘導性の ……【regulatory】 353
regulatory, modulatory, inducible

24. 増加する ……【increasing】 354
increasing, growing

25. 独特な／特有の ……【unique】 355
unique, characteristic

26. 異常な ……【abnormal】 356
abnormal, aberrant, unusual

27. 急速な ……【rapid】 357
rapid, fast, prompt

28. 同時の ……【simultaneous】 359
simultaneous, concomitant, concurrent, coincident, synchronous

29. 高齢の／〜歳の ……【elderly】 360
elderly, old

【Ⅰ-C. 関係】関係を表す形容詞

30. 責任ある／原因である ……【responsible】 362
responsible for, due to, causative, causal

31. 関連する ……【related to】 363
related to, relevant

32. 類似する／一致する ……【similar】 364
similar, analogous, same, equal, identical, homologous, fit, consistent

33. 異なる／別々の／独立した ……【different】 367
different, distinct, separate, discrete, independent

34. 逆の ……【opposite】 369
opposite, inverse

【I-D. 頻度・時】頻度・時に関係する形容詞

35. 連続の ……【continuous】 370
continuous, serial

36. 次の／引き続いた ……【following】 371
following, next, subsequent

37. まれな ……【rare】 372
rare, uncommon, infrequent

38. 新しい ……【new】 373
new, novel, initial, latest

39. 以前の／最近の ……【previous】 374
previous, recent

40. 現在の ……【present】 375
present, current

副詞編

I）論文でよく使われる副詞

【I-A. 程度】程度を表す副詞

1. 著しく／非常に ……【markedly】 378
significantly, markedly, remarkably, strikingly, dramatically, drastically, very, extremely, unusually, highly

2. 強く／大きく ……【strongly】 382
strongly, potently, greatly

3. 主に ……【predominantly】 383
predominantly, mainly, mostly, primarily, principally, exclusively, largely

4. 完全に／十分に ……【completely】 385
completely, entirely, fully, sufficiently, enough, quite

5. 明らかに／明確に ……【clearly】 388
clearly, apparently, distinctly, unambiguously, evidently, obviously, definitively, conclusively

6. 広く／普遍的に ……【widely】 390
widely, broadly, extensively, universally, ubiquitously, constitutively

7. はるかに／もっと ……【far】 393
far, much

8. ほとんど ……【nearly】 394
nearly, almost, virtually

9. かなり ……【substantially】 396
substantially, considerably, fairly, reasonably, appreciably, increasingly

10. 中程度に／部分的に ……【moderately】 398
moderately, partially, somewhat

11. 比較的／選択的に ……【relatively】 399
relatively, comparatively, comparably, selectively

12. わずかに／単に ……【slightly】 401
slightly, modestly, marginally, poorly, weakly, simply, only

13. 通常／一般に ……【commonly】 403
commonly, generally, in general, usually

14. 特に／特異的に ……【particularly】 405
particularly, especially, in particular, notably, specifically

15. 実際に ……【actually】 407
actually, practically

16. およそ ……【approximately】 408
approximately, about, ca., roughly

【I-B. 様態】様子・状態・関係などを表す副詞

17. おそらく／もしかしたら ……【probably】 410
probably, presumably, likely, perhaps, possibly, potentially

18. 容易に／うまく ……【readily】 413
readily, easily, successfully

19. 効率的に／活発に ……【efficiently】 414
efficiently, effectively, actively

20. 適切に ……【properly】 415
properly, adequately, appropriately, correctly

21. 一様に ……【uniformly】 416
uniformly, evenly

22. 緊密に／直接 ……【closely】 417
closely, tightly, directly

23. 速く／手短かに ……【rapidly】 418
rapidly, quickly, briefly

24. 同時に ……【simultaneously】 419
simultaneously, concomitantly, concurrently

25. 同様に／同等に ……【similarly】 421
similarly, identically, as well as, not only ～ but also, equally, equivalently

26. 異なって／別々に／独立に ……【differently】 423
differently, separately, individually, independently

27. 逆に ……【inversely】 424
inversely, oppositely, adversely

【I-C. 頻度・時】 頻度・時を表す副詞

28. 連続的に ……【continuously】 426
continuously, serially, progressively, persistently, subsequently

29. しばしば ……【frequently】 427
frequently, often

30. まれに ……【rarely】 428
rarely, infrequently

31. 早期に／最初に／すぐに ……【early】 429
early, initially, originally, newly, immediately, soon, readily

32. 以前に ……【previously】 432
previously, before, formerly, already, recently, to date

33. 現在 ……【currently】 434
currently, presently, at present, now, today, still

34. 将来 ……【prospectively】 436
prospectively, hereafter, thereafter

接続詞・接続語編

I）前後の内容を接続するために使われる語句

【I-A. 逆説・比較】 逆説や比較の意味を持つ接続詞，副詞，前置詞，句

1. しかし／だが一方／にもかかわらず／それどころか ……【however】 440
however, but, whereas, while, nevertheless, nonetheless, on the contrary

2. 対照的に／一方では／代わりに ……【in contrast】 442
in contrast, by contrast, conversely, in contrast to, as opposed to, on the other hand, meanwhile, instead, alternatively, instead of

3. ～だけれども／～にもかかわらず／～と違って ……【although】 446
although, though, even though, while, despite, in spite of, albeit, unlike, contrary to

4. ～と比較して ……【compared with】 449
compared with，compared to，in comparison with，
in comparison to，relative to

【Ⅰ-B. 肯定】肯定の意味を持つ接続詞，副詞，前置詞，句

5. ～なので／～のせいで ……【because】 452
because，since，as，due to，because of，owing to

6. 従って／それゆえ ……【therefore】 454
therefore，hence，accordingly，consequently，thus

7. それから／それによって ……【then】 455
then，in turn，and，thereby，so that

8. さらに／そのうえ／加えて ……【furthermore】 457
furthermore，further，moreover，in addition，additionally，
in addition to，besides

9. ～に一致して／～に従って ……【in agreement with】 460
in agreement with，coincident with，in accordance with，
in accord with，according to

10. 実際に ……【indeed】 461
indeed，in fact

11. たとえば／すなわち ……【for example】 462
for example，for instance，e.g.，namely，i.e.

12. 同様に ……【similarly】 464
similarly，likewise，correspondingly

13. まとめると ……【taken together】 465
taken together，collectively，in summary，in conclusion

【Ⅰ-C. 条件】条件節に使われる接続詞

14. もし～なら／たとえ～だとしても ……【if】 467
if，Given，if any，even if，unless，once

第1章

動詞編

動詞は文の中心となるもので，論文を書く際にはその使い方に特に注意が必要である．文を構成する基本要素は，主語（S），動詞（V），目的語（O），補語（C）であり，動詞の文型には，①S＋V，②S＋V＋C，③S＋V＋O，④S＋V＋O＋C，⑤S＋V＋O＋Oの5つがある．そのうち①②が自動詞，③④⑤が他動詞の文型である．論文で使われる他動詞の数は自動詞より圧倒的に多く，また，そのほとんどが③のS＋V＋Oの文型で用いられる．S＋V＋OのOの部分になりうるものとしては，「名詞（句）」「that節」「疑問詞節」「to不定詞」「原形不定詞」「ing形」などさまざまなものがある．動詞によってどれが使われるのかが異なるので注意が必要であろう．また，論文では受動態の用例が非常に多く，実際にはS＋V＋Oは，S＋be動詞＋過去分詞の形になることが多い．このときの過去分詞には，さらに特定の前置詞が続く場合があるので注意を要する．動詞編には過去分詞＋前置詞を含め，たくさんの共起表現が収録されているので，その文型とともに留意して論文執筆に活用しよう．

【I-A. 知見・仮説・計画】 過去の知見や仮説・計画を述べるときに使う動詞

1. 知られている／認められている 【know】

知る	認める	理解する
know learn	appreciate accept	understand

使い分け

- **know** は，受動態で「知られている」という意味に用いられることが非常に多い．
- **learn** は「学習する」という意味だが，受動態で「知られている」という意味にも使われる．
- **known** about は否定的な文で，**learned** about は肯定的な文で用いられることが多い．
- **known** that は肯定文で，**known** whether は否定文で使われることが多い．
- **appreciate** は「（真価を）認める／理解する」，**accept** は「受け入れる／認める」という意味で用いられる．
- **understand** は「（メカニズムなどを）理解する」という意味だが，否定的な意味の用例がかなり多い．

頻度分析

	用例数		用例数
know	17,705	❶ (be) known to	4,189
		❷ little is known about	1,698
		❸ it is not known whether	229
		❹ it is known that	164
learn	678	❶ (be) learned about	42
appreciate	196	❶ than previously appreciated	29
		❷ (be) appreciated that	17
accept	945	❶ (be) accepted that	107
understand	6,542	❶ to understand	1,934
		❷ (be) poorly understood	1,582
		❸ little is understood about	31

共起・頻度分析

（数字：用例数）

直前の単語				直後の単語				
be動詞	to			to	about	that	whether	how
5,600	278	known	17,443	4,189	2,024	280	270	83
5	61	know (s)	197	3	18	13	14	13
97	5	learned	369	30	42	6	0	1
0	193	learn (s)	309	36	27	4	10	19
43	0	appreciated	173	3	0	17	1	0
0	13	appreciate (s)	19	0	0	1	0	3
67	5	accepted	626	13	0	107	0	0
0	73	accept (s)	192	0	0	1	0	0
216	0	understood	3,715	12	48	8	5	30
0	1,934	understand (s)	2,827	2	0	5	9	359

解説

◆ **know** は，受動態でto不定詞またはaboutを伴う用例が多い．また，that節，whether節を伴う用例もかなり多い．

◆ **learn** は，受動態でaboutを伴う用例が多い．また，能動態でto不定詞を伴う用例も多い．

◆ **appreciate**, **accept** は，受動態でthat節を伴う用例が多い．

◆ **understand** は，“to understand how”の用例が多い．また，受動態でaboutを伴う用例も多い．

★★ know（〜を知る）

他動詞．受動態の用例が非常に多い．

❶ **known to 〜（〜すると知られている）**

Gadd45a is known to be involved in a G2 checkpoint and may be involved in the normal progression from G2 to M and its coordination with S phase events.（*Oncogene. 2002 21:6228*）

訳 Gadd45aはG2チェックポイントに関与すると知られている

❷ **little is known about 〜（〜についてはほとんど知られていない）**

However, little is known about the spatial distribution of this phospholipid in neurons and its dynamics.（*J Cell Biol. 2001 154:355*）

訳 しかし，…におけるこのリン脂質の空間的分布についてはほとんど知られていない

❸ **it is not known whether 〜（〜かどうかは知られていない）**

However, it is not known whether such mechanisms are operative *in vivo*.（*Proc Natl Acad Sci USA. 2001 98:1288*）

訳 しかし，…かどうかは知られていない

I

A

❹ it is known that ～（～ということが知られている）

It is known that the extracellular matrix regulates normal cell proliferation, and it is assumed that anchorage-independent malignant cells escape this regulatory function.（*Proc Natl Acad Sci USA. 2000 97:10026*）

訳 …ということが知られている

learn（～を知る／～を学習する）

他動詞.

❶ learned about ～（～について知られている）

Although much has been learned about the physiological significance of this receptor tyrosine kinase, its catalytic mechanism remains poorly understood.（*Biochemistry. 2000 39:9786*）

訳 …の生理学的重要性については，多くのことが知られているけれども

appreciate（～の真価を認める／～を理解する）

他動詞．受動態の用例が多い．

❶ ～ than previously appreciated（以前，理解されていたより～な）

These data indicate that Cdc25A turnover is more complex than previously appreciated and suggest roles for an additional kinase(s) in Chk1-dependent Cdc25A turnover.（*Genes Dev. 2003 17:3062*）

訳 Cdc25A の代謝回転は，以前，理解されていたよりもっと複雑である

❷ appreciated that ～（～ということが認められている）

It is now appreciated that insulin resistance can result from a defect in the insulin receptor signaling system, at a site post binding of insulin to its receptor.（*J Med Chem. 2000 43:995*）

訳 ということが，今，認められている

accept（～を受け入れる／～を認める）

他動詞.

❶ accepted that ～（～ということが受け入れられている）

It is generally accepted that the ability of cocaine to inhibit the dopamine transporter (DAT) is directly related to its reinforcing actions.（*Proc Natl Acad Sci USA. 2004 101:372*）

訳 …ということは一般に受け入れられている

★ understand（～を理解する）

他動詞.

❶ to understand ～（～を理解するために）

To understand how CycD/Cdk4 promotes growth, we performed a screen for modifiers of CycD/Cdk4-driven overgrowth in the eye.（*Dev Cell. 2004 6:241*）

訳 どのようにCycD/Cdk4が増殖を促進するかを理解するために

❷ poorly understood〔十分には理解されていない（よく理解されていない）〕

Skeletal muscle perfusion during exercise is impaired in heart failure, but the underlying mechanisms are poorly understood.（*Circ Res. 2001 88:816*）

訳 しかし，根底にあるメカニズムは，十分には理解されていない

❸ little is understood about ～（～についてはほとんど理解されていない）

Attention powerfully influences auditory perception, but little is understood about the mechanisms whereby attention sharpens responses to unattended sounds.（*Nat Neurosci. 2004 7:658*）

訳 しかし，それによって…である機構についてはほとんど理解されていない

2. 考えられる／思われる 【think】

考える	思う	思われる	考慮に入れる
consider regard	believe think suspect	appear seem	take into account

使い分け

◆ **consider, regard** は「考える／みなす」，**believe** は「思う／信じる」，**think** は「考える」，**suspect** は「推測する／疑う」という意味だが，いずれも受動態で「考えられる／思われる」ことを表現するために使われる．

◆ **appear, seem** は，現在形の自動詞として用いられることが多い．

◆ **take into accout** は，「考慮に入れる」という意味で用いられる．

頻度分析

	用例数		用例数
consider	3,905	❶ (be) considered to be	344
		❷ (be) considered as	197
regard	968	❶ (be) regarded as	181
believe	1,822	❶ (be) believed to	1,239
think	3,875	❶ (be) thought to	3,276
suspect	633	❶ (be) suspected to	65
appear	13,005	❶ appears to be	2,785
seem	2,034	❶ seems to be	430
take into account	290	❶ (be) taken into account	115

解説

- **consider**, **believe**, **think**, **suspect** は，受動態で後に to 不定詞を伴う用例が多い．
- **regard**, **consider** は，受動態で後に as を伴う用例が多い．
- **appear** to, **seem** to はどちらも「～すると思われる／～するらしい」という意味だが，用例数は **appear** to が圧倒的に多い．

★ consider（～と考える）

他動詞．受動態で“considered to [*do*]”“considered as”の用例が多い．

❶ **considered to be ～**（～であると考えられる）

Xenotransplantation is considered to be a solution for the human donor shortage.（*J Virol. 2001 75:2825*）

訳 異種移植は，ヒトのドナー不足の解決策であると考えられる

❷ **considered as ～**（～として考えられる）

Judicious retransplantation should be considered as a therapeutic option in the management of polyoma virus induced graft failure.（*Transplantation. 2002 73:1166*）

訳 賢明な再移植は，治療上の選択肢として考えられるべきである

regard（～と考える／～とみなす）

名詞，特に“with regard to”の用例が多いが，他動詞としても用いられる．“regarded as”の用例が多い．

❶ **regarded as ～**（～であると考えられる）

This model can be regarded as a general explanation for the activity of hyperthermophilic enzymes.（*Proc Natl Acad Sci USA. 2004 101:14379*）

訳 このモデルは，…に対する一般的な説明であると考えられうる

★ believe（～を信じる／～と思う）

他動詞．受動態で“believed to［do］”の用例が非常に多い．

❶ believed to ～（～すると信じられる）

Oxidative stress is believed to be a major factor in the development of this disease and peroxides are suspected to be prominent stressing agents.（*FASEB J. 2004 18:480*）

訳 酸化ストレスはこの疾患の発症の主な要因であると信じられている

★ think（～と考える／～と思う）

他動詞．受動態で“thought to［do］”の用例が非常に多い．

❶ thought to ～（～すると考えられる）

MMP-2 is thought to be involved in cancer cell invasiveness.（*J Biol Chem. 2002 277:20919*）

訳 MMP-2は癌細胞の浸潤性に関与すると考えられる

suspect（～だと推測する／～と思う）

他動詞．

❶ suspected to ～（～するのではないかと思われる／～すると推測される）

Langerhans cells（LCs）are suspected to be initial targets for HIV after sexual exposure（by becoming infected or by capturing virus).（*Proc Natl Acad Sci USA. 2003 100:8401*）

訳 ランゲルハンス細胞（LCs）はHIVの初期の標的ではないかと思われる

★★ appear（～と思われる／～らしい）

自動詞．“appear to［do］”の用例が非常に多い．

❶ appear to be ～（～であると思われる）

The decrease of STAT4 protein appears to be due to specific degradation of phospho-STAT4, possibly through the proteasome degradation pathway.（*Blood. 2001 97:3860*）

訳 STAT4タンパク質の減少はリン酸化STAT4の特異的な分解によると思われる

★ seem（～と思われる／～らしい）

自動詞．“seem to［do］”の用例が非常に多い．

❶ **seem to be ～**（～であると思われる）

Overexpression of the splice variant therefore seems to be a general characteristic of insulinomas and is estimated to contribute about 90% to insulin synthesis by these tumours.（*Lancet. 2004 363:363*）

訳 それゆえ，スプライスバリアントの過剰発現は膵島細胞腺腫の一般的な特徴であると思われる

take into account（～を考慮に入れる）

“take into account” の形で他動詞として用いられる．

❶ **taken into account**（考慮に入れられる）

These large-scale, naturally occurring variations must be taken into account when considering human-induced climate change and the management of ocean living resources.（*Science. 2003 299:217*）

訳 これらの大規模で自然に起こる変動は，…を考えるときに考慮に入れられなければならない

3. 予想する／推定する 【predict】

予想する	推定する
predict expect	assume presume estimate

使い分け

◆ **predict**, **expect** は，「予想する／予期する」という意味で用いられる．

◆ **assume**, **presume** は，「推定する」という意味で使われる．

◆ **estimate** は「（数や量を）見積もる」という意味で用いられる．名詞として使われることもある．

共起・頻度分析　（数字：用例数）

直前の単語					直後の単語	
be動詞	we	to			that	to
1,439	46	27	predicted	6,180	142	1,323
0	129	829	predict(s)	2,928	558	2
959	11	22	expected	3,148	54	709
0	71	15	expect(s)	168	58	15

直前の単語					直後の単語	
be動詞	we	to			that	to
328	10	1	assumed	697	192	259
0	34	69	assume (s)	443	133	0
175	1	7	presumed	528	19	199
0	5	1	presume (s)	11	7	0
1,022	179	4	estimated	2,723	83	335
0	239	795	estimate (s)	3,485	179	21

解説

◆いずれの語も，to 不定詞を伴う受動態の用例が多い．
◆**assume** は，"it has been assumed that" あるいは "it is assumed that" の用例も多い．
◆**predict**, **assume**, **estimate** は，能動態で that 節を伴う用例も多い．
◆能動態の **estimate** は，we を主語にする用例が多い．
◆能動態の **predict** は，we 以外が主語の場合も多い．

★★ predict（～を予想する／～を予測する）

他動詞．現在形の用例が多い．

❶ **predicted to ～**（～すると予想される）

The Pawn protein is predicted to be a large cell adhesion molecule with a single transmembrane domain, a short cytoplasmic tail and two extracellular epidermal growth factor (EGF)-like repeats.（*Gene. 2003 310:169*）

訳 Pawn タンパク質は，…を持つ大きな細胞接着分子であると予想される

❷ **predict that ～**（～ということを予測する）

The model predicts that optimal N:P ratios will vary from 8.2 to 45.0, depending on the ecological conditions.（*Nature. 2004 429:171*）

訳 モデルは，…ということを予測する

★ expect（～を予期する／～を予想する）

他動詞．受動態の用例が多い．

❶ **expected to ～**（～すると予期される）

Thus, telomerase is expected to be a very strong candidate for targeted therapy of prostate cancer.（*Oncogene. 2000 19:2205*）

訳 テロメラーゼは，…の非常に強力な候補であると予期される

★ assume（～と推定する）

他動詞．受動態の用例が多い．

❶ **assumed to be ～**（～であると推定される）

The proper intracellular distribution of mitochondria is assumed to be critical for normal physiology of neuronal cells, but direct evidence for this idea is lacking.（*Cell. 2004 119:873*）

訳 ミトコンドリアの適当な細胞内分布が，神経細胞の正常な生理機能にとって決定的に重要であると推定される

❷ **assumed that ～**（～ということが推定される／～ということを推定した）

It has been assumed that cleavage of the N-terminal propeptide domain of membrane type-1 matrix metalloproteinase（MT1-MMP）is required for enzyme function.（*J Biol Chem. 2000 275:29648*）

訳 …ということが推定されている

presume（～と推定する）

他動詞．

❶ **presumed to ～**（～すると推定される）

These regions are presumed to be required for the DNA binding activity of the repressor.（*Mol Microbiol. 2000 35:1394*）

訳 これらの領域は，…のDNA結合活性に必要とされると推定される

★ estimate（～と見積もる／～と推定する／見積もり）

他動詞の用例が多いが，名詞としても用いられる．

❶ **estimated to be ～**（～であると見積もられる）

The affinity was estimated to be approximately 10^9 M^{-1}.（*Biochemistry. 2004 43:13875*）

訳 親和性はおよそ10^9 M^{-1}であると見積もられた

❷ **we estimate that ～**（われわれは，～ということを見積もる）

We estimate that approximately 6–8% of the total photoreceptor pool in each eye is removed by this mechanism.（*Development. 2004 131:2409*）

訳 われわれは，…ということを見積もる

4. 仮定する／提唱する 【hypothesize】

仮定する	提唱する
hypothesize postulate	propose

使い分け

◆ **hypothesize** は，「仮説を立てる／仮定する」という意味で用いられる．

◆ **postulate** は，「（議論の前提として）仮定する」ときに使われることが多い．

◆ **propose** は，「（結論としてモデルなどを）提唱する」場合に用いられる．

頻度分析

	用例数		用例数
hypothesize	3,468	❶ we hypothesized that	1,589
postulate	1,220	❶ (be) postulated to	349
		❷ we postulate that	190
propose	10,456	❶ we propose that	3,245
		❷ (be) proposed to	1,357

解説

◆いずれの語も we を主語として，that 節を目的語とする用例が多い．

◆ **postulate**, **propose** は，to 不定詞を伴う受動態の用例も多い．

★ hypothesize（～という仮説を立てる／～と仮定する）

❶ we hypothesized that ～（われわれは～という仮説を立てた）

We hypothesized that the control of DC survival is regulated by the antiapoptotic factor bcl-x_L. (*J Immunol. 2004 173:4425*)

訳 われわれは…という仮説を立てた

★ postulate（～と仮定する）

❶ postulated to ～（～すると仮定される）

SOD-2 has been postulated to be a tumor suppressor. (*Oncogene. 2003 22:1024*)

訳 SOD-2は，癌抑制因子であると仮定されてきた

❷ we postulate that ～（われわれは～ということを仮定する）

We postulate that ZEB likely plays an important role in regulating the

life cycle of EBV.（*J Virol. 2003 77:199*）
訳 われわれは…ということを仮定する

★★ propose（～を提唱する／～を提案する）

❶ we propose that ～（われわれは～ということを提唱する）

Based on these results, we propose that BMP-15 is an important determinant of FSH action through its ability to inhibit FSH receptor expression.（*J Biol Chem. 2001 276:11387*）
訳 これらの結果に基づいて，われわれは…ということを提唱する

❷ proposed to ～（～するために提唱される）

Several models have been proposed to explain the mechanism of cyst formation.（*Nat Genet. 2000 25:143*）
訳 いくつかのモデルが，…の機構を説明するために提唱されてきた

5. 計画する／予定する 【design】

計画する	予定する
design plan	schedule

使い分け
◆ **design** は，「（目的のために）計画する／設計する」という意味で用いられる．
◆ **plan**, **schedule** は，「（予定を）計画する」という意味で用いられる．

頻度分析

	用例数		用例数
design	10,456	❶（be）designed to	1,746
plan	6,867	❶ to plan	18
schedule	515	❶（be）scheduled for	32

★ design（～を計画する／～を設計する／設計）

名詞としても使われるが，他動詞の用例も多い．

❶ designed to ～（～するために計画された）

This study was designed to determine whether a muscle defect could

explain this gallbladder dysfunction.（*Gastroenterology. 2001 120:506*）
訳 この研究は，…かどうかを決定するために計画された

★ plan（～を計画する／計画）

名詞の用例が多いが，他動詞として使われることもある．

❶ to plan ～（～を計画するために）

Our prediction rule can be used to plan surveillance of new leprosy patients.（*Lancet. 2000 355:1603*）
訳 われわれの予測の法則は，…の調査を計画するために使われうる

schedule（～を予定する／～を計画する／設計）

名詞の用例が多いが，他動詞としても用いられる．

❶ scheduled for ～（～を予定される）

A phase I dose-escalation study was conducted in 15 end-stage renal disease patients scheduled for renal allografts from living donors.（*Transplantation. 2000 70:1707*）
訳 第一相投与量逐次漸増試験が，腎臓の同種移植を予定されている15人の末期の腎疾患患者において行われた

6. ～しようとする／試みる／取り組む 【attempt】

試みる	努める	取り組む／着手する	目的とする
attempt try	seek	address undertake	aim

使い分け

◆ **attempt**, **try** は「試みる」，**seek** は「努める」という意味だが，いずれの語も to 不定詞を後に伴って「～しようとする」という意味に用いられる．

◆ **attempt** は，「（困難なことを）試みる」場合に使われることが多い．

◆ **try** は口語的表現で，**attempt** に比べると用例数はかなり少ない．

◆ **address** は「取り組む」，**undertake** は「着手する」，**aim** は「目的とする」という意味で使われる．

頻度分析

	用例数		用例数
attempt	1,747	❶ attempted to	227
try	145	❶ try to	35
seek	1,835	❶ sought to	1,097
address	3,152	❶ to address	1,479
undertake	1,106	❶ (be) undertaken to	495
aim	2,785	❶ aimed to	332

★ attempt（～を試みる／試み）

名詞の用例の方が多いが，他動詞としても用いられる．“we（have）attempted to” の用例が多く，受動態の用例は少ない．

❶ **attempted to ～（～しようと試みた）**

In the present study, we attempted to identify the minimal Gag sequences required for the formation of VLP.（*J Virol. 2000 74:5395*）

訳 われわれは，…を同定しようと試みた

try（～しようとする／～を試みる）

他動詞．

❶ **try to ～（～しようとする）**

We did a randomised multicentre trial to try to resolve this issue.（*Lancet. 2001 357:979*）

訳 われわれはこの問題を解決しようとするために無作為化した多施設治験を行った

★ seek（～に努める／～を求める）

他動詞．“we sought to” の用例が非常に多く，また，受動態の用例は少ない．

❶ **sought to ～（～しようと努めた／～しようとする）**

In this study, we sought to determine whether MMP-9 is critical for SMC migration and for the formation of a neointima by using mice in which the gene was deleted（MMP-9 $^{-/-}$ mice）.（*Circ Res. 2002 91:845*）

訳 われわれは，…かどうかを決定しようと努めた

★ address（～に取り組む／住所）

名詞としても用いられるが，他動詞の用例の方が多い．

❶ to address ～（～に取り組むために）

Two approaches were used to address these issues.（*J Biol Chem. 2002 277:32243*）

訳 2つのアプローチがこれらの問題に取り組むために使われた

★ undertake（～に着手する／～を企てる）

他動詞．受動態で後にto不定詞を伴う用例が多い．

❶ undertaken to ～（～するために着手される）

The present study was undertaken to determine whether endogenous sphingolipids are involved in the TGF-β signaling pathway.（*J Biol Chem. 2003 278:9276*）

訳 現在の研究は，…かどうかを決定するために着手された

★ aim（～を目的とする／目的）

名詞の用例が多いが，他動詞としても用いられる．後にto不定詞を伴う用例が多い．

❶ aimed to ～（～することを目的とした／～しようとした）

We aimed to assess the effects of family planning services on abortion rates in two similar areas.（*Lancet. 2001 358:1051*）

訳 われわれは，…の効果を評価することを目的とした

【Ⅰ-B. 提示・報告・証明】研究内容の提示や証明を述べるときに使う動詞

7. 示す／意味する 【show】

示す	示唆・意味する
show	suggest
indicate	imply
exhibit	mean
display	
represent	

使い分け

◆いずれの語も能動態で用いられることが多い.

◆ **show**, **indicate** は,「（結論などを）示す」という意味で用いられる.

◆ **exhibit**, **display** は,「（形質などを）示す」場合に用いられる.

◆ **represent** は,「～を表す／～である」という意味で使われる.

◆ **suggest** は「示唆する」, **imply** は「意味する／示唆する」, **mean** は「意味する」ときに用いられる.

◆ **show**, **indicate**, **suggest** は, 著者の主張を示す場合によく使われる.

頻度分析

	用例数		用例数
show	75,454	❶ show that	23,059
indicate	35,819	❶ indicate that	13,925
exhibit	13,627	❶ exhibited a	1,233
display	7,662	❶ displayed a	602
represent	8,569	❶ represent a	1,335
suggest	61,923	❶ suggest that	24,786
imply	2,576	❶ imply that	649
mean	11,349	❶ means that	90

共起・頻度分析

（数字：用例数）

直前の単語						直後の単語		
may	mice	results	we			that	a	the
18	228	2,418	**16,048**	show (s)	36,360	**25,424**	1,573	726
0	596	521	1,086	showed	21,263	**8,439**	2,492	560
0	1	5	2	shown	15,872	**4,501**	93	89

直前の単語						直後の単語		
may	mice	results	we			that	a	the
151	50	**5,480**	3	indicate (s)	20,431	**17,474**	933	625
0	48	314	0	indicated	6,350	**4,394**	317	187
0	161	23	0	indicating	9,038	**6,346**	842	506
46	**262**	0	2	exhibit (s)	6,863	0	**1,354**	171
0	**409**	10	0	exhibited	6,005	0	**1,233**	179
12	**161**	2	1	display (s)	4,040	8	**643**	145
0	**222**	0	2	displayed	3,242	0	**602**	106
945	21	64	7	represent (s)	6,226	1	**2,733**	**1,194**
0	1	0	4	represented	1,047	0	55	57
90	91	**8,600**	**1,626**	suggest (s)	40,278	**32,442**	3,674	1,032
0	24	194	19	suggested	5,830	**3,396**	337	188
0	215	17	0	suggesting	15,815	**10,892**	1,985	704
23	1	**260**	0	imply/implies	1,593	**1,127**	203	69
0	0	1	0	implied	132	38	6	7
0	11	2	0	implying	851	**551**	123	44
2	13	3	2	mean (s)	11,203	**130**	4	9
0	0	0	0	meant	27	10	0	2

解説

- **show**, **indicate**, **represent**, **suggest**, **imply**, **mean** は現在形，**exhibit**, **display** は過去形の用例が多い．
- **show** は，we が主語になる場合が多い．
- **indicate**, **suggest**, **imply** は，results, data などが主語になる場合が多い．
- **means** は，this を主語にする場合が多い．
- **represent** は，may を伴う用例が多い．
- **show**, **indicate**, **suggest**, **imply**, **mean** は，that 節を伴う用例が多い．
- **exhibit**, **display**, **represent** は that 節を伴わず，"a＋名詞" を伴う場合が多い．
- **exhibit**, **display** には，mice などを主語とする用例が多く，一方，we や results は主語にはならない．
- **mean** は名詞の用例が多い．

☆☆ show（～を示す）

他動詞．

❶ show that ～（～ということを示す）

Here we show that the expression of this small RNA is increased at a low temperature and in minimal medium.（*J Bacteriol. 2004 186:6689*）

訳 ここにわれわれは，…ということを示す

☆ indicate（～を示す）

他動詞.

❶ indicate that ～（～ということを示す）

Our results indicate that the expression of local organizing factors is controlled by combinatorial interaction between inductive and modulatory factors.（*Nat Neurosci. 2001 4:1175*）

訳 われわれの結果は，…ということを示す

☆ exhibit（～を示す）

❶ exhibited a ～（～を示した）

In these studies, Obx-mice exhibited a significant increase in 5-HT levels in the OT relative to sham-operated controls, but similar NE and DA concentrations.（*Brain Res. 2003 963:150*）

訳 Obx マウスは 5-HT レベルの有意な上昇を示した

★ display（～を示す）

❶ displayed a ～（～を示した）

However, mPGES1 −/− mice displayed a marked reduction in inflammatory responses compared with mPGES1 +/+ mice in multiple assays.（*Proc Natl Acad Sci USA. 2003 100:9044*）

訳 mPGES1 −/− マウスは，…の顕著な低下を示した

★ represent（～を表す／～である）

❶ represent a ～（～を表す／～である）

Thus, the up-regulation of PP2A may represent a novel mechanism for E1A-mediated sensitization to anticancer drug-induced apoptosis.（*Cancer Res. 2004 64:5938*）

訳 PP2A の上方制御は，…の新規の機構を表すかもしれない

☆ suggest（～を示唆する／～を示す）

他動詞.

❶ suggest that ～（～ということを示唆する）

Thus, these data suggest that the ability of $\alpha_v\beta_3$ to recognize osteopontin can be differentially regulated in a cell-specific manner.（*J Biol Chem. 2000 275:18337*）

訳 これらのデータは，…ということを示唆する

★ imply（～を意味する／～を示唆する）

他動詞.

❶ **imply that ～**（～ということを意味する）

These results imply that the inactivation of p53 function by Tax contributes to Tax suppression of DNA repair.（*J Biol Chem. 2000 275:35926*）

訳 これらの結果は，…ということを意味する

mean（～を意味する／平均／手段）

名詞の用例が多いが，他動詞としても用いられる.

❶ **mean that ～**（～ということを意味する）

This means that the complex zeros of the zeta function are associated with the irregularity of the distribution of the primes.（*Proc Natl Acad Sci USA. 2000 97:7697*）

訳 これは，…ということを意味する

8. 提示する／提供する 【present】

提示・提供する	提起する
present provide	raise

使い分け

◆ **present**, **provide** は，「（証拠などを）提示する／提供する」ときに用いられる.

頻度分析

	用例数		用例数
present	25,835	❶ present evidence	677
provide	28,032	❶ provide evidence	1,931
raise	2,677	❶ raise the possibility that	298

解説

◆いずれの語も能動態で用いられることが多い.

◆ **raise** は，“raise the possibility that（～という可能性を示唆する）”の用例が多い.

★★ present（～を提示する／現在の）

形容詞の用例が多いが，他動詞としても用いられる．

❶ present evidence（証拠を提示する）

Here we present evidence that JNK has a major role in promoting tumorigenesis both *in vivo* and *in vitro*.（*Oncogene. 2002 21:5038*）

訳 ここにわれわれは，…という証拠を提示する

★★ provide（～を提供する）

❶ provide evidence（証拠を提供する）

Here we provide evidence that p43 is primarily associated with the telomerase ribonucleoprotein *in vivo*.（*Biochemistry. 2003 42:5736*）

訳 ここにわれわれは，…という証拠を提供する

★ raise（～を提起する／～を上げる／～を産生する）

“raise the possibility that” の用例が多いが，「(抗体を) 産生する」という意味で用いられることも多い．

❶ raise the possibility that ～（～という可能性を示唆する）

These results raise the possibility that BCL6 may regulate apoptosis by means of its repressive effects on PDCD2.（*Proc Natl Acad Sci USA. 2002 99:2860*）

訳 これらの結果は，…という可能性を示唆する

9. 述べる／報告する／結論する 【describe】

述べる	報告する	結論する
describe state mention note	report	conclude

使い分け

◆ **describe**, **state**, **mention**, **note** は「述べる」という意味に用いられるが，**note** には「注目する」という意味もある．

◆ **report** は「報告する」，**conclude** は「結論する」という意味で使われる．

頻度分析

	用例数		用例数
describe	12,462	❶ we describe ❷ (be) described in	3,766 723
state	18,733	❶ stated that	22
mention	89	❶ mentioned above	19
note	1,637	❶ note that	78
report	27,022	❶ we report	9,235
conclude	5,884	❶ conclude that	4,936

解説

◆ **describe**, **note**, **report**, **conclude** は，we が主語になることが多い．
◆ **state** は，第三者が主語になる場合が多い．
◆ **mention** は，受動態の用例が多い．

★★ describe（〜について述べる／〜を記述する）

他動詞．

❶ **we describe 〜**（われわれは，〜について述べる）

Here we describe the identification of a novel olfactomedin-related gene, named optimedin, located on chromosome 1p21 in humans.（*Hum Mol Genet. 2002 11:1291*）

訳 ここにわれわれは，…の同定について述べる

❷ **described in 〜**（〜において述べられる）

An improved mammalian two-hybrid system designed for interaction trap screening is described in this paper.（*Proc Natl Acad Sci USA. 2000 97:5220*）

訳 相互作用トラップスクリーニングのために設計された改良哺乳類二重ハイブリッドシステムがこの論文において述べられる

★ state（〜を述べる／状態）

名詞の用例が多いが，他動詞としても使われる．

❶ **stated that 〜**（〜ということを述べた／〜ということが述べられる）

Dobzhansky stated that nothing in biology makes sense except in the light of evolution.（*Plant Mol Biol. 2000 42:45*）

訳 Dobzhansky は，…ということを述べた

mention（〜を述べる）

他動詞．

❶ ～ mentioned above（上述された～）

Gene disruption analysis shows that, unlike the three genes mentioned above, alp31^{+} is dispensable for cell growth and division.（*Genetics. 2000 156:93*）

訳 上述された3つの遺伝子と異なって

★ note（～に注目する／～を述べる）

名詞の用例もあるが，他動詞として用いられることの方が多い．

❶ note that ～（～ということに注目する）

Finally, we note that sequence diversity in chromo domains may lead to diverse functions in eukaryotic gene regulation.（*EMBO J. 2001 20:5232*）

訳 最後に，われわれは…ということに注目する

★★ report（～を報告する／報告）

他動詞として使われることが多いが，名詞の用例もある．

❶ we report ～（われわれは，～を報告する）

Here we report that expression of Wnt or Dishevelled（Dvl）increased Akt activity.（*J Biol Chem. 2001 276:17479*）

訳 ここにわれわれは，…ということを報告する

★ conclude（～と結論する）

他動詞．"we conclude that" の用例が非常に多い．

❶ conclude that ～（～ということを結論する）

We conclude that the ability of NK cells to directly recognize and respond to viral products is important in mounting effective antiviral responses.（*J Immunol. 2004 172:138*）

訳 われわれは，…ということを結論する

10. 説明する 【explain】

説明する	例証する
explain account for	illustrate

使い分け ◆ **explain**, **account for** は,「説明する」という意味に用いられる.
◆ **illustrate** は,「例証する」という意味で使われる.

頻度分析

	用例数		用例数
explain	4,761	❶ explain the	1,298
		❷ explain why	266
account for	4,060	❶ account for the	995
illustrate	1,538	❶ illustrate the	349

解説 ◆ **explain** は,あとに why や how が続く用例も多いが,**account for**, **illustrate** にはほとんどない.

★ explain(～を説明する)

❶ **explain the ～**(～を説明する)

Two models are proposed to explain the observed differential sequence requirement for the two distinct stages of the protein priming reaction.(*J Virol. 2002 76:5857*)

訳 2つのモデルが,観察された…を説明するために提案される

❷ **explain why ～**(なぜ～かを説明する)

These findings may explain why caveolin-1 levels are normally down-regulated during lactation.(*J Biol Chem. 2001 276:48389*)

訳 これらの知見は,なぜ…かを説明するかもしれない

★ account for(～を説明する/～を占める/～を計算する)

"account for" という熟語で,他動詞として用いられる.

❶ **account for the ～**(～を説明する)

Incorporation of these changes into a model of neuronogenesis indicates that they are sufficient to account for the observed delay in radial expansion.(*J Neurosci. 2000 20:4156*)

訳 それらは,観察された…の遅延を説明するのに十分である

★ illustrate(～を例証する/～を説明する/～を図解する)

❶ **illustrate the ～**(～を例証する)

These results illustrate the importance of transmembrane helix 3 in CXCR4 signaling.(*J Biol Chem. 2002 277:24515*)

訳 これらの結果は膜貫通ヘリックス3の重要性を例証する

11. 証明する／確立する 【demonstrate】

証明する	確立する
demonstrate evidence document prove	establish

使い分け

◆ **demonstrate** は，we, results などを主語とし，that 節を伴って，「〜ということを実証する」という意味で用いられる．

◆ **evidence** は，as evidenced by の形で「〜によって証明されたように」という意味で使われることが多い．

◆ **document** は「〜を実証する／（実証したことを論文に）記述する」，**prove** は「〜であることが判明する」という意味に用いられる．

◆ **establish** は「確立する」という意味で使われる．

頻度分析

	用例数		用例数
demonstrate	40,699	❶ demonstrate that	13,339
evidence	17,099	❶ as evidenced by	528
document	1,989	❶（be）well documented	307
prove	2,044	❶ proved to be	253
establish	9,400	❶ established that	691

★★ demonstrate（〜を実証する）

他動詞．“demonstrate（s）／demonstrated that”の用例が圧倒的に多い．

❶ demonstrate that 〜（〜ということを実証する）

Here we demonstrate that the transcriptional coactivator p300 is a substrate of AMP-kinase.（*J Biol Chem. 2001 276:38341*）

訳 ここにわれわれは，…ということを実証する

★ evidence（〜を証明する／証拠）

名詞の用例が多いが，他動詞としても用いられる．

❶ as evidenced by 〜（〜によって証明されたように）

SLC-mediated antitumor responses were lymphocyte dependent as evidenced by the fact that this therapy did not alter tumor growth in

SCID mice. (*J Immunol. 2000 164:4558*)
訳 …という事実によって証明されたように

★ document（～を記述する／証拠を提供する／文書）

他動詞の用例が多いが，名詞としても用いられる．受動態の用例が多い．

❶ **well documented**（よく実証される／よく記述される）

Functional deficits in these cognitive processes have been well documented in patients with schizophrenia. (*Am J Psychiatry. 2004 161:1603*)
訳 これらの認知の過程における機能的な欠損は，統合失調症の患者においてよく記述されてきた

★ prove（判明する／～を証明する）

自動詞の用例が多いが，他動詞としても用いられる．

❶ **proved to be ～**（～であると判明した）

Prostacyclin has proved to be a beneficial treatment for patients with severe pulmonary hypertension. (*Circulation. 2000 102:3130*)
訳 プロスタサイクリンは，…の患者にとって有益な治療であると判明した

★ establish（～を確立する）

他動詞．

❶ **established that ～**（～ということが確立される／～ということを確立した）

Although it is well established that all four Notch genes can act as oncogenes, the mechanism by which Notch proteins transform cells remains unknown. (*Mol Cell Biol. 2001 21:5925*)
訳 …ということがよく確立されているけれども

12. 明らかにする 【reveal】

明らかにする		解く
reveal	clarify	solve
disclose	uncover	
elucidate	manifest	

使い分け
◆いずれの語も「明らかにする」という意味に用いられる．
◆ **solve** は「（X線解析によって三次元構造を）解く」という意味で用いられることが多いが，「（問題を）解決する」という意味でも用いられる．

共起・頻度分析

（数字：用例数）

直前の単語				直後の単語			
be動詞	to			by	that	the	a
347	0	revealed	14,139	**412**	**6,322**	819	**1,762**
0	311	reveal (s)	7,088	1	**2,379**	748	**1,514**
20	0	disclosed	128	3	**27**	8	13
0	7	disclose (s)	53	0	3	11	14
507	0	elucidated	751	31	2	50	10
0	**1,054**	elucidate (s)	1,427	1	3	**962**	59
56	0	clarified	91	5	2	8	0
0	**317**	clarify/clarifies	497	0	1	**334**	1
52	0	uncovered	303	5	6	20	**82**
0	**101**	uncover	247	0	2	46	**70**
150	0	manifested	461	**146**	0	8	18
60	26	manifest (s)	430	21	1	13	26
305	2	solved	587	**98**	1	**122**	5
0	**75**	solve (s)	131	0	0	**53**	13

解説
◆ **reveal**, **disclose** は，that 節を伴う用例が多い．
◆ **reveal**, **unconver** は，“a ＋名詞”を伴う用例も多い．
◆ **elucidate**, **clarify**, **uncover** は，“to elucidate ／ clarify ／ uncover the”の用例が多い．
◆ **manifest**, **solve** は，受動態で“manifested ／ solved by”の用例が多い．また，**manifest** は形容詞として使われることも多い．
◆ **reveal**, **disclose** は過去形，**elucidate**, **clarify**, **uncover** は to 不定詞の用例が多い．
◆ **reveal**, **disclose**, **elucidate**, **clarify**, **uncover** は，能動態の用例が多い．

★★ reveal（～を明らかにする）

他動詞．能動態の用例が非常に多く，受動態で用いられることはほとんどない．analysis, results, studies などを主語とし，人が主語となる用例は少ない．

❶ **revealed that ～**（～ということを明らかにした）

Sequence analysis revealed that the amplicons were identical to sequences of the *B. lonestari flaB* gene in GenBank.（*J Clin Microbiol.*

2003 41:5557)

訳 配列分析は，…ということを明らかにした

❷ **revealed a ～**（～を明らかにした）

Western blot analysis revealed a significant increase in both the 120-kDa (29%) and 80-kDa (69%) fragments in HF (P<0.05 versus control).（*Circ Res. 2003 92:897*）

訳 ウエスタンブロット分析は，…の有意な増加を明らかにした

disclose（～を明らかにする／～を開示する）

他動詞．

❶ **disclosed that ～**（～ということを明らかにした）

In vitro binding studies disclosed that Cer-1-P interacts directly with full-length $cPLA_2$ and with the CaLB domain in a calcium- and lipid-specific manner with a K_{Ca} of 1.54 microm.（*J Biol Chem. 2004 279:11320*）

訳 試験管内結合試験は，…ということを明らかにした

★ elucidate（～を解明する／～を明らかにする）

他動詞．to 不定詞の用例が非常に多い．

❶ **to elucidate ～**（～を解明するために／～を明らかにするために）

In the present study, MSA was used to elucidate the mechanisms of cell growth inhibition by selenium.（*Cancer Res. 2002 62:156*）

訳 MSA が，…の機構を解明するために使われた

clarify（～を明らかにする）

他動詞．to 不定詞の用例が非常に多い．

❶ **to clarify ～**（～を明らかにするために）

To clarify the role of JNK in tumorigenesis, we have investigated the role of JNK in a large panel of primary human brain tumors and tumor derived cell lines.（*Oncogene. 2002 21:5038*）

訳 腫瘍形成における JNK の役割を明らかにするために

uncover（～を明らかにする）

他動詞．

❶ **to uncover ～**（～を明らかにするために）

Recent work has begun to uncover the molecular mechanisms that underpin this process.（*Nat Cell Biol. 2001 3:E28*）

訳 最近の研究が，…する分子機構を明らかにするために始まった

❷ **uncovered a ～**（～を明らかにした）

Thus, we have uncovered a novel function of Cdc25B that serves as a steroid receptor coactivator in addition to its role as a regulator for cell cycle progression.（*Mol Cell Biol. 2001 21:8056*）

訳 われわれはの新規の機能を明らかにした

manifest（～を明らかにする／～を顕在化させる／著明な）……

他動詞として用いられることが多いが，形容詞の用例もある．

❶ **manifested by ～**（～によって明らかにされる）

Fibroblasts deficient in both c-Abl and p53 show reduced growth in culture, as manifested by reduction in the rate of proliferation, saturation density, and colony formation, compared with fibroblasts lacking p53 alone.（*Proc Natl Acad Sci USA. 2000 97:5486*）

訳 増殖速度の低下によって明らかにされたように

solve（～を解く／～を解決する）……………………………

他動詞．

❶ **solved by ～**（～によって解かれる）

The structure was solved by the multiple wavelength anomalous diffraction method using a 5-bromo-U DNA.（*Biochemistry. 2001 40:5587*）

訳 構造が多波長異常回折法によって解かれた

13. 確認する／検証する 【confirm】

確認する	
confirm	validate
verify	ascertain

使い分け

◆いずれの語も「（すでに知られていることを）確認する／立証する」場合に用いられる．

◆**validate** は，「（方法の精度などを）検証する／確認する」という意味に用いられることも多い．

共起・頻度分析 （数字：用例数）

直前の単語				直後の単語		
was/were	to			by	that	whether
1,900	2	confirmed	5,871	1,681	1,052	0
0	512	confirm (s)	2,253	4	801	7
237	0	verified	634	216	62	4
0	160	verify/verifies	248	1	50	0
237	0	validated	904	177	1	0
0	188	validate (s)	439	0	4	0
100	0	ascertained	269	45	4	2
0	254	ascertain (s)	264	0	2	77

解説

◆いずれの語も受動態でbyを伴う用例が多い．

◆**confirm**は，能動態でthat節を伴う用例も多い．

◆**ascertain**は，“to ascertain whether”の用例も多い．

★ confirm（～を確認する）

他動詞．

❶ **confirmed by ～**（～によって確認される）

This hypothesis was confirmed by the finding that the leader sequences are transcribed as parts of small RNAs encoded by genes located in the 5S rRNA clusters of Hydra.（*Proc Natl Acad Sci USA. 2001 98: 5693*）

訳 この仮説は，…という知見によって確認された

❷ **confirmed that ～**（～ということを確認した）

In the current study, we confirmed that NDGA induces a decrease in the fluorescence of thioflavin T associated with Aβ（1-40）fibrils and extended this observation to Aβ（1-40）protofibrils.（*Mol Pharmacol. 2004 66:592*）

訳 現在の研究において，われわれは…ということを確認した

verify（～を立証する／～を確認する）

他動詞．

❶ **verified by ～**（～によって立証される）

The chicken growth hormone（GH）structural peptide was identified, and the specific interaction was verified by coimmunoprecipitation.（*Proc Natl Acad Sci USA. 2001 98:9203*）

訳 特異的な相互作用が免疫共沈降によって立証された

★ validate（～を検証する／～を確認する）

他動詞.

❶ **validated by ～**（～によって検証される）

The reliability of the method is validated by comparison of anisotropic chemical shift and heteronuclear dipolar interactions from single site labeled samples.（*Biophys J. 2000 79:767*）

訳 その方法の信頼性は，…の比較によって検証される

ascertain（～を確認する／～を調べる）

他動詞．"to ascertain" の用例が非常に多い.

❶ **to ascertain whether ～**（～かどうか確認するために）

To ascertain whether ATP-dependent factors play a role in this process, we quantified virus-like particle（VLP）production by ATP-depleted cells.（*J Virol. 2001 75:5473*）

訳 ATP 依存性因子がこの過程において役割を果たすかどうかを確認するために

14. 支持する／強化する 【support】

支持する	強化する
support favor	reinforce strengthen

使い分け

◆ **support** は，「（仮説などを）支持する」という意味の他動詞として用いられる.

◆ **favor** は「好む」という意味の他動詞で，「（仮説などを）支持する」という意味で用いられることも多い.

◆ **reinforce**, **strengthen** は，「（仮説などを）強化する」ときに使われる.

頻度分析

	用例数		用例数
support	13,946	❶ support the	3,020
favor	1,455	❶ favor the	80
reinforce	391	❶ reinforce the	70
strengthen	364	❶ strengthen the	60

★★ support（～を支持する／支持）

名詞の用例もあるが，他動詞として使われることの方が多い．

❶ support the ～（～を支持する）

These results support the hypothesis that cellular enzyme (s) may catalyze the late steps of retroviral DNA integration.（*J Biol Chem. 2000 275:39287*）

訳 これらの結果は，…という仮説を支持する

★ favor（～を支持する／～を好む／支持）

名詞の用例もあるが，他動詞として使われることの方が多い．

❶ favor the ～（～を支持する）

The results favor the hypothesis that weakly polar pi–pi interactions exist between the aromatic group and the receptor.（*J Med Chem. 1999 42:3004*）

訳 結果は，…という仮説を支持する

reinforce（～を強化する）

❶ reinforce the ～（～を強化する）

These results reinforce the notion that mutant BACH1 participates in breast cancer development.（*Proc Natl Acad Sci USA. 2004 101:2357*）

訳 これらの結果は，…という考えを強化する

strengthen（～を強化する）

❶ strengthen the ～（～を強化する）

These data strengthen the hypothesis that the HCV–associated lymphomas are derived from clonally expanded B cells stimulated by HCV.（*Blood. 2001 97:1023*）

訳 これらのデータは，…という仮説を強化する

【I-C. 研究・発見・評価・達成】 研究・発見・評価・達成に関する動詞

15. 研究する／調査する／調べる 【study】

調査・研究する	調べる／分析する	問う
study	examine	ask
investigate	test	
survey	analyze	
explore	dissect	
search	look at	

使い分け

◆ **study**, **investigate** は，どちらも「研究する／調べる」という意味だが，**investigate** の方がより詳細に行うという意味合いがある．

◆ **survey** は「調査する」，**explore**, **search** は「探索する」という意味で使われる．

◆ **examine** は「調べる」，**test** は「テストする／検定する」，**analyze** は「分析する／解析する」，**look at** は「調べる」という意味に用いられる．

◆ **dissect** は本来「解剖する」という意味だが，「精査する／分析する」という意味で用いられることが多い．

◆ **ask** は「尋ねる」という意味だが，「問う／求める」という意味でも用いられる．

共起・頻度分析

（数字：用例数）

直前の単語					直後の単語		
were/was	we	to			by	for	whether
2,609	1,966	3	studied	9,007	592	174	58
3	162	3,276	study/studies	68,829	193	160	56
2,195	3,684	1	investigated	9,598	626	128	857
0	556	4,578	investigate (s)	5,597	1	1	445
50	53	0	surveyed	224	10	10	0
0	11	41	survey (s)	1,304	3	17	0
238	322	0	explored	1,388	71	35	57
0	216	1,146	explore (s)	1,690	0	0	51
57	128	0	searched	306	7	155	0
0	4	248	search (s)	2,202	14	688	0
3,348	4,356	1	examined	13,502	752	433	605
0	678	3,389	examine	4,829	1	3	281

直前の単語					直後の単語		
were/was	we	to			by	for	whether
1,808	1,898	0	tested	8,285	364	898	406
0	201	3,213	test (s)	10,582	14	646	504
2,606	1,299	0	analyzed	6,705	1,026	542	14
0	195	1,019	analyze (s)	1,481	2	4	12
71	18	0	dissected	230	7	5	0
0	14	260	dissect (s)	339	0	0	0
0	14	0	looked at	29	0	0	0
0	7	30	look (s) at	70	0	0	1
76	235	0	asked	443	1	4	216
0	35	47	ask (s)	154	0	1	55

解説

◆ **study**, **investigate**, **explore**, **examine**, **test**, **analyze**, **dissect**, **look at** は，to 不定詞として用いられることが多い．

◆ **studied**, **investigated**, **explored**, **examined**, **tested**, **analyzed**, **looked at**, **asked** は，we を主語とする用例が多い．

◆ **study**, **investigate**, **search**, **examine**, **test**, **analyze** は，受動態で後に by や for を伴う用例も多い．

◆ **study**, **survey**, **search** は名詞の用例が多い．

☆ study（～を研究する／～を調べる／研究）

名詞の用例が多いが，他動詞としても用いられる．

❶ to study ～（～を研究するために）

To study the role of Oct-1 in these processes, the lymphoid compartment of RAG-$1^{-/-}$ animals was reconstituted with Oct-1-deficient fetal liver hematopoietic cells. (*Proc Natl Acad Sci USA. 2004 101:2005*)

訳 これらの過程における Oct-1 の役割を研究するために

❷ we studied ～（われわれは，～を研究した）

Here we studied the effects of PKC-mediated phosphorylation on purified recombinant wild-type Cx43 and a PKC-unresponsive mutant (S368A). (*J Biol Chem. 2004 279:20058*)

訳 ここでわれわれは，…に対する PKC に仲介されるリン酸化の効果を研究した

☆ investigate（～を精査する／～を研究する／～を調べる）

他動詞．

❶ **to investigate ～**（～を精査するために）

To investigate whether this phenotype involves cell-cell interaction defects, we performed analysis of genetically mosaic animals.（*J Neurosci. 2001 21:6745*）

訳 …かどうかを精査するために

❷ **we investigated ～**（われわれは，～を精査した）

In the current study, we investigated the role of trans-heterozygous mutations in mouse models of polycystic kidney disease.（*Hum Mol Genet. 2002 11:1845*）

訳 現在の研究において，われわれは…の役割を精査した

★ survey（～を調査する／調査）

名詞の用例が多いが，他動詞としても用いられる．

❶ **we surveyed ～**（われわれは，～を調査した）

We surveyed the expression patterns of 13,977 mouse genes in male and female hypothalamus, kidney, liver, and reproductive tissues.（*Dev Cell. 2004 6:791*）

訳 われわれは13,977個のマウスの遺伝子の発現パターンを調査した

★ explore（～を探索する／～を調査する）

他動詞．

❶ **to explore ～**（～を探索するために）

The experiments here were undertaken to explore the role of Kv1 α-subunits in the generation of voltage-gated K^+ currents in SCG neurons.（*J Neurosci. 2001 21:8004*）

訳 実験が，…の役割を探索するために着手された

★ search（～を探索する／探す／探索）

名詞の用例が多いが，動詞としても用いられる．“search（ed） for” などの自動詞の用例が多い．

❶ **searched for ～**（～を探索した）

Because hyperdiploid ALL samples also show high-level expression of FLT3, we searched for the presence of FLT3 mutations in leukemic blasts from 71 patients with ALL.（*Blood. 2004 103:3544*）

訳 われわれは白血病芽球におけるFLT3変異の存在を検索した

★★ examine（～を調べる）

他動詞．

❶ we examined ～（われわれは，～を調べた）

In this study, we examined the effect of UV irradiation on the development of lymphoid malignancies in mice with no or only one functional copy of p53.（*Proc Natl Acad Sci USA. 2001 98:9790*）

訳 この研究において，われわれは…の発症に対するUV照射の影響を調べた

❷ to examine ～（～を調べるために）

To examine the role of the VP1 N terminus in infection, we altered that sequence in CPV, and some of those changes made the capsids inefficient at cell infection.（*J Virol. 2002 76:1884*）

訳 感染におけるVP1のN末端の役割を調べるために

★★ test（～をテストする／～を検定する／～を検査する／テスト）…

他動詞の用例が多いが，名詞としても用いられる．

❶ to test ～（～をテストすること／～をテストするために）

The purpose of this study was to test the hypothesis that AA could protect RGCs from glutamate neurotoxicity.（*Invest Ophthalmol Vis Sci. 2002 43:1835*）

訳 この研究の目的は，…という仮説をテストすることであった

❷ tested for ～（～についてテストされる）

The mutations were tested for their ability to rescue faulty *N*-linked glycosylation of carboxypeptidase Y in an ALG6-deficient *Saccharomyces cerevisiae* strain.（*Am J Pathol. 2000 157:1917*）

訳 変異体は，…するそれらの能力についてテストされた

★ analyze（～を分析する）

他動詞．

❶ we analyzed ～（われわれは，～を分析した）

In this study, we analyzed the effects of SIV infection on apoptotic pathways in thymic tissue from newborn macaques infected with SIV.（*J Immunol. 2000 165:3461*）

訳 この研究において，われわれは…に対するSIV感染の影響を分析した

❷ analyzed by ～（～によって分析される）

These data were analyzed by using time series methods.（*Am J*

Epidemiol. 2000 152:558）

訳 これらのデータは，時系列法を使うことによって分析された

❸ **to analyze ～**（～を分析するために）

The method was used to analyze 34 bacterial and archaeal genomes, and yielded more than 7600 pairs of genes that are highly likely（$P >/= 0.98$）to belong to the same operon.（*Nucleic Acids Res. 2001 29:1216*）

訳 その方法が，34の細菌と古細菌のゲノムを分析するために使われた

❹ **analyzed for ～**（～について分析される）

The blood samples were analyzed for IL-1A ＋4845 and IL-1B ＋3954 polymorphisms using polymerase chain reaction（PCR）-based methods.（*J Periodontol. 2000 71:164*）

訳 血液サンプルはIL-1A ＋4845とIL-1B ＋3954の多型性について分析された

dissect（～を精査する／～を分析する）

他動詞.

❶ **to dissect ～**（～を精査するために）

This assay system was also used to dissect the molecular mechanisms of lysosome exocytosis.（*Blood. 2000 96:1782*）

訳 このアッセイシステムは，…の分子機構を精査するためにも使われた

look at（～を調べる）

"look at"という熟語で，他動詞として用いられる.

❶ **look at ～**（～を調べる）

With the advent of microarray technology, it has become possible to look at changes in gene expression profiles in a biological process on an unprecedented scale.（*Genomics. 2003 82:109*）

訳 生物学的過程における遺伝子発現プロファイルの変化を調べることが可能になってきた

ask（～を問う／～を求める）

他動詞.

❶ **we asked ～**（われわれは，～を問うた）

In the present study, we asked whether this protective effect was attributable to the generation of one of the catabolic products of HO-1, carbon monoxide（CO）.（*J Immunol. 2001 166:4185*）

訳 現在の研究において，われわれは…かどうかを問うた

16. 見つける／観察する 【find】

見つける	観察する	注目する
find identify detect discover	observe see	note

使い分け

◆ **find** は「見つける」という意味で，もっとも広い範囲に用いられる．

◆ **identify**, **detect** は，「（現象などを）同定する／検出する」という意味で使われる．

◆ **discover** は「（重要な現象を）発見した」ときに使われるが，用例の数は多くない．

◆ **observe**, **see** は，主に受動態で「観察される」という意味で用いられる．

◆ **note** は，主に受動態で「（観察された内容が）注目される／（現象などが）認められる」ときに使われる．

共起・頻度分析 （数字：用例数）

直前の単語					直後の単語			
was/were	we	to			that	in	to	as
10,247	8,246	1	found	31,229	9,576	6,901	7,235	68
0	3,517	409	find (s)	4,755	3,346	19	3	3
5,533	2,795	3	identified	24,000	447	2,789	179	2,072
0	910	6,232	identify/identifies	10,519	14	19	1	11
4,811	358	1	detected	10,948	20	4,273	19	129
0	64	1,752	detect (s)	2,616	0	18	0	16
280	273	0	discovered	1,539	347	168	37	38
0	10	165	discover (s)	197	6	1	0	0
8,704	1,756	20	observed	24,938	1,072	7,610	517	131
0	384	165	observe (s)	728	127	20	0	0
1,734	1	1	seen	5756	11	3,012	18	84
0	40	131	see (s)	356	7	3	1	1
797	96	0	noted	1353	101	373	27	11
0	56	20	note (s)	264	78	7	1	1

解説
◆いずれの語も受動態でinを伴う用例が多い．
◆**find**, **discover**, **note**は，that節を目的語にする用例も多い．
◆**find**は，受動態でto不定詞を伴う用例も多い．
◆**identify**は，受動態でasを伴う用例も多い．

⁑ find（～を見つける）

他動詞．“found that”“found to［*do*］”“found in”の用例が多い．

❶ found that ～（～ということを見つけた）

We found that all classes of Golgi components are dynamically associated with this organelle, contrary to the prediction of the stable organelle model.（*J Cell Biol. 2001 155:557*）

訳 われわれは，…ということを見つけた

❷ found to ～（～することが見つけられる）

In these assays, R391 was found to have a stronger effect on SXT stability than vice versa.（*J Bacteriol. 2001 183:1124*）

訳 R391はSXTに対してより強い効果を持つことが見つけられた

❸ found in ～（～において見つけられる）

No significant differences were found in VEGF levels by different phases of CML（*P*=.1）.（*Blood. 2002 99:2265*）

訳 有意な違いは，…には見つけられなかった

⁑ identify（～を同定する）

他動詞．“identified in”“identified as”の用例が多い．

❶ identified in ～（～において同定される）

Seven novel mutations were identified in 10 families, with one additional family found to harbor one of the two previously described mutations.（*Blood. 1999 93:2261*）

訳 7つの新規の変異が10家系において同定された

❷ identified as ～（～として同定される）

In a search for novel non-TZD ligands for PPARγ, T0070907 was identified as a potent and selective PPARγ, antagonist.（*J Biol Chem. 2002 277:19649*）

訳 T0070907は強力で選択的なPPARγ拮抗物質として同定された

⁑ detect（～を検出する）

他動詞．“detected in”の用例が多い．

❶ **detected in ～**（～において検出される）

AHR mRNA expression was detected in all tissue types tested: adductor muscle, digestive gland, foot, gill, gonad, mantle, and siphon. (*Gene. 2001 278:223*)

訳 AHR メッセンジャー RNA 発現は調べられたすべての組織タイプにおいて検出された

★ discover（～を発見する）

他動詞．"we（have）discovered that" の用例が多い．

❶ **discovered that ～**（～ということを発見した）

In addition, we discovered that human CD14 is highly expressed in hepatocytes. (*J Biol Chem. 2000 275:36430*)

訳 われわれは，…ということを発見した

❷ **discovered in ～**（～において発見される）

A new cataract mutation was discovered in an ongoing program to identify new mouse models of hereditary eye disease. (*Genomics. 2000 63:314*)

訳 新しい白内障の変異が進行中のプログラムにおいて発見された

★★ observe（～を観察する）

他動詞．受動態の用例が非常に多い．"observed in" の用例が多い．

❶ **observed in ～**（～において観察される）

None of these effects was observed in the presence of 7E-peptide or 7Cya-peptide. (*Invest Ophthalmol Vis Sci. 2001 42:1439*)

訳 これらの効果のどれも，…の存在下では観察されなかった

★ see（～を見る）

他動詞．受動態の用例が非常に多い．"seen in" の用例が非常に多い．

❶ **seen in ～**（～において見られる）

A similar effect was seen in rats after chronic self-administration of heroin. (*Proc Natl Acad Sci USA. 2000 97:7579*)

訳 同じような効果が，…の後でラットにおいて見られた

★ note（～を認める／～に注目する／～を述べる）

他動詞．"noted in" の用例が多い．

❶ noted in ～（～において認められる／～において注目される）

Increased ET-1 peptide expression was noted in the renal vasculature and in the cortical tubular epithelium of kidneys exposed to I/R.（*Transplantation. 2001 71:211*）

訳 上昇したET-1ペプチド発現が腎臓の脈管構造において認められた

17. 評価する／比較する　【assess】

評価する	比較する
assess evaluate estimate	compare

使い分け

- ◆ **assess**, **evaluate** は，「評価する」という意味で用いられる．
- ◆ **estimate** は「推定する／見積もる」という意味だが，「評価する」という意味合いも持つ．
- ◆ **compare** は，「比較する」という意味で使われる．

頻度分析

	用例数		用例数
assess	9,879	❶ to assess	2,946
evaluate	8,891	❶ to evaluate	2,327
estimate	6,251	❶（be）estimated by	244
compare	28,566	❶ compared with	15,619
		❷ compared to	5,508

★★ assess（～を評価する）

他動詞．

❶ to assess ～（～を評価するために）

To assess the role of TRAF6 in p75 signaling, we analyzed mice with this gene deleted.（*J Neurosci. 2004 24:10521*）

訳 …におけるTRAF6の役割を評価するために

★ evaluate（～を評価する）

他動詞．

❶ **to evaluate ～（～を評価するために）**

To evaluate the potential role of these factors in the loss of secretory function of exocrine tissues, a panel of monoclonal and polyclonal antibodies was developed for passive transfer into the NOD animal model.（*Arthritis Rheum. 2000 43:2297*）

訳 …におけるこれらの因子の潜在的な役割を評価するために

★ estimate（～を評価する／～を推定する／～を見積もる）

他動詞．

❶ **estimated by ～（～によって評価される／～によって推定される）**

Subject lung function was estimated by calculating the ratio of forced expiratory volume（FEV）after 1 second（FEV1）/forced vital capacity（FVC）.（*J Periodontol. 2001 72:50*）

訳 被検者の肺機能は，…の割合を算出することによって評価された

★★ compare（～を比較する）

他動詞．受動態の用例が非常に多い．“compared with”“compared to”の用例が特に多い．

❶ **compared with ～（～と比較して）**

Results were compared with those in normal control subjects and patients with Wolman disease（WD）.（*Invest Ophthalmol Vis Sci. 2001 42:1707*）

訳 結果はコントロール被検者のそれと比較された

18. 決定する／特徴づける 【determine】

決定する	特徴づける
determine define	characterize mark feature

使い分け

◆ **determine** は，「決定する」という意味の他動詞として用いられる．
◆ **define** は，「決定する／定義する」という意味で使われる．
◆ **characterize**, **mark** は「～を特徴づける」，**feature** は「～を特徴とする」という意味の他動詞として用いられる．

頻度分析

	用例数		用例数
determine	27,883	❶ to determine	9,957
		❷ (be) determined by	4,005
define	10,669	❶ (be) defined by	976
		❷ (be) defined as	912
characterize	13,659	❶ (be) characterized by	4,074
mark	3,572	❶ (be) marked by	178
feature	7,417	❶ features a	51

⁂ determine（～を決定する）

他動詞．

❶ to determine ～（～を決定するために）

To determine whether the presence of substrates affects the monomer/dimer equilibrium, further ultracentrifugation studies were performed.（*J Biol Chem. 2001 276:7727*）

訳 …かどうかを決定するために

❷ determined by ～（～によって決定される）

ACVR2 expression was determined by immunohistochemistry using an antibody targeting an epitope beyond the predicted truncated protein.（*Gastroenterology. 2004 126:654*）

訳 ACVR2発現が…を使う免疫組織化学によって決定された

⁂ define（～を決定する／～を定義する／～を明らかにする）

他動詞．"defined by" "defined as" の用例が多い．

❶ defined by ～（～によって決定される）

These specimens belong to the same adenocarcinoma subgroup as defined by clustering of gene expression data.（*Cancer Res. 2002 62:7001*）

訳 遺伝子発現データのクラスタリングによって決定されるように

❷ defined as ～（～として定義される）

Progression of periodontal disease was defined as the percentage of teeth per decade that increased ABL by > or = 40%, and the percentage of teeth per decade that developed CAL > or = 5 mm.（*Periodontol. 2003 74:161*）

訳 歯周病の進行は，…の割合として定義された

★★ characterize（～を特徴づける）

他動詞．

❶ characterized by ～（～によって特徴づけられる）

Mutations in CREBBP cause Rubinstein-Taybi syndrome, which is characterized by mental retardation, skeletal abnormalities and congenital cardiac defects.（*Nat Genet. 2001 29:469*）

訳 そして，それは精神遅滞，骨格異常および先天的心臓欠陥によって特徴づけられる

★ mark（～を特徴づける／～を標識する）

他動詞．"marked（顕著な）"の形容詞的用法が非常に多い．

❶ marked by ～（～によって特徴づけられる）

Alzheimer's disease is marked by progressive accumulation of amyloid β-peptide（Aβ）which appears to trigger neurotoxic and inflammatory cascades.（*J Immunol. 2003 171:2216*）

訳 アルツハイマー病は，アミロイドβペプチド（Aβ）の進行性の蓄積によって特徴づけられる

★ feature（～を特徴とする／特徴）

名詞の用例が多いが，他動詞としても用いられる．

❶ feature a ～（～を特徴とする）

The microchannel plate features a novel injector for uniform sieving matrix loading as well as high resolution, tapered turns that provide an effective separation length of 15.9 cm on a compact 150-mm diameter wafer.（*Proc Natl Acad Sci USA. 2002 99:574*）

訳 マイクロチャネルプレートは，…のための新規のインジェクターを特徴とする

19. 促す／可能にする／許す　【allow】

促す	可能にする／許す	承認する
lead prompt	enable allow permit make it possible	approve

使い分け

◆ **lead**, **prompt**, **enable**, **allow** は，「（人が）～することを促す／～することを可能にする」という意味に用いられる．

◆ **permit** は，能動態で「許す／可能にする」という意味で使われる．

◆ **make it possible to** は，「～することを可能にする」という意味に用いられる．

◆ **approve** は受動態の用例が圧倒的に多く，「（薬剤が）承認される」という意味で使われることが多い．

頻度分析

	用例数		用例数
lead	18,824	❶ led us to	238
prompt	510	❶ prompted us to	127
enable	2,685	❶ enabled us to	138
allow	9,859	❶ allows for	508
		❷ allowed us to	436
permit	1,918	❶ permits the	141
make it possible to	212	❶ makes it possible to	73
approve	286	❶ (be) approved for	71

★★ lead（～を仕向ける／つながる／導く）

“lead to（～につながる）”の形で自動詞として用いられることが多いが，他動詞としても使われる．他動詞としては“lead us to［*do*］”の用例が多い．lead／led／led の語形変化をする．

❶ led us to ～（われわれに～させた）

This result led us to hypothesize that carnosine may modulate the neurotoxic effects of zinc and copper as well.（*Brain Res. 2000 852:56*）

訳 この結果は，われわれに…ということを仮定させた

prompt（～を促す／迅速な）

他動詞．“prompt us to［*do*］”の用例が多い．

❶ prompted us to ～（われわれが～することを促した）

These observations prompted us to investigate possible interactions between the ErbBs and the TGF-α precursors in CHO cells.（*Oncogene. 2000 19:3172*）

訳 これらの知見は，われわれが…の間のありうる相互作用を精査することを促した

★ enable（～を可能にする）

他動詞．"enable us／them to［*do*］" の用例が多い．

❶ enabled us to ～（われわれが～することを可能にした）

This approach enabled us to identify previously unknown mutations in the receptor tyrosine kinase gene EPHB2.（*Nat Genet. 2004 36:979*）

訳 このアプローチはわれわれが以前に知られていない変異を同定することを可能にした

★★ allow（～を許す／～を可能にする）

他動詞および自動詞の両方で用いられる．"allow us／them／one／it to［*do*］" "allows for" の用例が多い．

❶ allow for ～（～を可能にする）

We have developed a new chemical inducible genetic system that allows for the isolation of any cDNA molecule from *in vitro* generated genomic transgenes in transgenic plants.（*Plant J. 2002 32:615*）

訳 …の単離を可能にするシステム

❷ allowed us to ～（われわれが～することを可能にした）

This approach has allowed us to identify homologs to ymf58（nad4L）, ymf62（nad6） and ymf60（rpl6）.（*Nucleic Acids Res. 2003 31:1673*）

訳 このアプローチは，われわれが…を同定することを可能にした

★ permit（～を許す／～を可能にする）

❶ permit the ～（～を可能にする）

A simple modification of this strategy that permits the generation of conventional gene knockout, conditional gene knockout and conditional gene repair alleles using one targeting construct is discussed.（*Nucleic Acids Res. 2001 29:E10*）

訳 …の産生を可能にするこの戦略の単純な修正

make it possible to（～することを可能にする）

❶ make it possible to ～（～することを可能にする）

This technique makes it possible to examine the expression of thousands of genes simultaneously.（*Oncogene. 2003 22:6497*）

訳 この技法は同時に数千の遺伝子の発現を調べることを可能にする

approve（～を承認する／～を認可する）

他動詞.

❶ approved for ～（～に対して承認される）

Etanercept and infliximab are tumor necrosis factor（TNF）antagonists that have been recently approved for the treatment of rheumatoid arthritis（RA）and Crohn's disease（CD）.（*Arthritis Rheum. 2002 46:3151*）

訳 エタネルセプトとインフリキシマブは，…の治療に対して最近承認された腫瘍壊死因子（TNF）拮抗薬である

20. 役立つ／助ける 【help】

役立つ	助ける
help **serve**	**help** **assist** **aid**

使い分け

◆ **help** to［*do*］は，「～するのに役立つ／～するのを助ける」という意味で用いられることが多い．また，to は省略されることの方が多い．

◆ **serve** as は，「～として役立つ／～として働く」という意味で使われる．

◆ **assist**, **aid** は「助ける」という意味で，**assist** は自動詞・他動詞の両方，**aid** は他動詞として用いられる．

頻度分析

	用例数		用例数
help	2,849	❶ help to	528
		❷ help explain	126
serve	4,760	❶ serve as	1,916
		❷ serve to	332
assist	1,446	❶ assist in	172
aid	923	❶ to aid	128

★ help（～に役立つ／～を助ける／援助）

名詞の用例もあるが，他動詞として使われることが多い．“help to［*do*］”の用例が

多いが，to はしばしば省略される．

❶ help to ～（～するのに役立つ）

These results may help to explain why tissue damage in Buruli ulcer is not accompanied by an acute inflammatory response.（*Infect Immun. 2000 68:877*）

訳 これらの結果は，…を説明するのに役立つかもしれない

❷ help explain ～（～を説明するのに役立つ）

These findings may help explain the formation of gingival pockets between cementum and periodontal epithelium, a hallmark of periodontitis.（*Infect Immun. 2002 70:5846*）

訳 これらの知見は，…を説明するのに役立つかもしれない

★ serve（役立つ／役目をする／働く）

自動詞．

❶ serve as ～（～として役立つ／～として働く）

This study may serve as a model for the development of sensitive and "nonradioactive" immunoassays for peptides, including polypeptide tumor markers.（*Anal Chem. 2002 74:5507*）

訳 この研究は，…の開発のためのモデルとして役立つかもしれない

❷ serve to ～（～する役目をする）

We suggest that a calcium gradient may serve to regulate the timing of vesicle uncoating.（*J Biol Chem. 2001 276:34148*）

訳 カルシウム濃度勾配は，…のタイミングを調節する役目をするかもしれない

★ assist（助ける／援助）

自動詞・他動詞の両方で用いられるが，自動詞の用例の方が多い．名詞としても用いられる．

❶ assist in ～（～の助けになる）

The formula and look-up tables based on the formula, can be used to assist in the design of microarray experiments.（*Bioinformatics. 2004 20:2821*）

訳 …は，マイクロアレイ実験の設計を助けるために使われうる

aid（～を助ける／援助）

名詞の用例が多いが，他動詞としても用いられる．

❶ **to aid ～**（～を助けるために）

We did a pilot study to establish whether it can be used to aid diagnosis of necrotising enterocolitis in preterm infants.（*Lancet. 2003 361:310*）

訳 われわれは，それが…の診断を助けるために使われうるかどうかを確立するために予備研究を行った

21. 達成する／成す 【achieve】

達成する	成す／つくる
achieve accomplish	make

使い分け
◆ **achieve**, **accomplish** は，「達成する」という意味に用いられる．
◆ **make** は，「～を成す／つくる」という意味で使われる．

頻度分析

	用例数		用例数
achieve	4,848	❶ to achieve	1,007
		❷ (be) achieved by	771
accomplish	858	❶ (be) accomplished by	267
make	8,319	❶ (be) made in	417

★ achieve（～を達成する／～を成し遂げる）

他動詞．

❶ **to achieve ～**（～を達成するために）

Only moderate exposure times and laser powers were required to achieve efficient dissociation.（*Anal Biochem. 2004 326:200*）

訳 レーザーの出力が，…を達成するために必要とされた

❷ **achieved by ～**（～によって達成される）

More specific blockage of CD38 expression was achieved by using morpholino antisense oligonucleotides targeting its mRNA, which produced a corresponding inhibition of differentiation as well.（*J Biol Chem. 2002 277:49453*）

訳 CD38 発現のより特異的な遮断が，…を使うことによって達成された

accomplish（～を達成する／～を成し遂げる）

他動詞．受動態の用例が多い．

❶ accomplished by ～（～によって達成される）

Cloning of the sal1 gene was accomplished by using Mu tagging, and the identity of the cloned gene was confirmed by isolating an independent sal1-2 allele by reverse genetics.（*Proc Natl Acad Sci USA. 2003 100:6552*）

訳 sal1 遺伝子のクローニングが，…を使うことによって達成された

★ make（～を成す／～をつくる／～を行う／～にする）

他動詞．「つくる」に関係するかなり広い意味で用いられる．

❶ made in ～（～において成される）

Considerable progress has been made in understanding the molecular basis of vernalization in *Arabidopsis*.（*Development. 2004 131:3829*）

訳 シロイヌナズナの春化の分子基盤の理解におけるかなりの進歩が成された

【Ⅱ-A. 発生・由来】 事象の発生・由来・獲得に関係する動詞

1. 現れる 【emerge】

現れる	発現する
emerge appear	express

使い分け
- emerge は「（姿が）現れる」という意味で用いられるが，「（世の中に）登場する」という意味もある．
- appear は「～のように思われる」の意味で使われることが多いが，「現れる」という意味にも用いられる．
- expressed は，「（遺伝子などが）発現する」場合に使われる．
- emerge, appear は自動詞，express は他動詞として用いられる．

頻度分析

	用例数		用例数
emerge	1,943	❶ emerged as	283
appear	13,005	❶ appeared in	148
express	39,161	❶ (be) expressed in	9,029

★ emerge（出現する／現れる）

自動詞．“emerged as”の用例が多い．

❶ emerged as ～（～として現れた）

Interleukin (IL)-13 has emerged as a central mediator of T helper cell (Th) 2-dominant immune responses, exhibiting a diverse array of functional activities including regulation of airway hyperreactivity, resistance to nematode parasites, and tissue remodeling and fibrosis. (*J Exp Med. 2003 197:687*)

訳 インターロイキン（IL）-13は，…の中心的なメディエーターとして現れた

★★ appear（思われる／現れる）

自動詞．“appear to be（～であるように思われる）”の用例が非常に多い（**動詞編** Ⅰ-A. 2.参照）．

❶ appeared in ～（～に現れた）

During myoblast fusion this level strongly increased, p204 became phosphorylated, and the bulk of p204 appeared in the cytoplasm of the myotubes.（*Mol Cell Biol. 2000 20:7024*）

訳 大量のp204が筋管の細胞質に現れた

★★ express（～を発現させる）

他動詞．受動態で“(be) expressed in”の用例が多い．

❶ expressed in ～（～において発現する）

We show that Hes6 is expressed in the murine embryonic myotome and is induced on C2C12 myoblast differentiation *in vitro*.（*Development. 2002 129:2195*）

訳 Hes6はマウスの胚性筋分節において発現する

2. 起こる／～が生じる／発生する【occur】

起こる／生じる	発生する
occur take place arise	develop

使い分け

◆ **occur**, **take place** は，「(現象が) 起こる」場合に使われる．

◆ **arise** は，「(組織や腫瘍などが) 発生する」という意味に用いられることが多い．

◆ **develop** は，「(生物が) 発生する」という意味で使われる．

頻度分析

	用例数		用例数
occur	20,310	❶ occurs in	1,763
take place	638	❶ takes place in	76
arise	3,172	❶ arise in	137
develop	18,466	❶ mice developed	410

解説

◆いずれの語も後にinを伴う自動詞的な用法が多い．

★★ occur（起こる／生じる）

自動詞．“occur in”の用例が多い．

❶ occur in ～（～において起こる）

The first step occurs in the inner or sensorial layer of the non-neural ectoderm where a subset of cells are chosen to differentiate into ciliated-cell precursors.（*Development. 1999 126:4715*）

訳 最初のステップは非神経性外胚葉の内側あるいは感覚層において起こる

★ take place（起こる）

❶ take place in ～（～において起こる）

This process takes place in all eukaryotic cells.（*Science. 2000 290:1717*）

訳 この過程はすべての真核生物細胞において起こる

★ arise（生じる）

自動詞. "arise from（～から生じる）" の用例が非常に多い（Ⅱ-A. 3. 参照）.

❶ arise in ～（～において生じる／～において発生する）

The earliest erythroblasts arise in yolk sac blood islands and subsequently enter the embryo proper to initiate circulation.（*Circ Res. 2003 92:133*）

訳 もっとも初期の赤芽球は卵黄嚢血島において発生する

★★ develop（発生する／～を発症する／～を開発する）

「発生する」の意味では自動詞として,「発症する／開発する」の意味では他動詞として用いられることが多い.

❶ mice developed（マウスが発生した）

Transgenic mice developed normally with no increased mortality and displayed normal body weight, blood glucose levels, and islet architecture.（*Proc Natl Acad Sci USA. 2002 99:16992*）

訳 トランスジェニックマウスは死亡率の上昇なしに正常に発生した

3. ～に由来する／～から生じる　【derive】

生じる／由来する	引き出す／由来する	出現する／由来する
derive	derive	emerge
originate	draw	
stem		
arise		
come from		

使い分け

◆ (be) **derived** from ～, **originate** from ～は「～に由来する」という意味で，細胞・組織などの発生に関して使われることが多い．

◆ **stem** from, **arise** from, **come from**, **emerge** from は，「(原因となる現象や根拠となる研究) から生じる／に由来する」という意味で用いられることが多い．

◆ (be) **drawn** from は，「(研究対象や材料などが) ～に由来する」という意味で使われる．

◆ **emerge** は「現れる」という意味だが，**emerge** from ～で「～から生じる／～に由来する」という意味にもなる．

◆ **originate**, **stem**, **arise**, **come**, **emerge** は自動詞として，**derive** は他動詞あるいは自動詞として使われる．

頻度分析

	用例数		用例数
derive	13,667	❶ (be) derived from	5,257
		❷ derive from	181
originate	1,058	❶ originate from	181
stem	5,933	❶ stem from	51
arise	3,172	❶ arise from	617
come from	300	❶ comes from	116
draw	498	❶ (be) drawn from	112
emerge	1,943	❶ emerged from	74

解説 ◆いずれの語も後に from を伴う場合が多い．

★ derive（～を引き出す／由来する）

“[*be*] derived from”の形の受動態で用いられることが非常に多い．他動詞として使われることが多いが，自動詞の用例もある．

❶ **derived from 〜（〜に由来する）**

Three peptide epitopes derived from LCMV have been shown to bind the mouse class I molecule H-2 Db and to stimulate CTL responses in LCMV-infected mice.（*J Exp Med. 1998 187:1647*）

訳 LCMVに由来する3つのペプチド抗原決定基は，…に結合すると示されている

❷ **derive from 〜（〜に由来する）**

We now demonstrate that CD8low cells derive from a proliferative compartment, but do not divide *in vivo*.（*J Immunol. 1999 163:155*）

訳 CD8low細胞は増殖性コンパートメントに由来する

★ originate（由来する／生じる／始まる）

自動詞．"originate from" の用例が多い．

❶ **originate from 〜（〜に由来する／〜から生じる）**

We thus conclude that murine Merkel cells originate from the neural crest.（*Dev Biol. 2003 253:258*）

訳 マウスのメルケル細胞は神経堤に由来する

★ stem（生じる／由来する／幹）

名詞の用例が多いが，自動詞としても使われる．

❶ **stem from 〜（〜から生じる／〜に由来する）**

Amyloid and prion diseases appear to stem from the conversion of normally folded proteins into insoluble, fiber-like assemblies.（*Biochemistry. 2001 40:9089*）

訳 アミロイドとプリオンの疾患は，正常に折りたたまれたタンパク質の…への転換から生じるように思われる

★ arise（生じる／由来する）

自動詞．"arise from" の用例が非常に多い．

❶ **arise from 〜（〜から生じる／〜に由来する）**

These phenotypes arise from a defect in ventral folding morphogenesis that occurs normally around E8.0.（*Genes Dev. 1999 13:1475*）

訳 これらの表現型は，…の欠陥から生じる

come from（～に由来する／～から生じる）

❶ come from ～（～に由来する／～から生じる）

Evidence of this fact comes from several studies that document a lack of nucleotide diversity in the *Y. pestis* genome.（*J Clin Microbiol. 2001 39:3179*）

訳 この事実の証拠はいくつかの研究に由来する

draw（～を引き出す）

他動詞.

❶ drawn from ～（～から得られた）

Data were drawn from a 21-year longitudinal birth cohort study (N=1,265).（*Am J Psychiatry. 2004 161:88*）

訳 データは21年の長期出生コホートの研究から得られた

★ emerge（出現する／現れる）

自動詞. "emerged as" の用例の方が多い.

❶ emerged from ～（～から出現した／～に由来した）

Insights into end-stage renal disease have emerged from many investigations but less is known about the epidemiology of chronic renal insufficiency (CRI) and its relationship to cardiovascular disease (CVD).（*J Am Soc Nephrol. 2003 14:S148*）

訳 最終ステージの腎疾患に対する洞察は多くの研究に由来した

4. ～に起因する／帰する 【result from】

起因する	帰する
result from	attribute ascribe

使い分け

◆動詞編II-A. 3. の「～に由来する／～から生じる」に意味が近いが，原因を強調したいときに用いられる.

◆**result from** は，「～に起因する／～の結果生ずる」という意味で用いられる.

◆(be) **attributed** to, (be) **ascribed** to は，「～に起因すると考えられる」という意味で使われる.

◆ **result** は自動詞，**attribute**, **ascribe** は他動詞として使われる．

頻度分析	用例数		用例数
result from	5,071	❶ result from	1,296
attribute	2,020	❶（be）attributed to	1,439
ascribe	321	❶（be）ascribed to	237

★ result from（～に起因する）

自動詞として用いられるが，名詞の用例も多い．

❶ result from ～（～に起因する）

These defects result from a failure to accumulate Gurken protein, which is required to initiate dorsoventral patterning during oogenesis.（*Nat Cell Biol. 1999 1:354*）

訳 これらの欠陥はGurkenタンパク質を蓄積することの不全に起因する

★ attribute（…を～に起因すると考える／…を～に帰する）

他動詞．“［*be*］attributed to”の用例が多い．

❶ attributed to ～（～に起因すると考えられる）

The increase in affinity is attributed to greater stability in the mismatched site associated with stacking by the heterocyclic aromatic ligand.（*Proc Natl Acad Sci USA. 2003 100:3737*）

訳 親和性の上昇は，…におけるより大きな安定性に起因すると考えられる

ascribe（…を～に起因すると考える／…を～に帰する）

他動詞．“［*be*］ascribed to”の用例が多い．

❶ ascribed to ～（～に起因すると考えられる）

These effects were ascribed to stimulation of adenylyl cyclase by increased intracellular free tubulin.（*Circ Res. 2001 88:E32*）

訳 これらの効果はアデニリルシクラーゼの刺激に起因すると考えられた

5. 得る／得られる 【obtain】

得る	
obtain	gain
acquire	get

使い分け
- ◆ **obtain** は，「得る」という意味で広く用いられる．
- ◆ **acquire** は，「獲得する」という意味で用いられる．
- ◆ **gain** は，洞察やアクセスなど抽象的なものに対して用いられる．

頻度分析

	用例数		用例数
obtain	10,054	❶ (be) obtained from	2,294
acquire	2,526	❶ (be) acquired by	79
gain	3,136	❶ gain insight into	292
get	151	❶ to get	40

解説
- ◆ **obtain**, **acquire** は受動態で，**gain**, **get** は能動態で使われることが多い．
- ◆ **get** の用例は非常に少ない．

★★ obtain（～を得る）

他動詞．“obtained from” の用例が非常に多い．

❶ **obtained from ～**（～から得られる）

Plasma samples were obtained from 105 women and 25 infants enrolled in a Ugandan clinical trial.（*J. Clin Microbiol. 2001 39:4323*）

訳 血漿サンプルは 105 人の女性と 25 人の乳児から得られた

★ acquire（～を獲得する／～を得る）

他動詞．

❶ **acquired by ～**（～によって得られる）

The data were acquired by means of a Charge-Coupled Device（CCD）camera and image-processing software.（*J Dent Res. 2000 79:1584*）

訳 データは電荷結合素子（CCD）カメラによって得られた

★ gain（～を得る）

他動詞の用例が多いが，名詞としても使われる．“gain insight into”の用例が多い．

❶ gain insight into ～（～に対する洞察を得る）

To gain insight into a possible role of AP-4 in intracellular trafficking, we constructed a Tac chimera bearing a mu4-specific YXXphi signal.（*J Biol Chem. 2001 276:13145*）

訳 …における AP-4 の可能な役割に対する洞察を得るために

get（～を得る）

他動詞．

❶ to get ～（～を得るために）

To get a better understanding of its function, we sought to identify the proteins that interact with the BHV-1 U_L3.5 protein.（*J Virol. 2000 74:2876*）

訳 それの機能についてのよりよい理解を得るために

6. 引き起こす／～を生じる 【cause】

引き起こす	もたらす	一因になる
cause result in lead to give rise to produce	yield bring about	contribute

使い分け

◆ **cause**, **result in**, **lead to**, **give rise to** は，「（結果として何かを）生じる／引き起こす」場合に用いられる．

◆ **produce** は「～を産生する」という意味だが，「～を引き起こす」という意味にも使われる．

◆ **yield**, **bring about** は，「（結果や現象を）生じる／もたらす」という意味に使われる．

◆ **contribute** to は「～に寄与する」という意味だが，「（病気の発症など）の一因になる」という意味で使われることも多い．

頻度分析

	用例数		用例数
cause	22,775	❶ (be) caused by	4,119
result in	23,833	❶ resulted in	9,722
lead to	16,199	❶ leads to	4,830
give rise to	1,180	❶ gives rise to	530
produce	18,157	❶ (be) produced by	2,305
yield	5,525	❶ yielded a	364
bring about	222	❶ (be) brought about by	94
		❷ bring about	60
contribute	11,977	❶ contribute to	6,357

解説

◆ **cause** は，受動態の用例が多い．

◆ **yield** は，能動態の用例が多い．

★★ cause（～を引き起こす／～の原因となる／原因）

名詞として使われることもあるが，他動詞受動態の用例の方が多い．

❶ caused by ～（～によって引き起こされる）

Motor neurone disease is caused by mutations in Cu/Zn superoxide dismutase（SOD1）in 15–20% of familial cases, due to a toxic gain of function by the mutant enzyme.（*Hum Mol Genet. 2002 11:2061*）

訳 運動神経疾患はCu/Znスーパーオキシドジスムターゼの突然変異によって引き起こされる

★★ result in（～という結果になる／～という結果をもたらす）……

❶ resulted in ～（～という結果になった）

Coexpression of the cSHMT and DNcSHMT proteins in bacteria resulted in the formation of heterotetramers with a cSHMT/DNcSHMT subunit ratio of 1.（*J Biol Chem. 2003 278:10142*）

訳 細菌でのcSHMTとDNcSHMTタンパク質の共発現はヘテロ四量体の形成という結果になった

★★ lead to（～につながる／～という結果を導く）……………………

leadは「～を仕向ける」という意味の他動詞としても使われるが，“lead to”という形で自動詞として用いられることが圧倒的に多い．lead／led／ledの語形変化をする．

❶ lead to ～（～につながる）

TGF-β1 treatment leads to an increase in MLK3 activity.（*J Biol Chem. 2004 279:29478*）

訳 TGF-β1処理はMLK3活性の上昇につながる

★ give rise to（～を生じる／～を引き起こす）……………………

❶ give rise to ～（～を生じる）

Mutations in the human GCS1 gene give rise to the congenital disorder of glycosylation termed CDG IIb.（*J Biol Chem. 2004 279:49894*）

訳 ヒトGCS1遺伝子の変異はCDG IIbと名付けられたグリコシル化の先天性疾患を生じる

★★ produce（～を引き起こす／～を産生する）……………………

他動詞．

❶ produced by ～（～によって引き起こされる）

The effects produced by the Ca^{2+} ionophore ionomycin mimicked those produced by ACh.（*J Neurosci. 2000 20:5940*）

訳 Ca^{2+}イオノフォア-イオノマイシンによって引き起こされる効果は，アセチルコリンによって引き起こされるそれらを模倣した

★ yield（～を生じる／～をもたらす）

他動詞として使われるが，名詞の用例も多い．

❶ yielded a ～（～を生じた／～をもたらした）

The results show that induction of apoE yielded a 2-2.5-fold increase in the uptake of low density lipoprotein-cholesteryl ester（LDL-CE）but had little effect on high density lipoprotein-CE uptake.（*J Biol Chem. 1998 273:12140*）

訳 apoEの誘導は，…の2～2.5倍の上昇をもたらした

bring about（～をもたらす）

"bring about"の熟語で他動詞として用いられる．

❶ brought about by ～（～によってもたらされる）

These rapid changes were brought about by induction of apoptosis in the α-cell population.（*Diabetes. 2002 51:398*）

訳 これらの急速な変化が，…のアポトーシスの誘導によってもたらされた

❷ bring about ～（～をもたらす）

Our results demonstrate that even conservative point mutations can bring about dramatic changes in the kinetics of crystallization.（*J Mol Biol. 2001 314:663*）

訳 保存的な点突然変異でさえ，…の劇的な変化をもたらしうる

★★ contribute（一因になる／寄与する）

自動詞．"contribute to"の用例が非常に多い．

❶ contribute to ～（～の一因になる／～に寄与する）

Hepatitis C virus（HCV）infection may contribute to the development of diabetes mellitus.（*Ann Intern Med. 2000 133:592*）

訳 C型肝炎ウイルス（HCV）感染は真性糖尿病の発症の一因になるかもしれない

7. 誘導する／誘発する 【induce】

誘導する	誘発する
induce	elicit evoke provoke

使い分け
- ◆ **induce** は，「（遺伝子発現などを）誘導する」ときに使われる．
- ◆ **elicit**, **evoke**, **provoke** は，「（反応などを）誘発する」という意味に用いられる．
- ◆ **evoke** は，神経の反応に対して使われることが多い．

頻度分析

	用例数		用例数
induce	60,504	❶（be）induced by	6,800
elicit	3,802	❶（be）elicited by	651
evoke	2,818	❶（be）evoked by	495
provoke	257	❶（be）provoked by	24

解説
- ◆ **induce**, **elicit**, **evoke** は，受動態の用例が多い．

⁑ induce（～を誘導する）

他動詞．受動態の用例が多い．

❶ **induced by ～**（～によって誘導される）

In barley aleurone layers, the expression of genes encoding α-amylases and proteases is induced by GA but suppressed by ABA.（*Plant Cell. 2001 13:667*）

訳 α-アミラーゼやプロテアーゼをコードする遺伝子の発現はGAによって誘導される

* elicit（～を誘発する／～を引き出す）

他動詞．受動態の用例が多い．

❶ **elicited by ～**（～によって誘発される）

To maximize immune responses elicited by a DNA vaccine, therefore, it appears that the immune system should first be primed with a specific Ag and then amplified with cytokines.（*J Immunol. 1998*

161:1875)

訳 DNA ワクチンによって誘発される免疫応答を最大にするために

★ evoke（～を誘起する）

他動詞．受動態の用例が多い．

❶ evoked by ～（～によって誘起される）

In contrast, synaptic responses evoked by stimulation of CA3 pyramidal neurons are mediated by calcium-impermeable AMPA receptors.（*Nat Neurosci. 1998 1:572*）

訳 CA3 錐体路ニューロンの刺激によって誘起されるシナプス性の反応は，…によって仲介される

provoke（～を誘発する）

他動詞．

❶ provoked by ～（～によって誘発される）

These findings reveal a genetic cause of autoimmune disease provoked by a defect in the pathway of protein N-glycosylation.（*Proc Natl Acad Sci USA. 2001 98:1142*）

訳 これらの知見は，…の経路の欠陥によって誘発される自己免疫疾患の遺伝的原因を明らかにする

8. 産生する 【produce】

産生する		
produce	**generate**	**raise**

使い分け

◆いずれの語も「産生する」という意味に用いられるが，**produce** は「～を引き起こす」という場合にも使われる．

◆ **generate** は，「～を作製する」という意味で使われることの方が多い．

◆ **raise** は，「（抗体を）産生する」ときに使われることが多い．

頻度分析

	用例数		用例数
produce	18,157	❶（be）produced by	2,305
generate	14,114	❶（be）generated in	584
raise	2,677	❶（be）raised against	250

★★ produce（～を産生する／～を引き起こす）

他動詞．受動態の用例が多い．

❶ produced by ～（～によって産生される）

INF-γ is produced by $CD4^+CD28^{null}$ T cells, which are expanded in UA and distinctly low in SA and controls.（*Circulation. 1999 100:2135*）

訳 INF-γ は $CD4^+CD28^{null}$ T 細胞によって産生される

★★ generate（～を産生する／～を作製する）

他動詞．

❶ generated in ～（～において産生される）

Flow cytometry analysis revealed that reactive oxygen intermediates were generated in CPT-treated PLB-985 cells.（*J Clin Invest. 1998 102:1961*）

訳 活性酸素中間体が CPT 処理された PLB-985 細胞において産生された

★ raise（～を産生する／～を上げる／～を提起する）

他動詞．“raised against” や “raise the possibility that（～という可能性を示唆する）” の用例が多い．

❶ raised against ～（～に対して産生された）

Affinity-purified antibodies raised against the human SEC34 protein（hSec34p）recognized a cellular protein of 94 kDa in both soluble and membrane fractions.（*J Biol Chem. 2001 276:22810*）

訳 ヒト SEC34 タンパク質に対して産生された抗体

9. 上昇する／上昇させる／増加させる【increase】

上昇する	上昇・増加させる	誘導する
increase	increase elevate up-regulate raise	induce

使い分け

- **increase**, **elevate** は，「（測定できる数や量を）上昇させる／増加させる」ときに用いられる．
- **up-regulate** は，受動態で「（遺伝子発現などが）上昇する」という意味に用いられる．
- **raise** は，「（温度や pH などを）上げる」という意味で使われる．
- **induce** は「（遺伝子発現などを）誘導する」という意味だが，大きな上昇がある場合にも用いられる．
- **increase** は，他動詞・自動詞の両方で用いられる．

頻度分析

	用例数		用例数
increase	72,178	❶ increased the	2,487
		❷ increased in	2,431
		❸ was increased in	407
elevate	6,684	❶ (be) elevated in	829
up-regulate	2,262	❶ (be) up-regulated in	470
upregulate	1,362		
raise	2,677	❶ (be) raised from ～ to	20
induce	60,504	❶ (be) induced by	6,800

解説

- **elevate**, **up-regulate**, **induce** は他動詞として用いられ，受動態の用例が多い．

☆☆ increase（～を増加させる／～を上昇させる／～が上昇する／増加）

他動詞・自動詞の両方の用例がある．他動詞受動態と自動詞とは，ほぼ同じ意味になる場合が多いが，受動態の用例の方が多い．名詞の用例も多い．

❶ **increased the ～**（～を増加させる）

In addition, this injury significantly increased the number of apoptotic cells over that accruing from mechanical injury alone.（*FASEB J. 1999 13:1875*）

訳 この傷害はアポトーシスを起こす細胞の数を有意に増加させた

❷ **increased in ～**（～において上昇した）

Three days after a unilateral olfactory bulbectomy, Sema3A transcript levels increased in regenerating neurons.(*J Comp Neurol. 2000 423:565*)

訳 Sema3A 転写物のレベルは再生しているニューロンにおいて上昇した

❸ **was increased in ～**（…は，～において上昇した）

Expression of mRNA for TNF-α was increased in brains of muMT mice.（*J Immunol. 2000 164:2629*）

訳 TNF-α メッセンジャー RNA の発現は muMT マウスの脳において上昇した

★ elevate（～を上昇させる）

他動詞．受動態の用例が多い．

❶ **elevated in ～**（～において上昇した）

Plasma sCD25 levels were elevated in systemic mastocytosis; the highest levels were associated with extensive bone marrow involvement.（*Blood. 2000 96:1267*）

訳 血漿 sCD25 レベルは全身性肥満細胞症において上昇した

★ up-regulate, upregulate（～を上方制御する／～を上昇させる）

他動詞．受動態の用例が多い．"upregulate" "up-regulate" のどちらも使われる．

❶ **up-regulated in ～**（～において上方制御される／～において上昇する）

However, we find that P450scc mRNA is up-regulated in thymocytes on the initiation of positive selection.（*J Immunol. 1999 163:5781*）

訳 P450scc メッセンジャー RNA は胸腺細胞において上方制御される

★ raise（～を上げる／～を提起する／～を産生する）

他動詞．"raise the possibility that（～という可能性を示唆する）" の用例が多く（動詞編 I -B. 8. 参照），「（抗体を）産生する」という意味で用いられることも多いが，「上げる」という意味で使われることは比較的少ない．

❶ **raised from ～ to**（～から…に上げられる）

When the temperature was raised from 37 to 39 degrees C, the block in membrane formation persisted throughout the infection.（*J Virol. 2004 78:257*）

訳 温度が37から39℃に上げられたとき

★★ induce（～を誘導する）

他動詞．受動態の用例が多い．

❶ **induced by ～**（～によって誘導される）

Expression of Cdc25A mRNA and protein was induced by E_2 in control and p16[INK4a]-expressing MCF-7 cells; however, functional activity of Cdc25A was inhibited in cells expressing p16[INK4a].（*Mol Cell Biol. 2001 21:794*）

訳 Cdc25Aメッセンジャー RNAとタンパク質の発現はE_2によって誘導された

10. 増強する／強化する 【enhance】

増強・強化する	活性化する
enhance augment potentiate	activate

使い分け

◆ **enhance** は，「（能力や遺伝子発現などを）強化する／増強する」という意味に用いられる．

◆ **augment**, **potentiate** は，「（誘導物質などによって引き起こされる現象を）増強する」ときに使われる．

◆ **activate** は，「（タンパク質などを）活性化する」ときに使われる．

頻度分析

	用例数		用例数
enhance	17,099	❶ enhance the	915
augment	1,784	❶ (be) augmented by	191
potentiate	1,549	❶ (be) potentiated by	154
activate	29,833	❶ (be) activated by	2,502

解説

◆ **augment**, **potentiate**, **activate** は，受動態の用例が多い．

⁑ enhance（～を強化する／～を増強する）

❶ enhance the ～（～を強化する）

The goal of this series is to enhance the ability of radiologists to evaluate the literature competently and critically, not make them into statisticians.（*Radiology. 2002 225:318*）

訳 このシリーズの目的は，…を評価する放射線科医の能力を強化することである

★ augment（～を増強する／～を増大させる）

他動詞．受動態の用例が多い．

❶ augmented by ～（～によって増強される）

Finally, the transactivation capability of EKLF is augmented by co-transfection of CKIIα.（*J Biol Chem. 1998 273:23019*）

訳 EKLFのトランス活性化能はCKIIαの同時遺伝子導入によって増強される

★ potentiate（～を増強する／～を強化する）

他動詞．受動態の用例が多い．

❶ potentiated by ～（～によって増強される）

VEGI-induced cytotoxicity was potentiated by inhibitors of protein synthesis.（*Oncogene. 1999 18:6496*）

訳 VEGIに誘導される細胞傷害性はタンパク質合成の阻害剤によって増強された

⁑ activate（～を活性化する）

他動詞．受動態の用例が多い．

❶ activated by ～（～によって活性化される）

Protein kinase PKR is activated by double-stranded RNA（dsRNA）and phosphorylates translation initiation factor 2α to inhibit protein synthesis in virus-infected mammalian cells.（*J Biol Chem. 2001 276:24946*）

訳 タンパク質リン酸化酵素PKRは二本鎖RNAによって活性化される

11. 促進する 【promote】

促進する		
promote	accelerate	facilitate

使い分け
- ◆動詞編Ⅱ-C. 10. の「増強する」に近い意味になることがある.
- ◆ **promote**, **accelerate** は,「(現象を) 促進する」ときに用いられる.
- ◆ **accelerate** は,「～を加速させる」という意味合いが強い.
- ◆ **facilitate** は,「(開発など人の行為を) 促進する」ときに使われることが多い.

頻度分析

	用例数		用例数
promote	8,882	❶ promote the	497
accelerate	2,602	❶ accelerates the	143
facilitate	5,314	❶ facilitate the	670

解説
- ◆ **promote**, **facilitate** は,能動態の用例が多い.

★ promote (～を促進する) ……………………………………

❶ **promote the ～** (～を促進する)

These variants were designed to promote the formation of heterodimers and to destabilize the formation of inactive variant homodimers of HIV-1 protease through substitutions at Asp-25, Ile-49, and Gly-50. (*J Biol Chem. 2000 275:7080*)

訳 これらの変異体はヘテロ二量体の形成を促進するように設計された

★ accelerate (～を加速させる／～を促進する) …………………

❶ **accelerate the ～** (～を加速させる)

Preassembly of the coactivator complex accelerates the rate of transcription in a cell-free system depleted of TFIID and mediator. (*Genes Dev. 2002 16:1852*)

訳 コアクチベーター複合体の事前の構築は,…における転写の速度を加速させる

★ facilitate（～を促進する）

❶ **facilitate the ～**（～を促進する）

The modifications will facilitate the development of high throughput methods for whole blood folate.（*Anal Biochem. 2002 301:14*）

訳 その改変はハイスループットな方法の開発を促進するであろう

12. 進行する／進行させる／進歩させる 【proceed】

進行する	進行・進歩させる
proceed progress	advance

使い分け

◆ **proceed** は，「（物事が）進行する」という意味で使われる．

◆ **progress**，**advance** は，「（病気が）進行する」あるいは「進歩する／進歩させる」の意味に用いられる．

◆ **proceed**，**progress** は自動詞，**advance** は他動詞として用いられる．

頻度分析

	用例数		用例数
proceed	1,198	❶ proceeds through	105
progress	1,572	❶ progressed to	123
advance	2,961	❶ with advanced	297
		❷ to advance	59

★ proceed（進行する）

自動詞として使われることが多いが，名詞の用例もある．

❶ **proceed through ～**（～を経て進行する）

Thymocyte development proceeds through two critical checkpoints that involve signaling events through two different receptors, the TCR and the pre-TCR.（*J Immunol. 1999 163:2610*）

訳 胸腺細胞の発達は2つの決定的に重要なチェックポイントを経て進行する

★ progress（進行する／進歩）

名詞として用いられることが多いが，自動詞の用例も少なくない．

❶ **progressed to ～**（～へ進行した）

When papillomas further progressed to squamous cell carcinomas (SCC), both control and ΔβRII SCC showed similar BrdU labeling indices and percentages of S phase cells.（*Oncogene. 2000 19:3623*）

訳 パピローマがさらに扁平上皮癌へ進行したとき

★ advance（～を進歩させる／～を進行させる／進歩）

他動詞の用例が多いが，名詞としても使われる．

❶ **with advanced ～**（進行した～を持つ…）

In addition, a tendency for greater activation of CXCR4+CD4+ T cells in patients with advanced disease was observed.(*J Immunol. 1998 161:3195*)

訳 進行した疾患を持つ患者におけるCXCR4+CD4+ T細胞のより大きな活性化の傾向が観察された

❷ **to advance ～**（～を進歩させるために）

To advance our understanding of the mode of coordinated gene regulation in multicellular organisms, we performed a genome-wide analysis of the chromosomal distribution of co-expressed genes in *Drosophila*.（*Nature. 2002 420:666*）

訳 …に対するわれわれの理解を進歩させるために

13. 増殖する／増幅する 【grow】

増殖する	増殖させる	増幅する
grow proliferate replicate	propagate	amplify

使い分け

◆ **grow** は生物全般に使われるが，**proliferate**, **propagate** は細胞や微生物だけに用いられる．

◆ **replicate** は「(DNAが) 複製する」という意味で，「(ウイルスが) 増殖する」場合にも用いられる．

◆受動態の **grown**, **propagated** は，人が増殖させた場合に用いられることが多い．

◆ **propagate** は「（菌やウイルスを）増殖させる」だけでなく，「伝搬する」という意味でも用いられる．

◆ **amplify** は，「（PCR によって遺伝子を）増幅する」場合などに用いられる．

◆ **proliferate** と **replicate** は自動詞，**propagate** は他動詞，**grow** は他動詞・自動詞の両方で用いられる．

頻度分析	用例数		用例数
grow	5,098	❶ (be) grown in	673
		❷ grow in	154
proliferate	1,707	❶ proliferate in	126
replicate	1,940	❶ replicate in	163
propagate	617	❶ (be) propagated in	56
amplify	1,766	❶ (be) amplified by	141

解説

◆ **grown**, **grow**, **proliferate**, **replicate**, **propagated** は，後に in を伴うことが多い．

★ grow（～を生育する／増殖する／成長する）

他動詞・自動詞の両方で用いられる．

❶ grown in ～（～において生育される）

P. vivax cannot be grown in culture; the reason for its resistance to DHFR inhibitors is unknown.（*Proc Natl Acad Sci USA. 2002 99: 13137*）

訳 三日熱マラリア原虫は培養で生育させることができない

❷ grow in ～（～において増殖する）

After mechanical dissociation, the melanoma specimen cells' ability to grow *in vitro* was assessed.（*Ann Surg. 2000 231:664*）

訳 メラノーマ検体細胞の生体内で増殖する能力が評価された

★ proliferate（増殖する）

自動詞．

❶ proliferate in ～（～において増殖する）

Under most circumstances, CLL B cells do not proliferate in culture and express a limited repertoire of surface antigens, including CD19, CD20, CD23, CD27, CD40, and CD70.（*Blood. 1998 91:2689*）

訳 慢性リンパ性白血病 B 細胞は培養では増殖しない

動詞

★ replicate（増幅する／複製する／～を再現する）

自動詞の用例が多いが，「～を再現する」という意味では他動詞として使われる．

❶ replicate in ～（～において増幅する）

The ability of the virus to replicate *in vivo* is determined by virally encoded determinants contained within a defined region of the VP2 gene.（*J Virol. 1999 73:8713*）

訳 ウイルスが生体内で増幅する能力は，…によって決定される

propagate（～を増殖させる／伝搬する）

「～を増殖させる」の意味では他動詞，「伝搬する」の意味では自動詞で用いられる．

❶ propagated in ～（～において増殖させられる）

When *M. pulmonis* strain X1048 was propagated in laboratory culture medium, > 95% of colony-forming units (cfu) lacked R-M activity and produced the variable surface protein VsaA.（*Mol Microbiol. 2001 40:1037*）

訳 *M. pulmonis* 株 X1048 は研究室の培養液で増殖させられた

★ amplify（～を増幅する）

他動詞．

❶ amplified by ～（～によって増幅される）

Two genes encoding KPHMT and one for PtS were identified in the *Arabidopsis thaliana* genome, and cDNAs for all three genes were amplified by PCR.（*Plant J. 2004 37:61*）

訳 3 つすべての遺伝子の cDNA が PCR によって増幅された

14. 拡大する／延長させる【extend】

広げる	広がる	延長させる
extend expand enlarge	spread	prolong

使い分け

◆ **extend** は，「（一方向に）広げる／広がる」という意味に用いられる．

◆ **expand, enlarge** は，「～を拡大する」という意味で使われる．

II

C

◆ **expand** は，特定の細胞集団を増殖させる場合に用いられることも多い．

◆ **spread** は，病気（細菌やウイルス）の蔓延などに用いられる．

◆ **prolong** は，「～を延長させる」という意味を持つ．

頻度分析	用例数		用例数
extend	5,362	❶ (be) extended to	288
expand	2,233	❶ (be) expanded in	109
enlarge	360	❶ (be) significantly enlarged	11
spread	2,369	❶ spread to	90
prolong	2,950	❶ prolonged survival	139

★ extend（～を広げる／～を伸ばす／広がる／伸びる）

他動詞および自動詞として用いられる．

❶ extended to ～（～にまで広げられる／～するために拡大される）

This approach can be extended to more-general pedigree structures and quantitative traits.（*Am J Hum Genet. 2004 74:432*）

訳 このアプローチはもっと一般的な系統構造にまで広げられうる

★ expand（～を拡大する）

他動詞．

❶ expanded in ～（～において拡大される）

GAG-binding cells seem to be expanded in bone marrow of GAG-immunized mice.（*Proc Natl Acad Sci USA. 2002 99:14362*）

訳 GAG結合細胞が，…の骨髄において拡大されるらしい

enlarge（～を拡大する）

他動詞．

❶ significantly enlarged（有意に拡大される）

In Down syndrome, early endosomes were significantly enlarged in some pyramidal neurons as early as 28 weeks of gestation, decades before classical AD neuropathology develops.（*Am J Pathol. 2000 157:277*）

訳 初期のエンドソームはいくつかの錐体ニューロンにおいて有意に拡大された

★ spread（広がる／蔓延する／伝播）

名詞および自動詞として用いられる．

❶ spread to 〜（〜に広がる／〜に広がった）

Infection spread to the brain as it was cleared from the lung, again without leukocyte accumulation.（*J Immunol. 2004 173:4030*）

訳 感染が脳に広がった

★ prolong（〜を延長させる／〜を延ばす）

他動詞．

❶ prolonged survival（生存を延長させた）

We observed that four weekly treatments with 4 mg/kg HAT significantly prolonged survival of MET-1-bearing mice.（*Cancer Res. 2003 63:6453*）

訳 4 mg/kg の HAT による 4 週間の処理は，MET-1 を持つマウスの生存を有意に延長させた

【II-D. 低下・抑制・破壊】 低下・抑制・抑止・破壊を表す動詞

15. 低下させる／低下する／減少させる 【decrease】

低下・減少させる	低下する	短縮させる	最小化する
decrease reduce diminish down-regulate lower depress	decrease decline	shorten	minimize

使い分け

◆ **decrease** は「～を低下させる／～を減少させる」「低下する／減少する」の意味で，他動詞・自動詞の両方で使われる．

◆ **reduce**, **diminish**, **down-regulate**, **lower** は，「～を低下させる／～を減少させる」という意味の他動詞として用いられ，受動態で使われるときは「低下する／減少する」という意味になることが多い．

◆ **depress** は「（人を）憂鬱にさせる」という意味だが，「～を低下させる」ときにも用いられる．

◆ **decline** は，「低下する」という意味の自動詞として用いられる．

◆ **shorten** は「～を短縮する」，**minimize** は「～を最小化する」という意味の他動詞として用いられる．

頻度分析

	用例数		用例数
decrease	25,120	❶ decreased the	1,041
		❷ decreased in	966
		❸ was decreased in	161
reduce	29,270	❶ reduced the	2,036
		❷ (be) reduced in	1,766
diminish	2,766	❶ (be) diminished in	264
down-regulate	1,625	❶ (be) down-regulated in	246
downregulate	844		
lower	12,331	❶ lowered the	110
depress	909	❶ (be) depressed in	46
decline	3,010	❶ declined to	86
shorten	1,226	❶ to shorten	22

minimize	806	❶ to minimize	256

⁑ decrease（〜を低下させる／低下する／〜を減少させる／減少）

他動詞・自動詞の両方の用例がある．他動詞受動態と自動詞とは，ほぼ同じ意味になる場合が多いが，受動態の用例の方が多い．名詞の用例も多い．

❶ decreased the 〜（〜を低下させた）

These mutations significantly decreased the rate of primer synthesis, due primarily to a decreased rate of initiation, and the extent of impairment correlated with the severity of the mutation（A > Q > K）.（*Biochemistry. 1999 38:7727*）

訳 これらの変異がプライマー合成の速度を有意に低下させた

❷ decreased in 〜（〜において低下した）

During MD, mRNA levels for collagen, MMP-3, and TIMP-1 decreased in both the deprived and control eyes, compared with age-matched normal eyes.（*Invest Ophthalmol Vis Sci. 2002 43:2067*）

訳 コラーゲン，MMP-3およびTIMP-1のメッセンジャーRNAレベルは，…両者において低下した

❸ was decreased in 〜（…は，〜において低下した）

Epithelial cell Nos2 mRNA expression was decreased in adenomas compared with histologically normal $Apc^{Min/+}Nos2^{+/+}$ intestine.（*Gastroenterology. 2001 121:889*）

訳 上皮細胞Nos2メッセンジャーRNA発現は，…に比べて腺腫において低下した

⁑ reduce（〜を低下させる／〜を減少させる）

他動詞．

❶ reduced the 〜（〜を低下させた）

Mutation of all four serine residues reduced the ability of TFIIA to stimulate transcription in transient transfection assays with various activators and promoters, indicating that TFIIA phosphorylation is required globally for optimal function.（*J Biol Chem. 2001 276:15886*）

訳 4つのセリン残基すべての変異は，…するTFIIAの活性を低下させた

❷ reduced in 〜（〜において低下した）

Gsα expression was reduced in the lung and tracheae of albuterol-treated rats, and cholera toxin-induced cAMP accumulation was blunted.（*J Clin Invest. 2000 106:125*）

訳 Gsαの発現が，…の肺と気管において低下した

★ diminish（～を低下させる／～を減少させる）

他動詞．

❶ diminished in ～（～において低下した）

The anti-tumor effect was diminished in mice deficient in $CD4^+$ T-cells.（*J Clin Invest. 2000 105:1623*）

訳 抗腫瘍効果は$CD4^+$ T細胞を欠損しているマウスにおいて低下した

★ down-regulate, downregulate（～を下方制御する／～を低下させる）

他動詞．受動態が多い．

❶ down-regulated in ～（～において低下する）

We have found that Cdc6 expression is down-regulated in prostate cancer as detected by semiquantitative reverse transcriptase-PCR of prostate cell lines and laser-captured microdissected prostate tissues.（*J Biol Chem. 2002 277:25431*）

訳 Cdc6の発現は前立腺癌において低下している

★ lower（～を低下させる／より低い）

形容詞lowの比較級としての用例がほとんどだが，他動詞の用例もある．

❶ lowered the ～（～を低下させた）

Chemotherapy lowered the recurrence rate outside the abdomen or pelvis.（*Radiology. 1999 211:183*）

訳 化学療法は再発率を低下させた

depress（～を低下させる／～を抑制する）

他動詞．"depressed patients（うつ病の患者）"の用例が多い．

❶ depressed in ～（～において低下する）

Epimerase activity was depressed in the mutant, but increased upon restoration of 2-*O*-sulfotransferase, suggesting that their physical association was required for both epimerase stability and translocation to the Golgi.（*Proc Natl Acad Sci USA. 2001 98:12984*）

訳 エピメラーゼ活性は変異体において低下した

★ decline（低下する／減少）

名詞の用例が多いが，自動詞としても使われる．

❶ declined to ～（～に減少した）

Output declined to basal levels throughout the remainder of the night.（*J Neurosci. 1998 18:5045*）

訳 出力は基底レベルに低下した

★ shorten（～を短縮する）

他動詞．

❶ to shorten ～（～を短縮するために）

The timely evaluation of new drugs that can be used to shorten tuberculosis（TB）treatment will require surrogate markers for relapse.（*J Infect Dis. 2003 187:270*）

訳 結核（TB）の治療を短縮するために使われうる新しい薬

minimize（～を最小化する）

他動詞．

❶ to minimize ～（～を最小化するように／～を最小化するために）

For the latter, the spontaneous object recognition task was conducted in a modified apparatus designed to minimize the potentially confounding influence of spatial and contextual factors.（*J Neurosci. 2004 24:5901*）

訳 潜在的に混乱させる影響を最小化するように設計された装置

16. 抑制する／弱める／干渉する 【inhibit】

抑制する	弱める／軽減する	干渉する
inhibit	attenuate	interfere
suppress	weaken	
repress	relieve	
	alleviate	
	mitigate	

使い分け ◆動詞編II-D. 15. の「低下させる」，動詞編II-D. 17. の「抑止する」と近い意味で用いられることがある．

- **inhibit**, **suppress**, **repress** は，「～を抑制する」という意味に用いられる．
- **inhibit** は，「～を阻害する」という意味で使われることも多い．
- **attenuate**, **weaken** は，「～を弱める」という意味に用いられる．
- **relieve**, **alleviate**, **mitgate** は，「～を軽減する」という意味で使われる．
- **interfere** は，「干渉する」という意味の自動詞で，“**interfere with**”の用例が非常に多い．
- **suppress**, **repress** は，遺伝子発現に対して使われることが多い．
- **inhibit**, **suppress**, **repress**, **attenuate**, **weaken** は，他動詞として使われる．

頻度分析

	用例数		用例数
inhibit	26,453	❶ (be) inhibited by	3,565
suppress	6,583	❶ (be) suppressed by	730
repress	2,861	❶ (be) repressed by	279
		❷ repress transcription	161
attenuate	4,363	❶ (be) attenuated by	415
		❷ attenuated the	407
weaken	385	❶ weakens the	26
relieve	464	❶ (be) relieved by	84
alleviate	316	❶ (be) alleviated by	45
mitigate	168	❶ (be) mitigated by	24
interfere	2,550	❶ interfere with	812

★★ inhibit（～を抑制する／～を阻害する）

他動詞．

❶ inhibited by ～（～によって抑制される）

Apoptosis was inhibited by RB mutants with constitutive ABL binding, but ABL overexpression overcame the effect of the RB mutant constructs.（*J Biol Chem. 2002 277:44969*）

訳 アポトーシスはRB変異体によって抑制された

★ suppress（～を抑制する）

他動詞．

❶ **suppressed by ～**（～によって抑制される）

When AtFKBP13 expression was suppressed by RNA interference method, the level of Rieske protein was significantly increased in the transgenic plants.（*Proc Natl Acad Sci USA. 2002 99:15806*）

訳 AtFKBP13発現がRNA干渉法によって抑制された

★ repress（～を抑制する）

他動詞.

❶ **repressed by ～**（～によって抑制される）

In addition, *Hoxa-10* expression is repressed by estrogen in a protein synthesis-independent manner.（*Dev Biol. 1998 197:141*）

訳 *Hoxa-10*発現は，エストロゲンによって抑制される

❷ **repress transcription**（転写を抑制する）

EthR has been shown to repress transcription of the activator gene ethA by binding to this intergenic region, thus contributing to ethionamide resistance.（*J Mol Biol. 2004 340:1095*）

訳 EthRは活性化因子遺伝子ethAの転写を抑制することが示されている

★ attenuate（～を減弱させる／～を弱める）

他動詞.

❶ **attenuated by ～**（～によって減弱される）

JNK activation was attenuated by blocking antibodies to β2 integrins, the tyrosine kinase inhibitors, genistein, and tyrphostin A9, a Pyk2-specific inhibitor, and piceatannol, a Syk-specific inhibitor.（*J Biol Chem. 2001 276:2189*）

訳 JNKの活性化は，…に対するブロッキング抗体によって減弱された

❷ **attenuated the ～**（～を減弱させる）

The same MEK1/2 inhibitors dose-dependently attenuated the increase in venular permeability caused by histamine.（*J Physiol. 2005 563:95*）

訳 同じMEK1/2抑制剤は，…の上昇を用量依存的に減弱させた

weaken（～を弱める）

他動詞．能動態の用例が多い．

❶ **weaken the ～**（～を弱める）

Activated Cdc42 weakens the interaction between ACK2 and clathrin and thus reverses the ACK2-mediated inhibition of endocytosis.（*J Biol Chem. 2001 276:17468*）

訳 活性化されたCdc42はACK2とクラスリンの間の相互作用を弱める

relieve（～を軽減する）

他動詞．

❶ **relieved by ～**（～によって軽減される）

This repressive activity can be relieved by the HDAC inhibitor trichostatin A.（*Mol Cell Biol. 2001 21:3118*）

訳 この抑制活性はHDAC阻害剤trichostatin Aによって軽減されうる

alleviate（～を軽減する）

他動詞．

❶ **alleviated by ～**（～によって軽減される）

This inhibition is alleviated by phosphorylation by protein kinase A.（*Nature. 2002 419:947*）

訳 この抑制はプロテインキナーゼAによるリン酸化によって軽減される

mitigate（～を軽減する）

他動詞．

❶ **mitigated by ～**（～によって軽減される）

The effect of RepA was mitigated by over-expression of ZmRb1.（*Proc Natl Acad Sci USA. 2002 99:11975*）

訳 RepAの効果はZmRb1の過剰発現によって軽減された

★ interfere（干渉する／妨害する）

自動詞．"interfere with" の用例がほとんどである．

❶ **interfere with ～**（～に干渉する）

IL-4/IL-13 did not interfere with the expression or activity of iNOS but up-regulated arginase I (the liver isoform of arginase) in a Stat6-dependent manner.（*J Immunol. 2001 166:2173*）

訳 IL-4/IL-13はiNOSの発現や活性に干渉しなかった

17. 抑止する／阻止する／妨げる 【block】

抑止する	妨げる
block	prevent
abrogate	hinder
silence	occlude
	hamper

使い分け

◆動詞編Ⅱ-D. 16. の「抑制する」と近い意味になることがある．
◆**block**, **abrogate** は，「～を抑止する／～を阻止する」という意味に使われる．
◆**silence** は「～を沈黙化する」という意味で，「～を抑止する」とほぼ同様の意味で用いられる．
◆**silence** は，遺伝子発現に対して用いられることが多い．
◆**prevent**, **hinder**, **occlude**, **hamper** は，「～を妨げる／～を阻止する」という意味で使われる．

頻度分析

	用例数		用例数
block	16,092	❶ (be) blocked by	2,412
abrogate	1,765	❶ (be) abrogated by	294
		❷ abrogated the	256
silence	563	❶ (be) silenced in	45
prevent	9,938	❶ (be) prevented by	783
		❷ prevented the	684
hinder	400	❶ (be) hindered by	101
occlude	332	❶ (be) occluded by	38
hamper	324	❶ (be) hampered by	204

★★ block（～をブロックする／～を阻止する／遮断）

他動詞としての用例が多いが，名詞としても使われる．

❶ **blocked by ～**（～によってブロックされる）

ERK activation was blocked by inhibiting MEK, the upstream activator of ERK.（*J Neurosci. 2003 23:2348*）

訳 ERKの活性化はMEKを抑制することによってブロックされた

★ abrogate（～を抑止する／～を阻止する）

❶ abrogated by ～（～によって抑止される）

Apoptosis was abrogated by the competitive inhibitor of ceramide synthase, fumonisin B1.（*Cancer Res. 1999 59:5194*）

訳 アポトーシスは，…の競合阻害剤によって抑止された

❷ abrogated the ～（～を抑止した）

In contrast, mutation of the ITSM abrogated the ability of PD-1 to block cytokine synthesis and to limit T cell expansion.（*J Immunol. 2004 173:945*）

訳 ITSM の変異は，…する PD-1 の能力を抑止した

silence（～を沈黙化する／～を抑止する）

他動詞の用例が多いが，名詞としても用いられる．

❶ silenced in ～（～において抑止される）

We also show that TMS1 is aberrantly methylated and silenced in human breast cancer cells.（*Cancer Res. 2000 60:6236*）

訳 TMS1 は，ヒト乳癌細胞において異常にメチル化されて抑止されている

★★ prevent（～を妨げる／～を阻止する）

❶ prevented by ～（～によって妨げられる）

Apoptosis was prevented by pretreatment with HGF but not IGF-I.（*Am J Pathol. 2001 158:275*）

訳 アポトーシスは HGF による前処置によって妨げられた

❷ prevented the ～（～を妨げた）

Administration of Crry-Ig also prevented the development of AHR.（*Am J Respir Crit Care Med. 2004 169:726*）

訳 Crry-Ig の投与はまた，AHR の発生を妨げた

hinder（～を妨げる）

他動詞．

❶ hindered by ～（～によって妨げられる）

Developing gene therapy for cystic fibrosis has been hindered by limited binding and endocytosis of vectors by human airway epithelia.（*J Clin Invest. 2000 105:589*）

訳 嚢胞性線維症の遺伝子治療の開発は，ベクターの限られた結合とエンドサイトーシスによって妨げられてきた

occlude（～を妨げる／～を閉鎖する）

他動詞.

❶ **occluded by ～**（～によって妨げられる）

This inhibition was occluded by ADP, suggesting that α, β-methylene-ADP is an agonist at p2y1 receptors.（*J Physiol. 2002 540:843*）

訳 この抑制は ADP によって妨げられた

hamper（～を妨げる）

他動詞.

❶ **hampered by ～**（～によって妨げられる）

The detection of microquantities of glycosaminoglycans（GAGs）in biological samples has been hampered by the lack of sensitive methods.（*Anal Biochem. 2002 306:298*）

訳 生物試料における微量のグリコサミノグリカン（GAGs）の検出は，感度の高い方法の欠如によって妨げられてきた

18. 破壊する／損傷する 【disrupt】

破壊する	損傷する
disrupt	**impair**
destroy	**damage**
lesion	

使い分け

◆動詞編Ⅱ-D. 19. の「消滅させる」に近い意味で用いられることがある.
◆**disrupt** は，「壊す／分裂させる」場合などに使われる.
◆**destroy** の用例は，かなり少ない.
◆**lesion** は，「(臓器が) 傷害される」ときに受動態で用いられる.
◆**impair**, **damage** は，「～を損傷する」という意味で使われる.

頻度分析

	用例数		用例数
disrupt	4,055	❶ disrupt the	337
		❷ (be) disrupted by	292
destroy	379	❶ (be) destroyed by	35
lesion	8,508	❶ lesioned rats	115

impair	4,238	❶（be）impaired in	692
damage	8,080	❶（be）damaged by	39

★ disrupt（～を壊す／～を分裂させる／～を混乱させる）

❶ **disrupt the ～**（～を壊す）

Mutations in the PXQXT motif in CTAR-1 that disrupt the interaction between LMP-1 and TRAFs abolished the induction of IRF-7.（*J Virol. 2001 75:12393*）

訳 LMP-1 と TRAF の間の相互作用を壊す CTAR-1 の PXQXT モチーフの変異は IRF-7 の誘導を消滅させた

❷ **disrupted by ～**（～によって壊される）

The interaction between MDM2 and hsp90 is disrupted by the 2A10 antibody, which recognizes a site on MDM2 important for binding to alternative reading frame（ARF）.（*J Biol Chem. 2001 276:40583*）

訳 MDM2 と hsp90 の間の相互作用は 2A10 抗体によって壊される

destroy（～を破壊する）

他動詞.

❶ **destroyed by ～**（～によって破壊される）

In oxygenated and iron replete cells, HIF-α subunits are rapidly destroyed by a mechanism that involves ubiquitylation by the von Hippel-Lindau tumor suppressor（pVHL）E3 ligase complex.（*Science. 2001 292:468*）

訳 HIF-α サブユニットは，…によるユビキチン化を含む機構によって急速に破壊される

★ lesion（～を破壊する／～に傷害を起こさせる／病変）

名詞として用いられることが多いが，他動詞の用例もある.

❶ **lesioned rats**（傷害を受けたラット／～を破壊されたラット）

This deficit was ameliorated when the rats were tested with the small object boxes, although the performance of the hippocampal-lesioned rats was still below that of controls.（*Behav Neurosci. 2001 115:1193*）

訳 海馬を破壊されたラットの成績は，まだ対照群のそれよりも低かった

★ impair（～を損傷する）

他動詞.

❶ impaired in ～（～において損傷される）

This feedback regulation by ABA is impaired in the ABA-insensitive mutant abi1 but not in abi2.（*J Biol Chem. 2002 277:8588*）

訳 ABAによるこのフィードバック調節は，ABA非感受性の変異体abi1において損傷されている

★ damage（～を損傷する／損傷）

名詞として用いられることが多いが，他動詞の用例もある.

❶ damaged by ～（～によって損傷される）

When the template DNA is damaged by a carcinogen, the fidelity of DNA replication is sometimes compromized, allowing mispaired bases to persist and be incorporated into the DNA, resulting in a mutation.（*J Mol Biol. 2001 309:519*）

訳 鋳型DNAが発癌剤によって損傷されるとき

19. 消滅させる／消失する／欠失している 【abolish】

消滅・消失させる	消去・除去する	消失する	失う	欠失させる
abolish ablate	eliminate remove	disappear	lose	delete

使い分け

◆動詞編II-D. 18.の「破壊する」に近い意味で用いられることがある.

◆**abolish**, **ablate**は「消滅させる／消失させる」という意味の他動詞として，**eliminate**, **remove**は「除去する／消去する」という意味の他動詞として用いられる.

◆**disappear**は，「消失する」という意味の自動詞として用いられる.

◆**lose**は，「～を失う」という意味の他動詞として使われる.

◆**delete**は，受動態で「欠失している」という意味で用いられることが多い.

頻度分析

	用例数		用例数
abolish	3,770	❶ (be) abolished by	582
		❷ abolished the	548

ablate	447	❶ (be) ablated in	31
eliminate	2,973	❶ (be) eliminated by	219
		❷ eliminating the need for	32
remove	2,646	❶ (be) removed from	338
disappear	419	❶ disappeared from	26
lose	2,243	❶ (be) lost in	276
delete	1,855	❶ (be) deleted in	207

★ abolish（～を消滅させる）

❶ abolished by ～（～によって消滅させられる）

Induction of c-fos was abolished by the cAMP-dependent protein kinase inhibitor H-89, suggesting that the transient c-fos mRNA increase is mediated by cAMP.（*FASEB J. 1999 13:553*）

訳 c-fosの誘導は，cAMP依存性プロテインキナーゼ抑制剤H-89によって消滅させられた

❷ abolished the ～（～を消滅させた）

Indeed, blockade of NF-κB activity abolished the ability of Ref-1 to rescue TNF-induced apoptosis.（*Circ Res. 2001 88:1247*）

訳 NF-κB活性の遮断は，…するRef-1の能力を消滅させた

ablate（～を消失させる）

他動詞.

❶ ablated in ～（～において消失している）

The exchanger is completely ablated in 80% to 90% of the cardiomyocytes as determined by immunoblot, immunofluorescence, and exchange function.（*Circ Res. 2004 95:604*）

訳 交換体は心筋細胞の80%～90%で完全に消失している

★ eliminate（～を消去する／～を除去する）

他動詞.

❶ eliminated by ～（～によって消去される／～によって除去される）

This activity was eliminated by the deletion of the PAS subdomain, demonstrating that the PAS subdomain participates in signal reception.（*J Bacteriol. 2004 186:1694*）

訳 この活性はPASサブドメインの欠失によって消去された

❷ **eliminating the need for ～**（～の必要性を消去する／～の必要性を除去する）

Tyrosine from the same sample is determined by its UV absorption at 280 nm, thus eliminating the need for an internal standard.（*Anal Biochem. 2000 280:278*）

訳 このように内部標準の必要性を消去する

★ remove（～を除く／～を除去する）

他動詞.

❶ **removed from ～**（～から除かれる／～から除去される）

T-cell reactivity was also restored if MSCs were removed from the cultures.（*Blood. 2003 101:3722*）

訳 MSC が培養物から除かれた

disappear（消失する）

自動詞.

❶ **disappeared from ～**（～から消失した）

T cells also completely disappeared from V-2-infected spinal cords coincident with the absence of viral RNA.（*J Virol. 2000 74:7903*）

訳 T 細胞はまた，V-2 に感染した脊髄から完全に消失した

★ lose（～を失う）

他動詞.

❶ **lost in ～**（～において失われる）

ARHI is expressed in normal ovarian and breast epithelial cells, but ARHI expression is lost in a majority of ovarian and breast cancers.（*Cancer Res. 2003 63:4174*）

訳 ARHI 発現は大部分の卵巣癌および乳癌において失われている

★ delete（～を欠失させる）

他動詞．受動態の用例が多い.

❶ **deleted in ～**（～において欠失している）

The GPC3 gene is located at Xq26, a region frequently deleted in advanced ovarian cancers.（*Cancer Res. 1999 59:807*）

訳 進行した卵巣癌においてしばしば欠失している領域

【II-E. 変化・移動・影響】 変化・移動・影響を表す動詞

20. 変化する／変化させる 【change】

変化する	変換する	形質転換する	修飾する	変動する
change alter shift	convert	transform	modify	vary

使い分け

- **change**, **alter**, **shift** は，「（何かを）変化させる」ときに用いられる．
- **shift** は，「移す」という意味にも用いられる．
- **convert** は，「（構造などを）変換する」場合に使われる．
- **transform** は「〜を形質転換する」，**modify** は「〜を修飾する」という意味で使われる．
- **vary** は「変動する」という意味の自動詞の用例が多いが，「〜を変化させる」という意味の他動詞としても用いられる．
- **alter**, **shift**, **convert**, **transform**, **modify** は，他動詞として使われる．
- **change**, **vary** は，他動詞・自動詞の両方で用いられる．

頻度分析

	用例数		用例数
change	31,315	❶ not change	721
		❷ (be) changed to	190
alter	10,563	❶ not alter	990
		❷ (be) altered in	557
shift	6,246	❶ (be) shifted from 〜 to	49
convert	2,570	❶ (be) converted to	535
transform	4,586	❶ (be) transformed with	157
modifiy	5,844	❶ (be) modified by	440
vary	4,645	❶ varying the	171
		❷ varied from 〜 to	142

解説

- **change**, **alter** は，否定形の用例も多い．

★ change（〜を変化させる／変化）

名詞の用例が多いが，動詞としても使われる．動詞としては，他動詞・自動詞の両方の用例がある．

❶ **not change ～**（変化しない／～を変えない）

During 1 year of treatment, the intimal index increased in the placebo group by 8%（SE 2）but did not change significantly in the treatment group（0.8%［1］; p=0.008）.（*Lancet. 2002 359:1108*）

訳 しかし，治療群では有意には変化しなかった

❷ **changed to ～**（～に変えられた）

The tyrosine residues at positions 12, 24, 76, and 104 were changed to alanines by mutagenesis of an infectious FCV cDNA clone.（*J Virol. 2004 78:4931*）

訳 12, 24, 76 および 104 番目のチロシン残基が，…の突然変異誘発によってアラニンに換えられた

★★ alter（～を変化させる）

他動詞．

❶ **not alter ～**（～を変化させない）

IL-1β did not alter the expression of SOCS1, SOCS2, and CIS, indicating that they are not involved.（*J Biol Chem. 2000 275:3841*）

訳 IL-1βは，SOCS1, SOCS2 および CIS の発現を変化させなかった

❷ **altered in ～**（～において変化する）

These observations suggest that mdm2 expression is altered in invasive breast cancer and is associated with more aggressive disease.（*Cancer Res. 2001 61:3212*）

訳 mdm2 発現は浸潤性の乳癌において変化する

★ shift（～を変化させる／～を移す／変化）

名詞の用例が多いが，他動詞としても用いられる．

❶ **shifted from ～ to**（～から…に変化した）

Both groups exhibited contrast effects when the concentration was shifted from 32% to 4% within a session.（*Behav Neurosci. 1999 113:732*）

訳 濃度は 32%から 4 %に変化した

★ convert（～を変換する／変換）

他動詞として使われることが多いが，自動詞や名詞の用例もある．

❶ converted to ～（～に変換される）

About one-third of apoE2 was converted to apoE3, and the repair was stable through 12 passages.（*J Biol Chem. 2001 276:13226*）

訳 およそ3分の1のapoE2は，apoE3に変換された

★ transform（形質転換する／～を癌化させる）

他動詞の用例が多いが，自動詞としても用いられる．

❶ transformed with ～（～によって形質転換される）

Hence, cells transformed with Ki-ras tend to be more resistant to FTIs than Ha-ras-transformed cells.（*Cancer Res. 2001 61:8758*）

訳 Ki-rasによって形質転換された細胞は，Ha-rasによって形質転換された細胞よりもFTIsに抵抗性になる傾向がある

★ modify（～を修飾する）

他動詞．

❶ modified by ～（～によって修飾される）

HuCds1 was modified by phosphorylation and activated in response to ionizing radiation.（*Proc Natl Acad Sci USA. 1999 96:3745*）

訳 HuCds1はリン酸化によって修飾され活性化された

★ vary（変動する／異なる／～を変化させる）

自動詞の用例が多いが，他動詞としても使われる．

❶ varying the ～（～を変化させる）

We show that f:AR expression in E19 cells can be precisely modulated by varying the concentration of tetracycline or its chemical derivative doxycycline in the growth media.（*Anal Biochem. 2001 289:217*）

訳 E19細胞におけるf:ARの発現は，テトラサイクリンの濃度を変化させることによって正確に調節されうる

❷ varied from ～ to（～から…で変動した）

Concentrations of E2 in wastewater varied from 0.77 to 6.4 ng/L, while concentrations of E1 were greater (1.6–18 ng/L).（*Anal Chem. 2001 73:3890*）

訳 排水中のE2の濃度は，0.77から6.4 ng/Lで変動した

21. 移動する／移す 【transfer】

移動させる	輸送する	移動する	動員する	取り込む
transfer shift translocate	transport	migrate move	recruit mobilize	incorporate

使い分け

- **transfer** は，「(何かを) 移動させる」ときに広く用いられる.
- **shift** は「～を変化させる」という意味で用いられることも多いが，「～を移す」という意味でも使われる.
- **translocate** は，受動態で「(タンパク質が核へ) 移行する」ときなどに用いられる.
- **transport** は，受動態で「(タンパク質などが) 輸送される」ときに用いられる.
- **migrate** は，「(細胞などが) 移動する／遊走する」ときに用いられる.
- **move** は，「(いろいろなものが) 移動する」ときに広く用いられる.
- **recruit** は「(タンパク質因子など) を動員する」，**mobilize** は「(細胞内カルシウムなどを) 動員する」ときに用いられる.
- **incorporate** は，「取り込む」という意味に使われる.
- **transfer**, **shift**, **translocate**, **recruit**, **mobilize**, **incorporate** は，他動詞として使われる.
- **migrate**, **move** は，自動詞として使われることが多い.

頻度分析

	用例数		用例数
transfer	10,081	❶ (be) transferred to	278
shift	6,246	❶ (be) shifted to	155
translocate	1,243	❶ (be) translocated to	152
transport	9,279	❶ (be) transported to	177
migrate	1,512	❶ migrate to	133
move	1,692	❶ moved to	46
recruite	2,819	❶ (be) recruited to	580
mobilize	485	❶ to mobilize	46
incorporate	2,595	❶ (be) incorporated into	750

解説

- **transferred**, **shifted**, **translocated**, **transported**, **migrate**, **moved**, **recruited** は後に to を，**incorporated** は後に into を伴う用例が多い.

★ transfer（～を移動させる／転移）

名詞の用例が多いが，他動詞としても使われる．

❶ transferred to ～（～に移される）

Three embryos were transferred to the uterus on the fourth day after oocyte retrieval.（*JAMA. 1999 281:1701*）

訳 3つの胎仔が子宮に移された

★ shift（～を移す／～を変化させる／変化）

名詞の用例が多いが，他動詞としても使われる．

❶ shifted to ～（～に移動した）

Peak chemotactic activity was shifted to the right as receptor number decreased.（*J Immunol. 2000 165:4877*）

訳 走化活性のピークは右に移動した

★ translocate（～を移行させる）

他動詞．

❶ translocated to ～（～に移行する）

When activated, NF-κB is translocated to the nucleus, a process that involves the phosphorylation and proteasomal degradation of IκB proteins.（*J Immunol. 1999 163:269*）

訳 NF-κBは核に移行する

★ transport（～に輸送される）

名詞の用例が多いが，他動詞としても用いられる．

❶ transported to ～（～に輸送される）

Using green fluorescent protein（GFP）we show that a GFP-Ogg1 fusion protein is transported to mitochondria.（*Nucleic Acids Res. 2001 29:1381*）

訳 GFP-Ogg1融合タンパク質はミトコンドリアに輸送される

★ migrate（移動する／遊走する）

自動詞．

❶ migrate to ～（～に移動する／～に遊走する）

After activation, T cells can migrate to the liver and induce hepatocyte damage, and thereby serve as a model of autoimmune hepatitis.（*J Exp Med. 1998 188:1147*）

訳 T細胞は肝臓に遊走できる

★ move（移動する／～を移動させる／運動する）

自動詞の用例が多いが，他動詞としても使われる．他動詞受動態は，自動詞とほぼ同様の意味になる場合が多い．

❶ moved to ～（～に移動した）

Most CENP-E-free chromosomes moved to the spindle equator, but their kinetochores bound only half the normal number of microtubules.（*Dev Cell. 2002 3:351*）

訳 ほとんどのCENP-Eのない染色体は，紡錘体の赤道に移動した

★ recruit（～を動員する／～を新しく移す）

他動詞．

❶ recruited to ～（～に動員される）

We find that Cdc15 is recruited to both spindle pole bodies（SPBs）during anaphase.（*Mol Biol Cell. 2001 12:2961*）

訳 Cdc15は，分裂後期の間に両方の紡錘体極（SPBs）に動員される

mobilize（～を動員する）

他動詞．

❶ to mobilize ～（～を動員すること）

The full-length NmU-R2 cDNA was subsequently cloned, stably expressed in 293 cells, and shown to mobilize intracellular calcium in response to neuromedin U.（*J Biol Chem. 2000 275:39482*）

訳 全長のNmU-R2 cDNAが引き続いてクローニングされ，293細胞で安定に発現され，そしてニューロメディンUに応答して細胞内カルシウムを動員することが示された

★ incorporate（～を取り込む）

他動詞．

❶ incorporated into ～（～に取り込まれる）

SUMO-1 was incorporated into HIV-1 virions where it was protected within the virion membrane from digestion by exogenous protease.（*J Virol. 2005 79:910*）

訳 SUMO-1はHIV-1ビリオンに取り込まれた

22. 影響する 【affect】

影響する		
affect	influence	impact

使い分け ◆ **affect**, **influence**, **impact** は，「〜に影響を与える」という意味の他動詞として用いられる．

頻度分析

	用例数		用例数
affect	15,803	❶ not affect	2,230
		❷（be）affected by	1,929
influence	8,167	❶ influence the	1,047
		❷（be）influenced by	992
impact	3,246	❶（be）impacted by	18

解説 ◆ **affect** は，否定形の用例が多い．

★★ affect（〜に影響を与える）

他動詞．

❶ not affect 〜（〜に影響を与えない）

Mutation at L75 did not affect the ability of this protein to interact with pre-22a, as judged from the *in vitro* assay, but this mutation specified a lethal effect for virus growth and abolished the formation of any detectable assembled structure.（*J Virol. 2003 77:4043*）

訳 L75 の変異はこのタンパク質の能力に影響を与えなかった

❷ affected by 〜（〜によって影響を受ける）

In wild-type seeds, LeEXP4 mRNA accumulation was blocked by far-red light and decreased by low water potential but was not affected by abscisic acid.（*Plant Physiol. 2000 124:1265*）

訳 しかし，アブシジン酸によって影響を受けなかった

★ influence（〜に影響を与える／影響）

名詞の用例が多いが，他動詞としても使われる．

❶ **influence the ～**（～に影響を与える）

Genetic factors are believed to influence the development of arterial thromboses.（*J Clin Invest. 2000 105:793*）

訳 遺伝的因子は，動脈血栓症の発生に影響を与えると信じられている

❷ **influenced by ～**（～によって影響を受ける）

This paper investigates how binding kinetics is influenced by the folding of a protein.（*Proc Natl Acad Sci USA. 2000 97:8868*）

訳 この論文は，どのように結合のカイネティクスがタンパク質の折りたたみによって影響を受けるかを精査した

★ impact（～に影響を与える／衝撃）

名詞の用例が多いが，他動詞としても使われる．

❶ **impacted by ～**（～によって影響を受ける）

Phosphorylation and sequestration of the serotonin transporter（SERT）were substantially impacted by ligand occupancy.（*Science. 1999 285:763*）

訳 セロトニントランスポーター（SERT）のリン酸化と隔離現象は，リガンドの占有率によって実質的な影響を受けた

【Ⅱ-F. 疾患】 病気や治療に関係する動詞

23. 病気にかかる 【develop】

発症する	罹患する	罹患させる
develop	suffer	affect predispose infect

使い分け

◆ **develop** には「発生する／開発する」という意味もあるが，「（病気を）発症する」という意味で使われることも多い．

◆ **suffer** は from を伴って「（病気などで）苦しむ」という意味の自動詞だが，「（病気を）患う／（病気に）罹患する」という意味にも用いられる．

◆ **affect** は「影響を与える」という意味だが，受動態の **affected** with の形で「（病気に）冒された／罹患した」という意味で使われる．

◆ **predispose** は，「かかりやすくする」という意味で用いられる．

◆ **develop**，**affect** は他動詞として，**suffer** は自動詞として，**predispose** は自動詞・他動詞の両方で用いられることが多い．

◆ **infect** は，他動詞受動態で with を伴って「（細菌やウイルスに）感染した」という意味で用いられることが多い．

頻度分析

	用例数		用例数
develop	18,466	❶ mice developed	401
suffer	511	❶ suffer from	108
		❷ suffering from	78
affect	15,803	❶ (be) affected with	97
predispose	648	❶ predispose to	117
		❷ predispose ～ to	112
		❸ (be) predisposed to	88
infect	10,966	❶ (be) infected with	2,166

★★ develop（～を発症する／～を開発する／発生する）……………

「発症する／開発する」の意味では他動詞として，「発生する」の意味では自動詞として用いられることが多い．

❶ **mice developed ～**（マウスが～を発症した）

Unexpectedly, eNOS-deficient mice developed much smaller aortic lesions than did wild-type control mice（2544+/−1107 versus 7023+/−1569 microm2/section; P=0.03).（*Circulation. 2002 105:2078*）

訳 eNOS欠損マウスは，野生型のコントロールマウスよりずっと小さな大動脈病変を発症した

suffer（患う／罹患する／～を被る）

自動詞の用例が多いが，他動詞として用いられることもある．"suffer from" の用例が多い．

❶ **suffer from ～**（～を患う／～に罹患している）

Many diabetic patients suffer from cardiomyopathy, even in the absence of vascular disease.（*Diabetes. 2002 51:174*）

訳 多くの糖尿病患者は心筋症に罹患している

❷ **suffering from ～**（～を患う／～に罹患した）

Mutations in the DDB2 subunit have been detected in patients suffering from the repair deficiency disease xeroderma pigmentosum（group E).（*Mol Cell Biol. 2001 21:6738*）

訳 DDB2サブユニットの変異が，修復異常疾患色素性乾皮症に罹患した患者において検出された

⁂ affect（～を冒す／～を罹患させる／～に影響を与える）

他動詞．

❶ **affected with ～**（～に冒された／～に罹患した）

We studied 837 individuals affected with type 2 diabetes and 386 mostly unaffected spouse controls.（*Genome Res. 2001 11:1221*）

訳 われわれは2型糖尿病に罹患した837人を調査した

predispose（かかりやすくする）

他動詞と自動詞の両方の用例がある．"predispose (d) to" の用例が多い．

❶ **predisposed to ～**（～に罹患しやすい）

SCN patients with mutations in the G-CSF receptor gene are predisposed to acute myeloid leukemia.（*J Immunol. 2001 167:6447*）

訳 G-CSF受容体遺伝子に変異を持つSCN患者は，急性骨髄性白血病に罹患しやすい

❷ **predispose ～ to**（～を…にかかりやすくする）

Germline mutations in the tumor suppressor genes BRCA1 and BRCA2 predispose individuals to breast and ovarian cancers.（*Hum Mol Genet. 2001 10:705*）

訳 癌抑制遺伝子 BRCA1 と BRCA2 の生殖系列変異は，人を乳癌および卵巣癌に罹りやすくする

❸ **predispose to ～**（～の素因となる／～に罹患しやすくする）

It is difficult to identify genes that predispose to prostate cancer due to late age at diagnosis, presence of phenocopies within high-risk pedigrees and genetic complexity.（*Nat Genet. 2001 27:172*）

訳 前立腺癌の素因となる遺伝子の同定は困難である

★★ infect（～を感染させる）

他動詞．受動態の用例が非常に多い．

❶ **infected with ～**（～に感染した）

Decreased MHC class I expression was not observed in cells infected with any of the viruses.（*J Virol. 2002 76:6425*）

訳 MHC クラス I 発現の減少は，どのウイルスに感染した細胞にも観察されなかった

24. 治療する／治療を受ける 【treat】

治療する	投薬する	治療を受ける	予防接種する	入院させる	回復させる
treat cure	administer give	undergo receive take	vaccinate	admit	ameliorate

使い分け

◆ **treat** は，「治療する／処置する」という意味で用いられる．

◆ **cure** も「治療する」という意味で用いられるが，「治癒する」ことを前提としている場合が多い．

◆ **administer** は，「投薬する」という意味で使われる．

◆ **give** は，「(薬を) 投与する」という意味でも使われる．

◆ **undergo** は，「(患者が) 手術を受ける」「(細胞が) アポトーシスを起こす」などの意味で用いられることが多い．

◆ **receive** は，「(治療や薬の投与などを) 受ける」という意味で用いられことも多い．

◆ **take** は「(薬を) 服用する／(治療を) 受ける」，**vaccinate** は「ワクチンを接種する／予防接種する」という意味でも用いられる．

◆ **admit** は「認める」という意味だが,「入院させる」という意味でも用いられる.
◆ **ameliorate** は,「回復させる」という意味で使われる.

頻度分析

	用例数		用例数
treat	13,434	❶ (be) treated with	4,600
cure	486	❶ (be) cured of	24
administer	2,736	❶ (be) administered to	305
give	8,225	❶ (be) given to	178
undergo	8,609	❶ underwent transplantation	69
receive	8,261	❶ patients received	404
take	6,630	❶ patients taking	66
vaccinate	768	❶ (be) vaccinated with	161
admit	278	❶ (be) admitted to	161
ameliorate	471	❶ (be) ameliorated by	44

★★ treat（～を治療する／～を処置する）

他動詞．受動態の用例が多い．

❶ treated with ～（～で治療される）

Patients were treated with imatinib in daily oral doses of 400 mg or 600 mg.（*Blood. 2002 99:3530*）

訳 患者はイマチニブで治療された

cure（～を治療する／～を治す／治療）

動詞と名詞の両方の用例がある．動詞としては，“cure A of B〔A（患者）のB（病気）を治療する〕”の構文で用いられることもある．

❶ cured of ～（～が治癒した）

All 40 patients were cured of primary hyperparathyroidism.（*Ann Surg. 2000 231:732*）

訳 40人の患者すべては原発性副甲状腺機能亢進症が治癒した

★ administer（～を投薬する／～を投与する）

他動詞．受動態の用例が多い．

❶ administered to ～（～に投与される）

Thalidomide was administered to 83 patients with myelodysplastic syndrome (MDS), starting at 100 mg by mouth daily and increasing to 400 mg as tolerated. (*Blood. 2001 98:958*)

訳 サリドマイドが骨髄異形性症候群の83人の患者に投与された

★ give（～を与える）

他動詞.

❶ given to ～（～に投与される／～に与えられる）

Heparin was given to 23 patients (92%) for an average of 6 days (range, 8 hrs to 22 days); the mean dose was 588 units/hr, and the mean partial thromboplastin time was 37 secs. (*Crit Care Med. 2001 29:641*)

訳 ヘパリンが23人の患者に投与された

★ undergo（～を受ける／～を経験する）

他動詞.

❶ underwent transplantation（移植手術を受けた）

In a population of 980 patients who underwent transplantation in Oxford, we identified 68 patients with known fistula site who had developed cutaneous malignancies on the upper limbs. (*Transplantation. 2001 71:143*)

訳 オックスフォードで移植手術を受けた980人の患者集団の中で

★ receive（～を受ける）

他動詞.

❶ patients received ～（患者は～を受けた）

Patients received a single dose of either morphine or ketorolac as the first postoperative analgesic when the pain score indicated significant pain. (*Crit Care Med. 1999 27:2786*)

訳 患者はモルヒネかケトロラックのどちらかの1回投与を受けた

★ take（～を服用する／取る）

他動詞.

❶ patients taking ～（～を服用している患者／～を受けている患者）

The benefit was especially large in patients taking drug treatment for blood pressure. (*Lancet. 2002 359:204*)

訳 血圧のための投薬を受けている患者にとって利点は特に大きい

vaccinate（～にワクチン接種する／～に予防接種する）

他動詞．受動態の用例が多い．

❶ **vaccinated with ～（～を接種される）**

Fifty-five infants were vaccinated with measles vaccine at age 6（n=32）or 9（n=23）months, followed by measles-mumps-rubella（MMR）-II vaccine at age 12 months.（*J Infect Dis. 2004 190:83*）

訳 55人の乳児が，6あるいは9カ月で麻疹ワクチンを接種された

admit（～を入院させる／～を認める）

他動詞．受動態の用例が圧倒的に多い．

❶ **admitted to ～（～に入院した）**

Clinical and laboratory results were compared with those of 17 untreated patients who were admitted to the same hospital during this nonrandomized preliminary trial.（*J Infect Dis. 2004 190:1084*）

訳 臨床および検査室の結果は，…の間に同じ病院に入院した17名の未処置の患者のそれらと比較された

ameliorate（～を回復させる）

他動詞．

❶ **ameliorated by ～（～によって回復した）**

In contrast to the effect of inhibition, glomerulonephritis was ameliorated by systemic Slit2 administration.（*Am J Pathol. 2004 165:341*）

訳 糸球体腎炎はSlit2の全身投与によって回復した

【Ⅲ-A. 関連・異同・識別】 関連・異同・識別に関する動詞

1. 関与する 【involve】

関与させる	関与する	役割を果たす
involve implicate engage	participate take part in	play～role in play～part in

使い分け

◆ **involve**, **implicate**, **participate**, **engage**, **take part** は，後に in を伴う用例が多い．

◆ **involve**, **implicate** は「～を巻き込む」という意味の他動詞だが，（be）**involved** in,（be）**implicated** in の形で「～に関与する」という意味に用いられることが多い．

◆他動詞の **engage** は「～を従事させる」という意味だが，（be）**engaged** in の形で自動詞の **engage（s）** in と同様に「～に関与する」という意味で使われる．

◆ **engage** は，他動詞・自動詞の両方で用いられる．

◆ **participate** in, **take part in** は，「～に関与する／～に参加する」という意味で用いられる．

◆ **play ～ role（s） in**, **play ～ part in** は，「…において～な役割を果たす」という意味で用いられる．

頻度分析

	用例数		用例数
involve	24,415	❶（be）involved in	12,930
implicate	6,900	❶（be）implicated in	4,020
engage	654	❶（be）engaged in	125
		❷ engage in	83
participate	3,170	❶ participate in	1,607
take part in	43	❶ take part in	22
play ～ role in	13,949	❶ play a role in	2,124
		❷ play an important role in	1,441
play ～ part in	89	❶ play a part in	20

★★ involve（～を関与させる／～を含む／かかわる）

他動詞．“involved in”の用例が非常に多い．

❶ involved in ～（～に関与する）

The dopamine D2 receptor（D2）is involved in the regulation of pituitary hormone secretion.（*Brain Res. 2002 939:95*）

訳 ドパミンD2受容体（D2）は下垂体ホルモン分泌の調節に関与する

★ implicate（～を関連づける／～を関与させる／～を意味づける）

他動詞．“implicated in”の用例が圧倒的に多い．

❶ implicated in ～（～に関与する／～に関連づけられる）

Immune-mediated injury to the graft has been implicated in the pathogenesis of chronic rejection.（*Transplantation. 2002 74:1053*）

訳 移植に対する免疫介在性傷害は，慢性拒絶反応の病因に関与している

engage（～を関与させる／～に結合する）

他動詞・自動詞の両方の用例がある．“engage(d) in”の用例が多い．

❶ engaged in ～（～に関与する）

Here we report that Etk is engaged in phosphatidylinositol 3-kinase（PI3-kinase）pathway and plays a pivotal role in interleukin 6（IL-6）signaling in a prostate cancer cell line, LNCaP.（*Proc Natl Acad Sci USA. 1998 95:3644*）

訳 Etkはホスファチジルイノシトール-3-リン酸化酵素（PI3-kinase）経路に関与する

❷ engage in ～（～に関与する）

We have found that epithelial cells engage in a process of cadherin-mediated intercellular adhesion that utilizes calcium and actin polymerization in unexpected ways.（*Cell. 2000 100:209*）

訳 上皮細胞はカドヘリンで仲介される細胞間接着に関与する

★ participate（関与する／参加する）

自動詞．“participate(s) in”の用例が圧倒的に多い．

❶ participate in ～（～に関与する）

Thus, CdGAP is a novel GAP that is likely to participate in Cdc42- and Rac-induced signaling pathways leading to actin reorganization.（*J Biol Chem. 1998 273:29172*）

訳 CdGAPは，おそらくCdc42およびRac誘導性シグナル伝達経路に関与するであろう新規のGAPである

take part in（～に関与する／～に参加する）

❶ take part in ～（～に関与する）

Tumor necrosis factor α （TNFα）appears to take part in the pathogenesis of multiple sclerosis and to contribute to the degeneration of oligodendrocytes as well as neurons.（*Brain Res. 2000 864:213*）

訳 腫瘍壊死因子α（TNFα）は，多発性硬化症の病因に関与するように思われる

★★ play ～ role in（…において～な役割を果たす）

“play an important role in” など，「重要な」という意味の形容詞を伴う用例も多い．

❶ play a role in ～（～において役割を果たす／～に関与する）

Our studies suggest that mucin-type *O*-linked glycosylation may be required for normal development and that ppGaNTases may play a role in the regulation of apoptosis.（*J Biol Chem. 2004 279:50382*）

訳 ppGaNTases は，アポトーシスの調節において役割を果たすかもしれない

❷ play an important role in ～（～において重要な役割を果たす）

Phosphorylation of the p53 tumor suppressor protein is likely to play an important role in regulating its activity.（*J Biol Chem. 2000 275:20444*）

訳 p53 癌抑制タンパク質のリン酸化は，おそらくその活性を調節する際に重要な役割を果たしそうである

play ～ part in（…において～な役割を果たす）

“play an important part in” など，「重要な」という意味の形容詞を伴う用例も多い．

❶ play a part in ～（～において役割を果たす／～に関与する）

Together, these two effects may play a part in the pathogenesis of schizophrenia and other neuropsychiatric disorders.（*J Biol Chem. 2001 276:27753*）

訳 これらの２つの効果は，統合失調症の病因において役割を果たすかもしれない

2. 関連する／関連づける 【relate】

関連づける	相関させる
relate associate link connect couple concern	correlate

使い分け

◆いずれの語も「関連」を意味するが，(be) **related** to が「～に関連する」という日本語の意味にもっとも近い．

◆(be) **associated** with は，「～に付随する」という意味合いが強い．

◆(be) **associated** with, (be) **linked** to, (be) **coupled** to, (be) **connected** to は「～に関連する」だけでなく，「～に結合する」という意味も持つ．

◆**couple** は，「関連／併発／結合」などの密接に関係する事象を「共役させる／関連づける」ときに用いられる．

◆**concern** は「～を関連づける」という意味の他動詞で，(be) **concerned** with の形で「～に関係する」という意味で用いられる．

◆**correlate** は，相関性を強調したいときに使われる．

共起・頻度分析

(数字：用例数)

直前の単語				直後の単語	
be動詞	to			to	with
1,862	26	related	17,308	4,901	10
0	90	relate (s)	550	328	0
13,704	3	associated	37,848	25	26,756
1	364	associate (s)	2,716	62	2,038
1,663	5	linked	7,944	2,708	212
1	147	link (s)	3,625	235	47
165	1	connected	575	191	38
0	45	connect (s)	298	29	6
609	6	coupled	4,904	924	864
0	127	couple (s)	852	143	30
34	0	concerned	79	1	51
27	13	concern (s)	785	3	17
1,422	3	correlated	6,593	190	4,814
1	348	correlate (s)	3,545	36	2,405

解説

◆ to または with を伴う他動詞受動態の用例が多い．
◆ **relate**, **link**, **connect** は，後に to を伴うことが多い．
◆ **associate**, **correlate**, **concern** は，後に with を伴う用例が多い．
◆ **couple** の後には，to と with のどちらも用いられる．
◆ **associate**, **correlate** は，自動詞として用いられることも多い．

★★ relate（～を関連づける／関連する）

他動詞として用いられることが多いが，自動詞の用例もある．"related to" の用例が多い．

❶ **related to ～**（～に関連する／～と血縁関係にある）

This protective effect of eliprodil may be related to its reduction (by 78%) of NMDA-induced currents recorded under patch-clamp recording in these cells. (*Invest Ophthalmol Vis Sci. 1999 40:1170*)

訳 eliprodil のこの保護性効果は，…の減少に関連するかもしれない

❷ **relate to ～**（～に関連する）

These differing effects of GH may relate to the progressive increase of LV fibrosis in the CM hamster. (*Circulation. 1999 100:1734*)

訳 GH のこれらの異なる効果は，左心室線維症の進行性増加に関連するかもしれない

★★ associate（～を関連づける／～を付随させる／結合する）

他動詞受動態で，"associated with" の用例が非常に多い．自動詞として用いられることもある．

❶ **associated with ～**〔～に付随（関連）する〕

Epstein-Barr virus (EBV) is associated with the development of a variety of malignancies, including Hodgkin lymphoma. (*Blood. 2003 102:4166*)

訳 エプスタイン・バーウイルス（EBV）は，さまざまな悪性腫瘍の発症に関連している

★ link（～を関連づける／～を連結する／連結）

名詞として用いられることが多いが，他動詞の用例も少なくない．"linked to" の用例が多い．

❶ **linked to ～**（～に関連する／～に連結した）

The abnormal regulation of the Ke 6 gene has been linked to the development of recessive polycystic kidney disease in the mouse. (*J Biol Chem. 1998 273:22664*)

訳 Ke 6遺伝子の異常な調節は，劣性多発性嚢胞腎の発症と関連している

❷ **link ～ to**（～を…に関連づける／～を…に連結する）

The focal adhesion kinase (FAK) protein-tyrosine kinase (PTK) links transmembrane integrin receptors to intracellular signaling pathways.（*EMBO J. 1998 17:5933*）

訳 膜貫通型インテグリン受容体を細胞内シグナル伝達経路に関連づける

★ connect（～を関連づける／～を連結する）

自動詞の用例もわずかながらあるが，他動詞として用いられることが多い．

❶ **connected to ～**（～に関連する／～に連結した）

Their role in microbial infections is poorly defined, however, because no cell-surface HSPG has been clearly connected to the pathogenesis of a particular microbe.（*Nature. 2001 411:98*）

訳 細胞表面のどのHSPGも，これまではっきりとは特定微生物の病因に関連づけられていない

❷ **connect ～ to**（～を…に関連づける／～を…に連結する）

The identification of coding variants of NETs and SERTs would offer important opportunities to connect genotype to phenotype.（*J Neurosci. 2001 21:8319*）

訳 遺伝子型を表現型に関連づける重要な機会

★ couple（～を共役させる／～を関連づける／共役）

他動詞受動態として用いられることが多い．特に“coupled to”“coupled with”の用例が多い．“coupled to”はどのように共役するかを示すことも多い．“coupled with”は共役した結果どうなったかを示す意図がある場合が多い．名詞としても使われる．

❶ **coupled to ～**〔～と共役（関連）する〕

The GATA-1-mediated displacement of GATA-2 is tightly coupled to repression of GATA-2 transcription.（*J Biol Chem. 2005 280:1724*）

訳 GATA-1に仲介されるGATA-2の置換は，GATA-2転写の抑制と強く共役する

❷ **coupled with ～**〔～と共役（関連）する〕

This reduction in NAA was coupled with a marked rise in lactate.（*Brain Res. 2001 907:208*）

訳 このNAAの減少は乳酸の急激な上昇と共役した

concern（～を関連づける／懸念）

名詞の用例が多いが，他動詞としても用いられる．受動態で“concerned with”の用例が多い．

❶ **concerned with ～**（～に関係する）

This study is concerned with the role of impulse activity and synaptic transmission in early thalamocortical development.（*J Neurosci. 2002 22:10313*）

訳 この研究は活動電位活性の役割に関係する

⁑ correlate（～を相関させる／相関する）

他動詞・自動詞の両方の用例がある．“correlate(d) with”の用例が多い．

❶ **correlated with ～**（～と相関する／～と相関した）

We also demonstrate that the OM-induced morphological changes are correlated with increased cell motility in a STAT3-dependent manner.（*Oncogene. 2003 22:894*）

訳 OMに誘導される形態的変化は，細胞運動性の上昇と相関する

❷ **correlate with ～**（～と相関する）

Thus, the cell growth phenotype correlates with the inherent thermal stability of the guanylyltransferase.（*J Biol Chem. 2001 276:36116*）

訳 細胞増殖表現型は，グアニル酸転移酵素の固有の温度安定性と相関する

3. 伴う／付随する　【accompany】

伴う／付随する		
accompany	follow	associate

使い分け

◆ **accompanied** by は，物事が同時に起こる場合に用いられることが多い．

◆引き続いて何かが起こる場合は，**follwed** by が用いられる．

◆ **associated** with は，「～に付随（関連）する」という意味で用いられる．

頻度分析	用例数		用例数
accompany	3,487	❶ (be) accompanied by	2,531
follow	20,136	❶ (be) followed by	4,574
associate	40,738	❶ (be) associated with	26,765

★ accompany（～を伴う／～を付随させる）

他動詞．"accompanied by" の用例が圧倒的に多い．

❶ **accompanied by ～**〔～を伴う（～によって伴われる）〕

This reduction in the lesion volume has been accompanied by an increase in the intensity in the 18F-FDG signal per voxel.（*J Nucl Med. 2002 43:876*）

訳 この病変部の体積の減少は，18F-FDG シグナルの強度の上昇を伴っていた

★★ follow（～を伴う／～に続く）

他動詞．"followed by" の用例が多い．

❶ **followed by ～**〔～を伴う（～によって伴われる）〕

For many G protein-coupled receptors, agonist-induced activation is followed by desensitization, internalization, and resensitization.（*J Biol Chem. 2002 277:38524*）

訳 作用物質に誘導される活性化は脱感作を伴う

★★ associate（～を付随させる／～を関連づける／結合する）

他動詞受動態で，"associated with" の用例が非常に多い．自動詞として用いられることもある．

❶ **associated with ～**〔～に付随（関連）する〕

Mutation in the BRCA1 gene is associated with an increased risk of breast and ovarian cancer.（*J Biol Chem. 2002 277:33422*）

訳 BRCA1 遺伝子の変異は，乳癌および卵巣癌のリスクの上昇に付随（関連）している

4. 一致する／似ている 【correspond】

一致する	一致させる	似ている
correspond coincide agree	fit match	resemble represent

使い分け

◆ **correspond** to, **coincide** with, **agree** with は，「～に一致する」という意味の自動詞として使われる．

◆ **agreed** to は，「（患者が）～に同意した／～を承諾した」場合に用いられることが多い．

◆ **fit** は「一致させる／適合させる」という意味の他動詞で，（be）**fitted** to ／ with の形で「～に一致する／～に合致する」という意味に用いられる．

◆ **match** は，（be）**matched** to ／ with の形で，調査研究の対照群（control）の条件（年齢・性別など）が被験群と「一致する」ことを示すために用いられることが多い．

◆ **resemble** は「～に似ている」という意味の他動詞で，能動態で使われる．

◆ **represent** は「～を表す」という意味の他動詞として使われることが多いが，「～に相当する」という意味でも用いられる．

共起・頻度分析　　（数字：用例数）

直前の単語			直後の単語			
be動詞			to	with	by	a
0	corresponded	389	**288**	61	1	0
0	correspond (s)	1,143	**1,037**	29	1	0
0	coincided	277	0	**268**	0	0
0	coincide (s)	399	2	**364**	0	0
5	agreed	207	36	**72**	0	0
0	agree (s)	356	3	**188**	0	0
116	fitted	259	**62**	**53**	27	5
121	fit (s)	1,141	152	48	34	49
200	matched	2,398	**112**	**58**	**58**	6
0	match (es)	845	73	23	5	10
0	resembled	299	1	0	0	13
0	resemble (s)	1,256	0	0	0	90
288	represented	1,047	1	4	**286**	55
2	represent (s)	6,226	3	1	1	**2,733**

解説
- ◆**correspond** は自動詞で，後に to を伴うことが非常に多い．
- ◆**coincide**, **agree** は自動詞で，後に with を伴うことが多い．
- ◆他動詞の **fit**, **match** は受動態で使われる場合が多く，後に to または with を伴うことも多い．
- ◆**resemble** は，通常，能動態の他動詞として用いられ，後ろに to や with がくることはない．

★ correspond（一致する／相当する）

自動詞．“correspond to” の用例が圧倒的に多い．

❶ correspond to ～（～と一致する／～に相当する）

The processing of gp84 by the cellular protease furin generates gp43, which corresponds to the C-terminal part of gp84.（*J Virol. 2001 75:7078*）

訳 そして，それは gp84 の C 末端部分に相当する

coincide（一致する）

自動詞．“coincide with” の用例が非常に多い．

❶ coincided with ～（～と一致する）

Increased arginase activity coincided with decreased plasma arginine concentration.（*Ann Surg. 2001 233:393*）

訳 アルギナーゼ活性の上昇は，血漿アルギニン濃度の低下と一致した

agree（一致する／同意する）

自動詞．“agree with” の用例が非常に多い．

❶ agree with ～（～と一致する）

These results agree with previous results obtained in this laboratory using newly synthesized L subunits made in intact chloroplasts.（*Plant Physiol. 1993 101:523*）

訳 これらの結果は，…において得られた以前の結果と一致する

★ fit（～を一致させる／～を適合させる／一致する）

他動詞として用いられることが多いが，形容詞や名詞の用例も少なくない．

❶ fitted to ～（～に一致する／～に合致する）

Logistic regression models were fitted to cross-sectional data to estimate the effects of in utero exposure to maternal smoking and previous and current ETS exposure on the prevalence of wheezing and physician-diagnosed asthma.（*Am J Respir Crit Care Med. 2001 163:429*）

訳 ロジスティック回帰モデルは，横断的なデータに一致した

❷ **fitted with ～**（～と一致する／～と合致する）

The a-wave data were fitted with a model based on photopigment transduction to obtain values for log Rmax（maximum response）and log S（sensitivity）.（*Invest Ophthalmol Vis Sci. 2004 45:275*）

訳 a 波のデータは，…に基づくモデルと一致した

★ match（～を一致させる）

名詞としても用いられるが，他動詞の用例が多い．

❶ **matched to ～**（～に一致する／～に匹敵する）

In the case-control study, case patients were limited to theme park hotel visitors and controls were matched to case patients by age group and hotel check-in date.（*JAMA. 1998 280:1504*）

訳 コントロールは，年齢グループとホテルのチェックイン日について症例患者に一致した

❷ **matched with ～**（～と一致する／～に匹敵する）

Control subjects and their plasma samples were matched with case patients according to baseline $CD4^{+}$ T cell count, transfusion history, HIV risk factor, and follow-up time.（*J Infect Dis. 2002 186:114*）

訳 コントロール患者とその血漿サンプルは，…について症例患者と一致した

★ resemble（～に似ている）

他動詞．受動態にはならない．

❶ **closely resemble ～**（～と緊密に似ている）

The overall structure of HMP2 most closely resembles that of meprins, a subgroup of astacin metalloproteinases.（*Development. 2000 127:129*）

訳 HMP2 の全体の構造は，メプリンのそれともっとも緊密に似ている

★ represent（～に相当する／～を表す／～である）

他動詞．

❶ **represent a ～**（～に相当する）

These data suggest that C5 inhibition may represent a novel therapeutic strategy for preventing complement-mediated inflammation and tissue injury.（*Circulation. 1999 100:2499*）

訳 C5の抑制は，…のための新規の治療戦略であるかもしれない

5. 異なる 【differ】

異なる	
differ	vary

使い分け
◆ **differ** は，「異なる」という意味の自動詞として用いられる．
◆ **vary** は「変動する」という意味の自動詞として使われることが多いが，「異なる」という意味にも用いられる．

頻度分析

	用例数		用例数
differ	5,483	❶ differ in	728
		❷ differ from	497
vary	4,645	❶ varied in	121

★ differ（異なる）

自動詞．“differ in”“differ from”の用例が多い．

❶ **differ in ～**（～において異なる）

FGF-1 and FGF-2 differ in their ability to bind isoforms of the FGF receptor family as well as the heparin-like glycosaminoglycan (HLGAG) component of proteoglycans on the cell surface to initiate signaling in different cell types.（*Proc Natl Acad Sci USA. 1999 96:1892*）

訳 FGF-1とFGF-2は，…するそれらの能力において異なっている

❷ **differ from ～**（～と異なる）

Cell lines produced by these approaches are invariably transformed, genomically unstable and display cellular properties that differ from their normal counterpart.（*Hum Mol Genet. 2000 9:403*）

訳 それらの正常な対応物と異なる性質

★ vary（異なる／変動する）

自動詞．

❶ varied in ～（～において異なる）

Both *in vitro* and *in vivo*, individual RSV PZ escape mutants varied in their susceptibility to PZ.（*J Infect Dis. 2004 190:1941*）

訳 個々のRSV PZエスケープミュータントは，PZに対する感受性において異なっていた

6. 区別する 【distinguish】

区別する	識別する
distinguish **differentiate**	**discriminate**

使い分け

◆いずれの語も，後にbetweenあるいはamongを伴って，「～の間を区別する（識別する）」という意味の自動詞として用いられる．

◆betweenは主に二者を区別する場合に用い，三者以上の場合はamongを用いる．

◆**distinguish**／**discriminate**／**differentiate**～from … の形で，「～を…と区別する（識別する）」という意味の他動詞として使われることもある．

◆**discriminate**は，**distinguish**よりも微妙な差を区別する場合に用いられる．

◆**differentiate**は，「分化する」という意味の自動詞して用いられることが多いが，「区別する」という意味でも使われる．

頻度分析

	用例数		用例数
distinguish	2,160	❶ distinguish ～ from	320
		❷ distinguish between	311
differentiate	3,721	❶ differentiate between	88
		❷ differentiate ～ from	83
discriminate	896	❶ discriminate between	248
		❷ discriminate ～ from	67

★ distinguish（区別する）

自動詞として用いられることが多いが，他動詞の用例もある．

❶ **distinguish ～ from**（～を…と区別する）

An ANN was used to distinguish benign from malignant nodules on the basis of subjective or objective features.（*Radiology. 2000 214:823*）

訳 ANNが，良性の結節を悪性の結節と区別するために使われた

❷ **distinguish between ～**（～を区別する）

The pattern of joint involvement is often used to distinguish between rheumatoid and psoriatic arthritis.（*Arthritis Rheum. 2000 43:865*）

訳 …は，リウマチ様関節炎と乾癬性関節炎とを区別するためにしばしば使われる

★ differentiate（区別する／分化する）

自動詞として用いられることが多いが，他動詞の用例もある．

❶ **differentiate between ～**（～を区別する）

To differentiate between signals for proliferation and morphogenesis, we used a cloned mammary epithelial cell line that lacks epimorphin, an essential mammary morphogen.（*Development. 2001 128:3117*）

訳 増殖と形態形成のシグナルを区別するために

❷ **differentiate ～ from**（～を…と区別する）

It is possible to differentiate malignant from healthy cells and to classify diseases based on identification of specific gene expression profiles.（*Blood. 2004 104:4002*）

訳 悪性の細胞を健康な細胞と区別することは可能である

discriminate（識別する）

自動詞として用いられることが多いが，他動詞の用例もある．

❶ **discriminate between ～**（～を識別する）

By contrast, NBF-1 carrying the ΔF508 mutation loses the ability to discriminate between these two phospholipids.（*Biochemistry. 2002 41:11161*）

訳 ΔF508変異を持つNBF-1は，これら2つのリン脂質を識別する能力を失う

❷ **discriminate ～ from**（～を…と識別する）

The diagnostic potential of expression profiling is emphasized by its ability to discriminate primary lung adenocarcinomas from metastases of extra-pulmonary origin.（*Proc Natl Acad Sci USA. 2001 98:13790*）

訳 原発性肺腺癌を肺以外の由来の転移と区別する能力

【Ⅲ-B. 性質】 性質について述べるときに使う動詞

7. 持つ 【have】

持つ	
have	possess

使い分け
◆どちらも「～を持つ」という意味の他動詞として用いられる．
◆能動態の用例が多い．

頻度分析

	用例数		用例数
have	153,660	❶ had a	4,241
possess	3,063	❶ possesses a	261

★ have（～を持つ）

現在完了の用例が多いが，「～を持つ」という意味の他動詞としても使われる．

❶ had a ～（～を持っていた）

African-Americans had a significantly higher relative risk for graft loss than either Caucasians (1.57, *P*<0.0005) or Hispanics (2.01, *P*<0.0003). (*Transplantation. 2000 70:288*)

訳 アフリカ系アメリカ人は，…に関して有意により高い相対リスクを持っていた

★ possess（～を持つ）

❶ possesse a ～（～を持つ）

These data demonstrate that human brain endothelial cells possess a unique hematopoietic activity that increases the repopulating capacity of adult human bone marrow. (*Blood. 2002 100:4433*)

訳 ヒト脳内皮細胞はユニークな造血活性を持つ

8. 含む 【contain】

含む		
contain	include	involve

使い分け
- ◆いずれの語も現在分詞の用例が多い．
- ◆ **contain, include** は，「(実際に何かを) 含む」場合に用いられる．
- ◆ **involve** は，「(現象や機能など抽象的なものを) 含む」ときに用いられる．
- ◆ **involve** は，受動態で「～に関与する」の意味で使われることも多い．

頻度分析

	用例数		用例数
contain	33,276	❶ containing the	2,246
		❷ contains a	1,927
include	28,021	❶ including the	2,709
		❷ (be) included in	568
involve	24,485	❶ involving the	730

☆ contain（～を含む）

❶ containing the ～（～を含む）

GST-fusion proteins containing the DEP1 catalytic domain with a substrate-trapping D/A mutation were found to interact with p120 (ctn), a component of adherens junctions. (*Oncogene. 2002 21:7067*)

訳 DEP1 触媒ドメインを含む GST 融合タンパク質

❷ contain a ～（～を含む）

The predicted SPN-4 protein contains a single RNA recognition motif (RRM), and belongs to a small subfamily of RRM proteins that includes one *Drosophila* and two human family members. (*Development. 2001 128:4301*)

訳 予想される SPN-4 タンパク質は 1 つの RNA 認識モチーフを含む

☆ include（～を含む）

❶ including the ～（～を含む）

However, mu-calpain null platelets exhibit impaired tyrosine phosphorylation of several proteins including the β3 subunit of αⅡbβ3 integrin, correlating with the agonist-induced reduction in platelet aggregation.（*Mol Cell Biol. 2001 21:2213*）

訳 …のβ3サブユニットを含むいくつかのタンパク質

❷ **included in ～**（～に含まれる）

Of 50 identified studies, 12 trials（941 patients）were included in the analysis.（*JAMA. 2001 285:193*）

訳 12の治験（941人の患者）が解析に含まれた

★★ involve（～を含む／～を関与させる）

"involved in（～に関与する）" の用例が多いが，「～を含む」という意味でも用いられる．

❶ **involving the ～**（～を含む）

We conclude that kainate receptors present at presynaptic terminals in the rat hippocampus mediate the facilitation of glutamate release through a mechanism involving the activation of an adenylyl cyclase-second messenger cAMP-protein kinase A signalling cascade.（*J Physiol. 2004 557:733*）

訳 …の活性化を含む機構によって

9. 維持する／保持する 【maintain】

維持する	保持する
maintain sustain keep	hold retain

使い分け

◆ **maintain**, **sustain**, **keep** は，「維持する」という意味の他動詞として用いられる．

◆ **maintain** は「(ある好ましい状態を) 維持する」ときに，**sustain** は「(何かを高い状態で) 維持する」場合に用いられることが多い．

◆ **hold**, **retain** は，「保持する」という意味で用いられる．

頻度分析

	用例数		用例数
maintaine	6,700	❶（be）maintained in	481

sustaine	3,020	❶ (be) sustained for	60
keep	527	❶ (be) kept in	41
hold	1,206	❶ (be) held in	102
retain	3,355	❶ (be) retained in	328

★ maintain（～を維持する）

他動詞.

❶ maintained in ～（～で維持される）

After lens formation, Pax6 expression is maintained in the lens epithelium, whereas its level abruptly decreases in differentiated fiber cells.（*Invest Ophthalmol Vis Sci. 2004 45:3589*）

訳 Pax6 の発現はレンズ上皮で維持されている

★ sustain（～を維持する）

他動詞.

❶ sustained for ～（～の間維持される）

β_8 integrin mRNA expression increased within 3–6 h of Fas ligation due to enhanced mRNA stabilization, and mRNA increases were sustained for 48-72 h.（*J Biol Chem. 2002 277:47826*）

訳 メッセンジャー RNA の上昇は，48 ～ 72 時間の間維持された

keep（～を維持する）

他動詞.

❶ kept in ～（～で維持される）

C57BL/6 mice were kept in an atmosphere of >95% O_2 for 4 days followed by return to room air.（*J Immunol. 2003 171:955*）

訳 C57BL/6 マウスは，…の大気中で維持された

★ hold（～を保持する／～を維持する）

他動詞.

❶ held in ～（～に保持される）

Each heme is held in place by thioether bonds between the heme vinyl groups and Cys residues.（*Biochemistry. 2001 40:13483*）

訳 それぞれのヘムは，…によって適当な位置に保持されている

★ retain（～を保持する）

他動詞．

❶ retained in ～（～に保持される）

In resting T cells, NFATc is retained in the cytoplasm by a mechanism that depends on multiple phosphorylations in an N-terminal regulatory domain.（*Proc Natl Acad Sci USA. 2000 97:7130*）

訳 NFATc は，…する機構によって細胞質に保持されている

10. 続ける／持続する 【continue】

続ける	持続する
continue	persist last

使い分け

◆ **continue** は，「続ける」という意味の他動詞として用いられることが多い．

◆ **persist**, **last** は，「持続する」という意味の自動詞として使われる．

頻度分析

	用例数		用例数
continue	2,515	❶ continues to	274
persist	1,808	❶ persisted for	138
last	2,317	❶ long-lasting	348
		❷ lasted for	22

★ continue（～を続ける）

他動詞の用例が多いが，自動詞としても用いられる．“continue to ［*do*］”の用例が多い．

❶ continue to ～（～し続ける）

Human immunodeficiency virus（HIV）continues to be a major public health problem throughout the world, with high levels of mortality and morbidity associated with AIDS.（*J Virol. 2002 76:2835*）

訳 ヒト免疫不全症ウイルス（HIV）は，世界中で主要な公衆衛生上の問題であり続ける

★ persist（持続する）

自動詞.

❶ persisted for ～（～の間持続した）

After introduction into DU 145 prostate cancer cells inhibition of telomerase activity persisted for up to 7 days, equivalent to six population doublings.（*Nucleic Acids Res. 2001 29:1683*）

訳 テロメアーゼ活性の抑制は，最大7日間まで持続した

last（持続する／最後の）

形容詞の用例が多いが，自動詞としても用いられる.

❶ long-lasting ～（長く持続する～）

These results suggest that the Lp channel may play a critical role in the induction of long-lasting changes in synaptic strength.（*J Neurosci. 2000 20:4786*）

訳 Lp チャネルは，シナプス強度の長く持続する変化の誘導において決定的に重要な役割を果たすかもしれない

❷ lasted for ～（～の間持続した）

The attenuation lasted for more than 4 days.（*Brain Res. 2003 988:97*）

訳 減弱は4日間以上持続した

11. 保存する／貯蔵する 【conserve】

保存する	貯蔵する
conserve preserve	store

使い分け

◆ **conserve** は，受動態で「(種や進化を超えて) 保存されている」という意味で使われる.

◆ **preserve** は，「(一部分を失っても機能を) 保存している／保持している」場合などに用いられる.

◆ **store** は，「貯蔵する」という意味で用いられることが多い.

頻度分析

	用例数		用例数
conserve	10,524	❶ (be) conserved in	974

preserve	1,297	❶（be）preserved in	155
store	1,954	❶（be）stored in	159

★★ conserve（～を保存する）

他動詞．受動態の用例が多い．

❶ conserved in ～（～において保存される）

Because this domain is conserved in all proteins of the large Maf family, we hypothesized that NRL-MTD played an important role in assembling the transcription initiation complex.（*J Biol Chem. 2004 279:47233*）

訳 このドメインは，…のすべてのタンパク質において保存されている

★ preserve（～を保存する／～を保持する）

他動詞．

❶ preserved in ～（～において保存される）

This unique and essential function of the C terminus is preserved in the absence of the N-terminal catalytic domains, suggesting that the C terminus can interact with and recruit other DNA polymerases to the site of initiation.（*Mol Cell Biol. 2001 21:4495*）

訳 C末端のこのユニークで必須の機能は，N末端の触媒ドメインなしに保存されている

★ store（～を貯蔵する）

他動詞．

❶ stored in ～（～において貯蔵される）

The mature 169 amino acid protein is stored in the cytoplasmic granules of neutrophils and eosinophils but is absent from lymphocytes, monocytes, and platelets.（*J Biol Chem. 2002 277:19658*）

訳 成熟した169アミノ酸のタンパク質は，…の細胞質顆粒に貯蔵されている

12. 蓄積する／沈着する 【accumulate】

蓄積する	沈着する
accumulate	deposit

使い分け
◆ **accumulate** は，「蓄積する」という意味の自動詞として用いられる．
◆ **deposit** は，「沈着させる」という意味の他動詞として使われる．

頻度分析

	用例数		用例数
accumulate	3,254	❶ accumulate in	374
deposit	1,065	❶ (be) deposited in	108

★ accumulate（蓄積する）

自動詞．

❶ **accumulate in ～**（～に蓄積する）

In the absence of G_q signaling, eosinophils failed to accumulate in the lungs following allergen challenge.（*J Immunol. 2002 168:3543*）

訳 好酸球は肺に蓄積するのに失敗した

deposit（～を沈着させる／沈着）

他動詞として用いられるが，名詞の用例も多い．

❶ **deposited in ～**（～に沈着する）

The studies reported here clearly demonstrate that, in similar cultures, crystalline calcium phosphate material is deposited in the chondrocytic and fibrocytic matrices.（*Dev Biol. 1976 52:283*）

訳 結晶性のリン酸カルシウム物質は，軟骨細胞性および線維芽細胞性の基質に沈着する

13. 構成する／～から成る 【compose】

構成する	成す
compose comprise constitute	consist

使い分け

◆（be）**composed** of,（be）**comprised** of は，「～で構成される」という意味で用いられる．

◆ **constitute** は，「（抽象的なものを）構成する」という意味の他動詞として使われる．

◆ **consist** of は，「～から成る／～で構成される」という意味の自動詞として用いられる．

頻度分析

	用例数		用例数
compose	2,590	❶（be）composed of	2,329
comprise	2,997	❶（be）comprised of	539
		❷ comprise a	246
constitute	1,632	❶ constitute the	241
consist	4,975	❶ consists of	1,690

★ compose（～を構成する）

他動詞．“composed of” の用例が圧倒的に多い．

❶ **composed of ～**（～で構成される／～から成る）

The interferon α receptor is composed of two subunits: IFNaR1 and IFNaR2.（*Biochemistry. 2002 41:11261*）

訳 インターフェロンα受容体は，2つのサブユニットで構成されている

★ comprise（～を構成する）

他動詞．“comprised of” の用例が多い．

❶ **comprised of ～**（～で構成される／～から成る）

MEKK1 is comprised of a kinase domain and a long amino-terminal regulatory domain.（*J Biol Chem. 2004 279:1872*）

訳 MEKK1は，キナーゼドメインと長いアミノ末端調節ドメインで構成されている

② **comprise a ～**（～を構成する）

Hence, a CpG motif is necessary and sufficient to comprise a binding site for CpG-binding protein, although the immediate flanking sequence affects binding affinity.（*J Biol Chem. 2001 276:44669*）

訳 CpG モチーフは，CpG 結合タンパク質の結合部位を構成するのに必要かつ十分である

★ constitute（～を構成する）

❶ **constitute the ～**（～を構成する）

Our results constitute the first demonstration of an impairment in the phagocytosis of apoptotic cells by macrophages *in vivo* in a mammalian system.（*J Exp Med. 2000 192:359*）

訳 われわれの結果は，…の最初の証拠を構成する

★ consist（成る）

自動詞．"consist of" の用例が非常に多い．

❶ **consist of ～**（～から成る／～で構成される）

The asymmetrical channel structure consists of two major regions: a heart-shaped region connected at its widest end with a handle-shaped region.（*Proc Natl Acad Sci USA. 2002 99:10370*）

訳 非対称的なチャネル構造は主要な 2 つの領域から成る

14. 必要とする 【require】

必要とする		
require	need	necessitate

使い分け

◆ **require**, **need**, **necessitate** は，「～を必要とする」という意味の他動詞として用いられる．

◆ **require**, **need** は，受動態で後に for または to 不定詞を伴う用例が多い．

◆ **necessitate** の用例は，非常に少ない．

頻度分析

	用例数		用例数
require	32,946	❶ (be) required for	14,896
		❷ (be) required to	2,386
need	5,528	❶ (be) needed to	1,107
		❷ (be) needed for	619
necessitate	194	❶ necessitated the	15

★★ require（～を必要とする）

他動詞．"required for" の用例が多い．

❶ required for ～ （～のために必要とされる）

Later in flower development, the AP1 gene is required for normal development of sepals and petals.（*Plant J. 2001 26:385*）

訳 AP1 遺伝子は，…の正常な発生に必要とされる

❷ required to ～ （～するために必要とされる）

We also show that endogenous Ras is required to maintain normal levels of dMyc, but not dPI3K signaling during wing development.（*Genes Dev. 2002 16:2286*）

訳 内在性の Ras は，dMyc の正常レベルを維持するために必要とされる

★ need（～を必要とする／必要性）

名詞の用例が多いが，他動詞としても使われる．助動詞の用例はほとんどない．

❶ needed to ～ （～するために必要とされる）

Further studies are needed to determine whether an increased calcium intake has long-term benefits in Gambian children.（*Am J Clin Nutr. 2000 71:544*）

訳 さらなる研究が，…かどうかを決定するために必要とされる

❷ needed for ～ （～のために必要とされる）

Because MHC class II $^{-/-}$ (class II $^{-/-}$) mice generate efferent Tr cells following a.c. inoculation, we conclude that conventional CD4^{+} T cells are not needed for the development of efferent CD8^{+} T cells.（*J Immunol. 2003 171:1266*）

訳 通常の CD4^{+} T 細胞は，輸出 CD8^{+} T 細胞の発生のためには必要とされない

necessitate（～を必要とする）

❶ **necessitated the ～**（～を必要とした）

Functional GJIC has been shown in long-term primary rat hepatocyte cultures, which have been implemented widely to study various aspects of hepatocellular function; however, the onset of transgenic technology in murine species has necessitated the development of a primary mouse hepatocyte system.（*Hepatology. 2003 38:1125*）

訳 マウス種におけるトランスジェニックテクノロジーの開始は，初代マウス肝細胞システムの開発を必要とした

15. 分ける／隔てる／分離する 【separate】

分ける／隔てる	分離する	解離する
separate divide interrupt	segregate	dissociate

使い分け

◆ **separate** は，他動詞受動態で「隔てられる／離れている」という意味に用いられることが多い．

◆ **divide** は，受動態で「（人によって）分けられた」場合に使われることが多い．

◆ **interrupt** は，「（遺伝子がイントロンなどによって）分断されている」という意味で用いられることが多い．

◆ **segregate** は，自動詞としては「（遺伝的に）分離する」，他動詞としては「分ける／分類する」という意味で使われる．

◆ **dissociate** は，自動詞としては「解離する」，他動詞としては「分離する」という意味で用いられる．

頻度分析

	用例数		用例数
separate	4,277	❶ (be) separated by	672
		❷ (be) separated from	237
divide	1,554	❶ (be) divided into	547
interrupt	386	❶ (be) interrupted by	91
segregate	703	❶ (be) segregated into	39
		❷ segregates with	31

dissociate	1,091	❶ (be) dissociated from	116
		❷ dissociates from	76

★ separate（～を隔てる／～を分ける／～を分離する／別々の）…

形容詞の用例が多いが，動詞の用例もかなりある．

❶ **separated by ～**（～によって隔てられる）

The first and second transmembrane helices are separated by a short loop from residues 48 to 52.（*Biochemistry. 2002 41:12876*）

訳 1番目と2番目の膜貫通ヘリックスは，短いループによって隔てられている

❶ **separated by ～**（～ほど離れている）

Previous studies suggest that the function of these elements involves formation of an RNA stem structure, even though they are separated by more than 700 nucleotides.（*J Biol Chem. 2002 277:50143*）

訳 それらは700塩基以上離れている

❷ **separated from ～**（～から隔てられる）

The parasite is separated from the host cell by a unique electron-dense structure of unknown composition.（*Infect Immun. 2000 68:2315*）

訳 寄生体は，ユニークな高電子密度の構造によって宿主細胞から隔てられている

★ divide（～を分ける／分裂する）……………………………………

他動詞の用例が多いが，自動詞として使われることもある．"divided into" の用例が多い．

❶ **divided into ～**（～に分けられる）

Patients were divided into two groups based on the radiographic or clinical status of their cervical spine: cleared and noncleared.（*Crit Care Med. 2000 28:3436*）

訳 患者は，…に基づいて2つのグループに分けられた

interrupt（～を分断する／～を隔てる）……………………………

他動詞．

❶ **interrupted by ～**（～によって分断される）

The gene spans about 17.8 kb and contains 22 exons interrupted by 21 introns.（*Gene. 2002 291:123*）

訳 その遺伝子は，およそ17.8 kbに渡っており，21個のイントロンによ

って分断される22個のエクソンを含んでいる

segregate（分離する／～を分ける／～を分類する）……………

自動詞・他動詞の両方の用例がある.

❶ segregated into ～（～に分けられる）

Patients were segregated into four diagnostic categories: meconium ileus (MI), prenatal/neonatal screening (SCREEN), positive family history (FH), and symptoms other than meconium ileus (SYMPTOM).（*Am J Epidemiol. 2004 159:537*）

訳 患者は4つの診断上のカテゴリーに分けられた

❷ segregate with ～（～とともに分離する）

We have identified a mutation in the ANKH gene that segregates with the disease in a family with this condition.（*Am J Hum Genet. 2002 71:985*）

訳 われわれは，その疾患とともに分離するANKH遺伝子の変異を同定した

★ dissociate（解離する／～を分離する）…………………………

自動詞・他動詞の両方の用例がある.

❶ dissociated from ～（～と分離される）

These results demonstrate that the effects of RU-24969 on locomotor activity can be dissociated from its effects on immediate early gene expression within the striatum.（*Brain Res. 2000 852:247s*）

訳 運動活性に対するRU-24969の効果は，即時型遺伝子発現に対するそれの効果とは分離されうる

❷ dissociate from ～（～から解離する）

Upon phosphorylation of Cbl, the CAP-Cbl complex dissociates from the insulin receptor and moves to a caveolin-enriched, triton-insoluble membrane fraction.（*Nature. 2000 407:202*）

訳 CAP-Cbl複合体はインスリン受容体から解離する

16. 制限する／限られる 【limit】

制限する		
limit	restrict	confine

使い分け ◆（be）**limited** to,（be）**restricted** to,（be）**confined** toは,「～に限られる」という意味で用いられる.

頻度分析

	用例数		用例数
limit	10,166	❶（be）limited to	746
		❷（be）limited by	601
restrict	4,610	❶（be）restricted to	1,359
confine	783	❶（be）confined to	511

解説 ◆受動態の用例が非常に多い.

★★ limit（～を制限する／制限）

名詞の用例もあるが，他動詞として用いられることの方が多い.

❶ limited to ～（～に限られる）

Expression of this gene is limited to cells of hemopoietic origin, in keeping with the previously defined tissue expression of the HA-2 Ag.（*J Immunol. 2001 167:3223*）

訳 この遺伝子の発現は造血性由来の細胞に限られている

❷ limited by ～（～によって制限される）

Interleukin-12 can act as a potent adjuvant for T cell vaccines, but its clinical use is limited by toxicity.（*J Immunol. 2004 172:5159*）

訳 それの臨床応用は毒性によって制限されている

★ restrict（～を制限する）

他動詞.“restricted to”の用例が多い.

❶ restricted to ～（～に限られる）

In situ hybridization demonstrates that dDAT mRNA expression is restricted to dopaminergic cells in the fly nervous system.（*Mol Pharmacol. 2001 59:83*）

訳 dDAT メッセンジャー RNA の発現はドーパミン作動性細胞に限られている

confine（～を制限する）

他動詞．"confined to" の用例が非常に多い．

❶ confined to ～（～に限られる）

Confocal microscopy demonstrated that PSCA expression in TCC is confined to the cell surface.（*Cancer Res. 2001 61:4660*）

訳 移行上皮癌における前立腺幹細胞抗原の発現は，細胞表面に限られている

17. 避ける【avoid】

避ける		
avoid	circumvent	escape

使い分け

◆ **avoid**, **circumvent** は，「避ける」という意味の他動詞として用いられる．

◆ **escape** は，「避ける／逃れる」という意味で，他動詞および自動詞の両方で用いられる．

頻度分析

	用例数		用例数
avoid	975	❶ to avoid	345
circumvent	257	❶ to circumvent	116
escape	961	❶ to escape	108

avoid（～を避ける）

他動詞．

❶ to avoid ～（～を避ける…）

Efficient detoxification of bile acids is necessary to avoid pathological conditions such as cholestatic liver damage and colon cancer.（*Hepatology. 2005 41:168*）

訳 胆汁酸の効率的な解毒は，病的な状態を避けるために必要である

circumvent（～を回避する）

他動詞.

❶ to circumvent ～（～を避ける…）

One potential means to circumvent this liability would be to express an inhibitory neuroactive peptide and constitutively secrete the peptide from the transduced cell.（*Nat Med. 2003 9:1076*）

訳 この傾向を回避する１つの可能な手段は，抑制的な神経ペプチドを発現することであろう

escape（～を避ける／逃れる）

他動詞および自動詞の両方で使われる.

❶ to escape ～（～を避ける…）

The successful strain（variant）that will cause the next epidemic is selected from a reduced number of progenies that possess relatively high transmissibility and the ability to escape from the immune surveillance of the host.（*J Theor Biol. 2003 222:437*）

訳 宿主の免疫サーベイランスから逃れる能力

18. ～のままである／存在する 【remain】

～のままである	存在する	生き残る
remain	**exist**	**survive**

使い分け

◆ **remain** は，「～のままである」という意味に用いられることが多い.
◆ **exist** は，「存在する」という意味で使われる.
◆ **survive** は，「生き残る／生存する」という意味に用いられることが多い.
◆いずれの語も自動詞として使われる.

頻度分析

	用例数		用例数
remain	12,824	❶ remains unclear	672
exist	5,109	❶ exist in	651
survive	1,791	❶ survive in	86

★★ remain（～のままである）

自動詞．

❶ **remain unclear**（はっきりしないままである）

However, it remains unclear whether this immune response is protective, pathogenic, or both.（*J Virol. 2004 78:13104*）

訳 …かどうかは，はっきりしないままである

★ exist（存在する）

自動詞．"exist in" の用例が多い．

❶ **exist in ～**（～に存在する）

Over 30 inositol polyphosphates are known to exist in mammalian cells; however, the majority of them have uncharacterized functions.（*J Biol Chem. 2005 280:1156*）

訳 30以上のイノシトールポリリン酸塩が，哺乳類細胞に存在することが知られている

★ survive（生き残る／生存する）

自動詞．

❶ **survive in ～**（～において生き残る）

GBS also provides a model system for studying adaptation to different host environments due to its ability to survive in a variety of sites within the host.（*J Bacteriol. 2003 185:6592*）

訳 宿主内のさまざまな部位で生き残る能力

19. 局在する／位置する 【localize】

局在する／位置する	
localize	position
locate	map

使い分け

◆（be）**localized** to,（be）**localized** in は「～に局在する」という意味で，細胞内局在などを示すときに用いられる．

◆（be）**located** in は「～に位置する」という意味で，タンパク質や遺伝子の構造内での位置に関して用いられることが多い．

◆（be）**positioned** in は，「～に位置する」という意味で広く用いられ

る．

◆（be）**positioned** to［*do*］は，「～するように位置する」という意味で使われる．

◆（be）**mapped** to は「～にマップされる」という意味で，遺伝子の染色体上での位置などに関して用いられる．

◆ **localize** は，自動詞としても用いられる．

頻度分析

	用例数		用例数
localize	7,773	❶（be）localized to	1,876
		❷（be）localized in	667
		❸ localizes to	583
locate	6,027	❶（be）located in	1,762
position	10,366	❶（be）positioned to	100
		❷（be）positioned in	96
map	6,007	❶（be）mapped to	727

解説 ◆いずれの語も受動態の用例が多い．

★ localize（～を局在させる）

他動詞として用いられることが多いが，自動詞として使われることも少なくない．

❶ localized to ～（～に局在する）

Although the majority of PR is localized to the nucleus, biochemical partitioning resulted in a loosely bound（cytosolic）fraction, and a tightly bound（nuclear）fraction.（*J Biol Chem. 2004 279:15231*）

訳 プロゲステロン受容体の大部分は核に局在する

❷ localized in ～（～に局在する）

PIAS1 is localized in the nucleus as distinct nuclear dots.（*J Biol Chem. 2001 276:36624*）

訳 PIAS1 は核に局在する

❸ localize to ～（～に局在する）

The CED-1 protein localizes to cell membranes and clusters around neighboring cell corpses.（*Cell. 2001 104:43*）

訳 CED-1 タンパク質は細胞膜に局在する

★ locate（～を位置づける）

他動詞．受動態の用例が非常に多い．

❶ **located in ～**（～に位置する）

MelR binds to 18 bp target sites using two helix-turn-helix（HTH）motifs that are both located in its C-terminal domain.（*Mol Microbiol. 2004 51:1297*）

訳 MelR は，両方とも C 末端ドメインに位置する 2 つのヘリックス・ターン・ヘリックス（HTH）モチーフを使って 18 塩基対の標的部位に結合する

★ position（～を位置づける／位置）

名詞の用例が多いが，他動詞としても用いられる．受動態の用例が多い．

❶ **positioned to ～**（～するように位置した）

We propose a model for the interaction between RPB4/RPB7 and the core RNA polymerase in which the RNA binding face of RPB7 is positioned to interact with the nascent RNA transcript.（*Mol Cell. 2001 8:1137*）

訳 RPB7 の RNA 結合面は，新生 RNA 転写物と相互作用するように位置している

❷ **positioned in ～**（～に位置した）

In interphase *Schizosaccharomyces pombe* cells, the nucleus is positioned in the middle of the cylindrical cell in an active microtubule (MT)-dependent process.（*J Cell Biol. 2001 153:397*）

訳 核は円柱状細胞の中央に位置している

★ map（～をマップする）

名詞として用いられることが多いが，他動詞としても使われる．

❶ **mapped to ～**（～にマップされる）

The human UBQLN3 gene was mapped to the 11p15 region of chromosome 11.（*Gene. 2000 249:91*）

訳 ヒト UBQLN3 遺伝子は第 11 染色体の 11p15 にマップされた

【Ⅲ-C. 機能】 機能に関係する動詞

20. 応答する／反応する 【respond】

応答する	反応する
respond	**react**

使い分け
◆ **respond** は「応答する／反応する」，**react** は「反応する」という意味の自動詞として用いられる．

頻度分析

	用例数		用例数
respond	4,032	❶ respond to	1,487
react	1,548	❶ react with	327

解説
◆ **respond** は後に to を，**react** は後に with を伴うことが多い．
◆ A **responds** to B（A は B に応答する）の場合は，A と B の入れ替えは不可で B は A には応答しない．
◆ A **reacts** with B（A は B と反応する）は，A と B の入れ替えが可能で両者の関係は等価である．

★ respond（応答する／反応する）

自動詞．“respond to” の用例が非常に多い．

❶ respond to ～（～に応答する）

The ability to respond to chemical signals is essential for the survival and reproduction of most organisms.（*Brain Res. 2002 941:62*）

訳 化学シグナルに応答する能力は，…にとって必須である

★ react（反応する）

自動詞．“react with” の用例が非常に多い．

❶ react with ～（～と反応する）

Both were found to react with Mtb-infected, but not bacillus Calmette-Guerin-infected, targets.（*J Immunol. 2001 166:439*）

訳 両者は，…と反応することが見つけられた

21. 認識する／知覚する 【recognize】

認識する	知覚する	情報交換する／連絡する
recognize	perceive	communicate

使い分け

◆ **recognize** は「認識する」，**perceive** は「知覚する」という意味の他動詞として用いられる．

◆ **communicate** は「情報交換する／連絡する」という意味の自動詞として用いられ，後に with を伴う用例が多い．

頻度分析

	用例数		用例数
recognize	5,035	❶ (be) recognized by	888
perceive	397	❶ (be) perceived as	34
communicate	230	❶ communicate with	44

★ recognize（～を認識する）

他動詞．

❶ recognized by ～（～によって認識される）

Electrophoretic mobility shift assays found that the C/EBP site is recognized by C/EBPα and that both LF Sp1 binding sites bind the Sp1 transcription factor specifically in myeloid cells.（*Blood. 2000 95:3734*）

訳 C/EBP 部位は C/EBPα によって認識される

perceive（～を知覚する／～を認知する）

他動詞．

❶ perceived as ～（～として知覚される）

Here, we test the hypothesis that 5-aza-CdR treatment is perceived as DNA damage, as assessed by the activation of the tumor suppressor p53.（*Mol Pharmacol. 2001 59:751*）

訳 5-aza-CdR 処理は DNA 損傷として知覚される

communicate（情報交換する／連絡する）

自動詞．ものが主語になる場合も多い．

❶ **communicate with ～**（～と情報交換する）

Nerve cells communicate with each other through two mechanisms, referred to as fast and slow synaptic transmission.（*Science. 2001 294:1024*）

訳 神経細胞は，２つの機構によってお互いに情報交換する

22. 結合する／接着する 【bind】

結合する	接着する	連結する	相互作用する
bind associate engage couple bond attach	adhere	connect join	interact

使い分け

◆ **bind**, **associate** は他動詞あるいは自動詞として，「結合する／～を結合させる」の意味で用いられる．

◆ **engage** は「～を従事させる／～を関与させる」という意味だが，「～に結合する」という意味でも使われる．

◆ **couple** は他動詞受動態あるいは自動詞で，後に to を伴って「～に共役する／～に結合する」の意味で用いられる．

◆ **bond** は，水素結合や共有結合に用いられることが多い．

◆ **attach** は他動詞受動態あるいは自動詞で，後に to を伴って「～に結合する／～に付着する」の意味で用いられる．

◆ **adhere** は，「接着する」という意味の自動詞として用いられる．

◆ **connect**, **join** は，他動詞受動態で「連結される」の意味で用いられることが多い．

◆ **interact** は with を伴って，「～と相互作用する」という意味の自動詞として使われることが多い．

◆ **associate**, **connect**, **couple** は，「関連づける」という意味でも用いられる．

◆ **bind**, **associate**, **connect**, **engage**, **couple**, **attach** は，他動詞・自動詞の両方で用いられる．

◆ **adhere**, **interact** は，自動詞として使われる．

◆ **join**, **bond** は，他動詞として用いられる．

頻度分析

	用例数		用例数
bind	26,567	❶ (be) bound to	3,300
		❷ binds to	3,222
		❸ bind the	504
associate	40,738	❶ (be) associated with	26,761
		❷ associate with	898
engage	654	❶ engage the	41
couple	5,756	❶ (be) coupled to	924
		❷ couple to	96
bond	6,741	❶ (be) hydrogen-bonded	219
attach	1,667	❶ (be) attached to	751
		❷ attach to	78
adhere	454	❶ adhere to	154
connect	1,246	❶ (be) connected to	191
		❷ connecting the	130
		❸ connect to	16
join	959	❶ (be) joined to	43
interact	11,347	❶ interact with	3,250

解説　◆ to または with を伴う用例が多い．

★★ bind（～を結合させる／～に結合する／結合する）

他動詞・自動詞の両方の用例がある．他動詞には，「～を結合させる」と「～に結合する」の2つの用法がある．受動態の“［*be*］bound to”は，「～に結合する」という意味で用いられる．

❶ **bound to ～**（～に結合する）

The glucosyl group of the substrate is bound to the protein via the side-chain carboxamide groups of Asn 187 and Asn 207. (*Biochemistry. 2000 39:5691*)

訳 基質のグルコシル基は，…を通してタンパク質に結合する

❷ **bind to ～**（～に結合する）

ZBP-89 is a transcription factor that binds to the G/C-rich elements and mediates p53-independent apoptosis. (*Nucleic Acids Res. 2003 31:7264*)

訳 ZBP-89 は，G/C に富んだエレメントに結合する転写因子である

❸ **bind the ～**（～に結合する）

Both classes of compounds appear to bind the same site, a relatively small portion of the GFCC'C" face of the N-terminal V-set domain of human B7.1, not present in the homologous B7.2 or even mouse B7.1.（*J Biol Chem. 2002 277:7363*）

訳 両方のクラスの化合物が，同じ部位に結合するように思われる

⁑ associate（～を結合させる／結合する）

他動詞受動態の用例が多いが，自動詞として使われることもある．"associated with" の用例が多い．

❶ **associated with ～**（～に結合している）

Although CAL does not have any predicted transmembrane domains, CAL is associated with membranes mediated by a region containing the coiled-coil domains.（*J Biol Chem. 2002 277:3520*）

訳 CAL は膜に結合している

❷ **associate with ～**（～に結合する）

Exposure of plant tissues to hyperosmotic stress led to the rapid phosphorylation of Ssh1p, a modification that decreased its ability to associate with membranes.（*Plant Cell. 2001 13:1205*）

訳 膜に結合する能力

engage（～に結合する／～を関与させる）

他動詞として用いられることが多い．

❶ **engage the ～**（～に結合する）

These activities require IL-22 to engage the cell surface receptors IL-22R1 and the low-affinity signaling molecule IL-10R2.（*J Mol Biol. 2004 342:503*）

訳 これらの活性は，IL-22 が細胞表面受容体 IL-22R1 に結合することを必要とする

★ couple（～を共役させる／～を結合させる／共役する）

他動詞の用例が多いが，自動詞として使われることもある．

❶ **coupled to ～**（～に共役する／～に結合する）

In the present study, we show that PKD is expressed in human platelets and that it is rapidly activated by receptors coupled to heterotrimeric G-proteins or tyrosine kinases.（*Blood. 2003 101:1392*）

訳 それは，ヘテロ三量体のGタンパク質に共役する受容体によって急速

に活性化される

❷ **couple to ～**（～に共役する／～に結合する）

We propose that YakA acts downstream of G-proteins, because cAMP receptors still couple to G-proteins in the yakA mutant.（*J Biol Chem. 2001 276:30761*）

訳 cAMP 受容体は，yakA 変異体においてもなおGタンパク質に共役する

★ bond（～を結合させる／結合）

名詞の用例が多いが，他動詞の現在分詞や過去分詞が形容詞的に用いられる．

❶ **hydrogen-bonded**（水素結合する）

The model suggests that with the quinazoline-based inhibitors, the N3 atom is hydrogen-bonded to a water molecule which, in turn, interacts with Thr 830.（*J Med Chem. 2000 43:3244*）

訳 N3 原子は水分子に水素結合している

★ attach（～を結合させる／付着する）

他動詞受動態の用例が多いが，自動詞として使われることもある．

❶ **attached to ～**（～に結合する）

Heteronuclear multiple-quantum coherence spectroscopy analysis demonstrated that a nitrogen atom is attached to C-4 of the sugar residue.（*J Biol Chem. 2001 276:43132*）

訳 窒素原子は糖残基の C-4 に結合している

❷ **attach to ～**（～に付着する）

S. epidermidis has the ability to attach to indwelling materials coated with extracellular matrix proteins such as fibrinogen, fibronectin, vitronectin, and collagen.（*J Biol Chem. 2002 277:43017*）

訳 *S. epidermidis* は，…に付着する能力を持つ

adhere（接着する）

自動詞．

❶ **adhere to ～**（～に接着する）

PDLF were quantitatively examined for their ability to adhere to a variety of BRG materials fluorometrically.（*J Periodontol. 2001 72:990*）

訳 歯根膜細胞は，…に接着するそれらの能力について定量的に調べられた

★ connect（～を連結する／つながる）

他動詞の用例が多いが，自動詞として使われることもある．

❶ connected to ～（～に連結される）

This domain is connected to the N-terminal domain by an unstructured linker, which is proposed to confer a high degree of mobility on αCTD.（*EMBO J. 2000 19:1555*）

訳 このドメインは，…によってN末端ドメインに連結されている

❷ connecting the ～（～を連結する）

The protein contains one α-helix and two strands of antiparallel β-sheet, with a type IV β-turn connecting the two strands.（*Biochemistry. 2002 41:12284*）

訳 2つの鎖を連結する4型βターン

❸ connect to ～（～につながる）

Smad1 −/− embryos die by 10.5 dpc because they fail to connect to the placenta.（*Development. 2001 128:3609*）

訳 それらは胎盤につながるのに失敗する

join（～を連結する）

他動詞．

❶ joined to ～（～に連結される）

Helix 1 is joined to helix 2 by a flexible linker.（*Biochemistry. 2004 43:3111*）

訳 ヘリックス1は可動性のリンカーによってヘリックス2に連結されている

★★ interact（相互作用する）

自動詞．"interact with" の用例が多い．

❶ interact with ～（～と相互作用する）

This surface might interact with the C-terminal domain of SU or with an adjacent monomer in the Env oligomer.（*J Virol. 2002 76:10861*）

訳 この表面は，…のC末端ドメインと相互作用するかもしれない

23. 働く／機能する 【function】

働く／機能する	
function	operate
act	work
serve	
behave	

使い分け
◆ **function**, **act**, **serve**, **behave** は後に as を伴って，「〜として働く」という意味の自動詞として用いられることが多い．
◆ **function**, **operate**, **work** は後に in を伴って，「〜において機能する／〜において働く」という意味で使われることも多い．

頻度分析

	用例数		用例数
function	47,863	❶ function in	3,830
		❷ function as	1,892
act	9,819	❶ act as	1,581
serve	4,760	❶ serve as	1,916
behave	691	❶ behaves as	104
operate	1,939	❶ operate in	123
work	5,958	❶ to work	90

★ function（働く／機能する／機能）

名詞の用例が多いが，自動詞としても使われる．

❶ function in 〜（〜において機能する／〜において働く）

However, epistasis analysis argues that only HDA1 and TUP1 are likely to function in the same pathway.（*Mol Cell. 2001 7:117*）

訳 HDA1 と TUP1 だけが，同じ経路でおそらく機能しそうである

❷ function as 〜（〜として働く）

This is the first time that TopBP1 has been shown to function as a transcriptional coactivator and that E2 interacts with TopBP1.（*J Biol Chem. 2002 277:22297*）

訳 TopBP1 は，転写のコアクチベーターとして働くことが示されてきた

★★ act（働く／作用する／作用）

自動詞として使われることが多い．特に，“act as”の用例が多い．

❶ act as ～（～として働く）

These results demonstrate that AZ may act as a tumor suppressor gene stimulating apoptosis and restraining cell proliferation, thereby inhibiting forestomach tumor development.（*Cancer Res. 2003 63:3945*）

訳 AZ は癌抑制遺伝子として働くかもしれない

★ serve（役立つ／役目をする／働く）

自動詞．“serve as”の用例が多い．

❶ serve as ～（～として役立つ／～として働く）

This animal model may serve as a useful tool to further evaluate mechanisms of tumorigenesis by JCV T-antigen.（*Oncogene. 2000 19:4840*）

訳 この動物モデルは，…するための有用なツールとして役立つかもしれない

behave（振る舞う／働く）

自動詞．“behave as”の用例が多い．

❶ behave as ～（～として働く／～として振る舞う）

Steady-state analysis reveals that the substituted piperidine likewise behaves as a competitive inhibitor.（*J Biol Chem. 2002 277:28677*）

訳 置換されたピペリジンは同様に競合阻害剤として働く

★ operate（働く／機能する／～を操作する）

自動詞・他動詞の両方の用例がある．

❶ operate in ～（～において働く／～において機能する）

This control mechanism is likely to operate in other Gram-positive bacteria containing similar pyrG leader sequences.（*Proc Natl Acad Sci USA. 2004 101:10943*）

訳 この制御機構は，他のグラム陽性菌においておそらく働きそうである

★ work（働く／研究）

名詞の用例が多いが，自動詞としても使われる．

❶ **～ to work**（働く～）

This is the first example in which these SOS inducible polymerases are shown to work in concert during lesion bypass.（*Biochemistry. 2004 43:13621*）

訳 これらのSOS誘導性のポリメラーゼは，…の間に協調して働くことが示される

24. 調節する 【regulate】

調節する	駆動する	仲介する	調整する
regulate control modulate	drive	mediate	adjust

使い分け

◆ **regulate**, **control**, **modulate** は，「～を調節する／～を制御する」，**drive** は「～を駆動する／～を制御する」という意味で使われる．

◆ **mediate** は「～を仲介する」という意味に用いられる．

◆ **adjust** は，「（年齢や性別などの比較の条件を）調整する」場合に用いられることが多い．

◆いずれの語も他動詞として使われる．

頻度分析

	用例数		用例数
regulate	30,593	❶（be）regulated by	4,082
		❷ regulate the	1,525
control	36,854	❶（be）controlled by	1,360
		❷ control the	819
modulate	7,005	❶（be）modulated by	831
		❷ modulate the	746
drive	4,757	❶（be）driven by	924
mediate	36,568	❶（be）mediated by	5,544
adjust	2,860	❶（be）adjusted for	424

★★ regulate（～を調節する）

❶ **regulated by ～**（～によって調節される）

IEX-1 gene expression is regulated by a variety of factors such as x-irradiation, ultraviolet radiation, steroids, growth factors, and inflammatory stimuli.（*J Biol Chem. 2002 277:14612*）

訳 IEX-1 遺伝子発現は，…のようなさまざまな因子によって調節される

❷ **regulate the ～**（～を調節する）

The two-component regulatory system PhoPQ has been shown to regulate the expression of virulence factors in a number of bacterial species.（*Mol Microbiol. 2002 44:1637*）

訳 二成分制御系 PhoPQ は，病原性因子の発現を調節することが示されてきた

⁑ control（～を制御する／コントロール）

名詞の用例が多いが，他動詞としても用いられる．

❶ **controlled by ～**（～によって制御される）

In mammals, nuclear localization of the circadian regulators PER1-3 is controlled by multiple mechanisms, including multimerization with PER and CRY proteins.（*J Biol Chem. 2001 276:45921*）

訳 概日性調節因子 PER1-3 の核局在は，複数の機構によって制御されている

❷ **control the ～**（～を制御する）

Two-component signaling systems, in which a receptor-coupled kinase is used to control the phosphorylation level of a response regulator, are commonly used in bacteria to sense their environment.（*Proc Natl Acad Sci USA. 2004 101:17072*）

訳 受容体と共役したキナーゼは，…のリン酸化レベルを制御するために使われる

★ modulate（～を調節する）

❶ **modulated by ～**（～によって調節される）

The activity of MT1-MMP is regulated extensively at the post-translational level, and the current data support the hypothesis that MT1-MMP activity is modulated by glycosylation.（*J Biol Chem. 2004 279:8278*）

訳 MT1-MMP 活性はグリコシル化によって調節される

❷ **modulate the ～**（～を調節する）

These data support the model that one function of K-bZIP is to modulate the activity of the transcriptional transactivator K-Rta.（*J Virol. 2003 77:1441*）

訳 K-bZIP の 1 つの機能は，転写のトランス活性化因子 K-Rta の活性を調

節することである

★ drive（～を駆動する／～を制御する／～を動かす）

他動詞.

❶ **driven by ～**（～によって駆動される）

Expression of luciferase reporter gene driven by the Pi starvation-induced AtPT2 promoter was also suppressed by Phi.（*Plant Physiol. 2002 129:1232*）

訳 リン酸欠乏誘導性の AtPT2 プロモータによって駆動されるルシフェラーゼレポータ遺伝子の発現はリン酸類似体によっても抑制された

★★ mediate（～を仲介する／～を媒介する）

他動詞.

❶ **mediated by ～**（～によって仲介される）

Neurotrophin-induced neuronal death is mediated by activation of Jun kinase.（*J Neurosci. 2000 20:6340*）

訳 ニューロトロフィン誘導性の神経細胞死は Jun キナーゼの活性化によって仲介される

★ adjust（～を調整する）

他動詞.

❶ **adjusted for ～**（～について調整される）

All analyses were adjusted for age, sex, and study location, the latter being a proxy measure of socioeconomic status.（*Arthritis Rheum. 2003 48:1686*）

訳 すべての分析は，年齢，性別および調査場所について調整された

25. 形成する／集合する 【form】

形成する	集合する	凝集する
form	assemble	aggregate

使い分け

◆ **form** は，「形成する」という意味の他動詞として用いられる.

◆ **assemble** は，「構築する」という意味の他動詞として用いられるが，「集合する」という意味の自動詞として使われることも多い.

◆ **aggregate** は，「凝集する」という意味の自動詞として用いられる.

頻度分析	用例数		用例数
form	32,945	❶ form a	2,125
assemble	2,683	❶ assemble into	192
aggregate	2,257	❶ to aggregate	105

⋆⋆ form（～を形成する／形成）

名詞の用例が多いが，他動詞としても用いられる．

❶ form a ～（～を形成する）

BLM is known to form a complex with the RAD51 recombinase, and to act upon DNA intermediates that form during homologous recombination, including D-loops and Holliday junctions.（*J Biol Chem. 2003 278:48357*）

訳 BLMは，RAD51リコンビナーゼと複合体を形成することが知られている

⋆ assemble（集合する／～を構築する）

自動詞および他動詞として用いられる．

❶ assemble into ～（集合して～を構築する）

In contrast, MuB that bound outside of the A/T-rich regions failed to assemble into large oligomeric complexes.（*J Biol Chem. 2004 279:16736*）

訳 A/Tに富んだ領域の外側に結合したMuBは，集合して大きなオリゴマー複合体を構築することに失敗した

⋆ aggregate（凝集する／凝集物）

名詞の用例が多いが，自動詞として用いられる．

❶ ～ to aggregate（凝集する～）

Characterization of these proteins *in vitro* has been hampered by their relative insolubility and tendency to aggregate.（*J Bacteriol. 2003 185:1808*）

訳 試験管内でのこれらのタンパク質の性質決定は，相対的な難溶性と凝集する傾向によって妨害されてきた

26. 与える 【confer】

与える	供給する
confer give render	supply

使い分け ◆ **confer** は「(形質などを) 与える」, **give** は「与える」, **render** は「～に…を与える／～を…にする」, **supply** は「供給する」という意味の他動詞として用いられる.

頻度分析

	用例数		用例数
confer	3,501	❶ confer resistance to	108
give	8,225	❶ give insight into	24
render	1,247	❶ renders cells	24
supply	1,074	❶ (be) supplied by	66

★ confer (～を与える)

他動詞.

❶ **confer resistance to ～** (～に対する抵抗性を与える)

Viruses have developed diverse non-immune strategies to counteract host-mediated mechanisms that confer resistance to infection. (*Nature. 2002 418:646*)

訳 感染に対する抵抗性を与える機構

★ give (～を与える)

他動詞. "give rise to (～を生じる)" の用例が多い (動詞編Ⅱ-B. 6. 参照).

❶ **give insight into ～** (～への洞察を与える)

Models based on tropomyosin crystal structures give insight into possible effects of the mutations on the structure. (*Biochemistry. 2003 s42:14114*)

訳 トロポミオシン結晶構造に基づくモデルは, 構造に対するその変異のありうる効果への洞察を与える

★ render（～を…にする／～に…を与える）

他動詞．

❶ render cells ～（細胞を～にする／細胞に～を与える）

Furthermore, forced expression of FLIP renders cells resistant to Fas-mediated apoptosis.（*J Biol Chem. 2001 276:6893*）

訳 FLIPの強制発現は，細胞をFasに仲介されるアポトーシスに対して抵抗性にする

★ supply（～を供給する／供給）

名詞および他動詞として用いられる．

❶ supplied by ～（～によって供給される）

The energy for glycogenesis is supplied by oxidative phosphorylation.（*Proc Natl Acad Sci USA. 2001 98:457*）

訳 糖新生のためのエネルギーは酸化的リン酸化によって供給される

【Ⅳ-A. 方法・実施】 研究の方法・実施に関係する動詞

1. 選択する 【select】

選択する	
select	choose

使い分け ◆ **select**, **choose** は,「選択する」という意味の他動詞として用いられる. 論文では, **select** が好まれる傾向にある.

頻度分析

	用例数		用例数
select	4,608	❶ (be) selected for	429
choose	839	❶ (be) chosen for	97

★ select（～を選択する）

他動詞.

❶ **selected for ～**（～のために選択される）

Eight samples were selected for further analysis.（*J Clin Microbiol. 2001 39:3282*）

訳 8個のサンプルが，さらなる分析のために選択された

choose（～を選択する）

他動詞.

❶ **chosen for ～**（～のために選択される）

CD58 was chosen for further analysis because of its abundant and prevalent overexpression.（*Blood. 2001 97:2115*）

訳 CD58 が，さらなる分析のために選択された

2. 使う／利用する 【use】

使う	利用する	適用する
use employ	utilize take advantage of exploit	apply

使い分け

◆ **use**, **employ** は「使う／用いる」, **utilize**, **take advantage of** は「利用する」, **exploit** は「活用する／利用する」, **apply** は「適用する」という意味の他動詞として用いられる．

頻度分析

	用例数		用例数
use	89,896	❶ (be) used to	12,706
		❷ by using	5,328
		❸ we used	4,509
employ	2,606	❶ (be) employed to	415
utilize	3,202	❶ (be) utilized to	240
take advantage of	344	❶ taking advantage of	102
exploit	977	❶ (be) exploited to	150
apply	4,506	❶ (be) applied to	1,639

解説

◆ **use**, **employ**, **utilize**, **exploit** は，受動態で to 不定詞を伴う用例が多い．

◆ **applied** to の後は，動詞ではなく名詞がくる場合が多い．

◆ **use** が使われる頻度は非常に高い．

★★ use（～を使う／～を使用する／使用）

名詞として用いられることもあるが，他動詞の用例が多い．

❶ **used to ～**（～するために使われる）

RT-PCR analysis was used to determine the pattern of expression of the larger isoform (I) of this receptor in a variety of tissues.（*Gene. 2000 257:307*）

訳 RT-PCR 分析が，…のパターンを決定するために使われた

❷ **by using ～**（～を使うことによって）

APE-1/Ref-1 function was assessed by using a luciferase-linked reporter construct containing 3 activator protein 1 binding sites.

(*Gastroenterology. 2004 127:845*)

訳 APE-1/Ref-1の機能は，ルシフェラーゼに連結されたレポータコンストラクトを使うことによって評価された

❸ **we used ～**（われわれは～を使った）

In this study, we used a combination of pharmacological and genetic approaches to determine which endogenous sphingolipid is the likely mediator of growth inhibition.（*J Biol Chem. 2001 276:35614*）

訳 この研究において，われわれは薬理学的および遺伝学的アプローチの組み合わせを使った

★employ（～を用いる／～を使用する）

他動詞.

❶ **employed to ～**（～するため用いられる）

A yeast two-hybrid assay was employed to identify androgen receptor（AR）protein partners in gonadotropin-releasing hormone neuronal cells.（*J Biol Chem. 2002 277:20702*）

訳 酵母ツーハイブリッドアッセイが，…におけるアンドロゲン受容体（AR）タンパク質のパートナーを同定するために用いられた

★utilize（～を利用する）

他動詞.

❶ **utilized to ～**（～するために利用される）

The approach described here may be utilized to identify penetrance modifiers in other autosomal dominant syndromes.（*Hum Mol Genet. 2002 11:1327*）

訳 ここで述べられる方法は，…における浸透度変更因子を同定するために利用されるかもしれない

take advantage of（～を利用する／～を活用する）

❶ **taking advantage of ～**（～を利用する）

By taking advantage of a novel phenomenon, we have developed a selectable assay to detect contractions of CTG/CAG triplets.（*Mol Cell Biol. 2003 23:4485*）

訳 新規の現象を利用することによって

exploit（～を活用する／～を利用する）

他動詞.

❶ **exploited to ～**（～するために活用される）

This study elucidates a mechanism that can be exploited to develop new strategies to improve AAV vector transduction efficiency.（*J Virol. 2004 78:13678*）

訳 この研究は，…を改善する新しいストラテジーを開発するために活用されうる機構を明らかにする

⋆ apply（～を適用する）

他動詞.

❶ **applied to ～**（～に適用される）

The method was applied to the analysis of 14 commercial purified proteins, yielding characteristic features of surface activity as a function of pH.（*Anal Chem. 2005 77:250*）

訳 その方法は，…の分析に適用された

3. 行う 【perform】

行う	完了する
perform carry out conduct do	complete

使い分け

◆ **perform, carry out** は，どちらも「行う／実行する」という意味だが，使用頻度は **perform** が圧倒的に高い.

◆ **conduct** には「指揮する」という意味もあるが，「行う」という意味で用いられることが非常に多い.

◆ **do** は助動詞としての用例が圧倒的に多いが，「行う」という意味の動詞としても用いられる.

◆ **complete** は「完了する／完成する」という意味で，「（患者が治療や検査を）完了した」場合などに用いられる.

共起・頻度分析　　（数字：用例数）

直前の単語					直後の単語			
were/was	we	to			a	the	in	to
4,335	1,115	0	performed	8,500	658	70	1,139	787
0	50	567	perform (s)	1,300	190	85	7	0

直前の単語					直後の単語			
were/was	we	to			a	the	in	to
560	162	0	carried out	1,358	120	9	176	148
0	5	139	carry/carries out	313	32	41	2	0
1,041	550	0	conducted	2,442	542	16	347	373
0	10	88	conduct (s)	240	20	9	0	0
289	0	0	done	622	3	2	89	97
138	8	2	completed	1,070	133	214	59	4
99	1	321	complete (s)	5,627	29	145	53	4

解説

◆ **perform**, **carry out**, **conduct**, **do** は，受動態で in や to 不定詞を伴う用例が多い．

◆ **perform**, **carry out**, **conduct** は，能動態で we を主語にし，a＋名詞を目的語にする用例も多い．

◆ **complete** は，形容詞の用例が多い．

★★ perform（～を行う／～を実行する）

他動詞．

❶ **performed in ～**（～において行われる）

Pharmacokinetic and pharmacodynamic studies were performed in 24 patients (at 1.45 to 2.0 mg/m^2). (*J Clin Oncol. 2004 22:2108*)

訳 薬物動態学的および薬力学的な研究が，24 人の患者において行われた

★ carry out（～を行う／～を実行する）

❶ **carried out in ～**（～において行われる）

82 MRI examinations were carried out in three groups: patients with healed myocardial infarction; patients with non-ischaemic cardiomyopathy; and healthy volunteers. (*Lancet. 2001 357:21*)

訳 82 人の MRI 検査が 3 つのグループにおいて行われた

★ conduct（～を行う）

名詞としても用いられるが，他動詞の用例が多い．

❶ **we conducted ～**（われわれは，～を行った）

We conducted a randomized, double-blind trial to determine the effect of aspirin on the incidence of colorectal adenomas. (*N Engl J Med. 2003 348:883*)

訳 われわれは無作為化二重盲検試験を行った

❷ **conducted to ～（～するために行われる）**

This study was conducted to determine whether smooth muscle cells (SMCs), a major source of arterial TF, could generate extracellular TF.（*Circ Res. 2000 87:126*）

訳 この研究は，…かどうかを決定するために行われた

★ do（～を行う）

助動詞の用例が圧倒的に多いが，他動詞としても用いられる．

❶ **done to ～（～するために行われる）**

The present analyses were, therefore, done to identify the extracellular matrix proteins with propensity to induce the release of intracellular galectin-3 from breast carcinoma cells.（*Cancer Res. 2001 61:1869*）

訳 現在の分析は，それゆえ，…を同定するために行われた

★ complete（～を完了する／完全な）

形容詞として用いられることが多いが，他動詞の用例もある．

❶ **completed the ～（～を完了した）**

Forty-six patients completed the study.（*Ann Intern Med. 2000 132:636*）

訳 46人の患者が調査を完了した

4. 切断する 【cleave】

切断する	
truncate	break
cleave	digest

使い分け ◆ **truncate**，**cleave** は「切断する」，**break** は「断ち切る／切断する」，**digest** は「消化する／切断する」という意味の他動詞として用いられる．

頻度分析

	用例数		用例数
truncate	2,486	❶ truncated form	206
cleave	1,984	❶ to cleave	199
break	2,207	❶ to break	47

★ truncate（～を切断する）

他動詞.

❶ truncated form（切断型）

Expression of a truncated form of the death receptor adaptor FADD (C-FADD) as a transgene in mice blocks T cell proliferation.（*J Biol Chem. 2003 278:41585*）

訳 …の切断型の発現

★ cleave（～を切断する）

他動詞.

❶ to cleave ～（～を切断する…）

A truncated protein lost the ability to cleave the core protein precursors.（*J Virol. 2003 77:11279*）

訳 切断型のタンパク質は，コアタンパク質前駆体を切断する能力を失った

★ break（～を断ち切る／～を切断する／切断／中断）

名詞の用例が多いが，他動詞としても用いられる.

❶ to break ～（～を断ち切る…）

Using this combination, it was possible to quantify the force required to break a single interaction between pilus and mannose groups linked to the SAM.（*Proc Natl Acad Sci USA. 2000 97:13092*）

訳 …の間の１つの相互作用を断ち切るために必要とされる力を定量することは可能であった

digest（～を消化する／～を切断する／消化）

名詞としても用いられるが，他動詞の用例の方が多い.

❶ digested with ～（～で消化される）

The proteins were digested with trypsin, and the released peptides were sequenced by tandem mass spectrometry.（*J Biol Chem. 2002 277:22010*）

訳 タンパク質はトリプシンで消化された

5. 置換する 【substitute】

置換する	
substitute	**replace**

頻度分析

	用例数		用例数
substitute	2,352	❶ (be) substituted for	188
		❷ (be) substituted with	175
replace	2,456	❶ (be) replaced by	656
		❷ (be) replaced with	465

解説

◆ **substitute** は，“**substituted** for”“**substituted** with”の用例が多い．

◆ **replace** は，“**replaced** by”“**replaced** with”の用例が非常に多い．

◆ “A（be）**substituted** with B”と“B（be）**substituted** for A”がほぼ同じ意味になり，AとB（A：置換前，B：置換後）の位置関係が逆であることに注意．

◆ “**substituted** with”“**replaced** with”“**replaced** by”は，ほぼ同じ意味に用いられる．

★ substitute（～を置換する／～を代わりに用いる）

他動詞．

❶ **substituted for ～**（～の代わりに用いられる）

Mutant human C9 in which Ala was substituted for Cys359/384 was found to express normal lytic activity and to be fully inhibited by CD59.（*Biochemistry. 1996 35:3263*）

訳 アラニンが，システイン359/384の代わりに用いられた

❷ **substituted with ～**（～によって置換される）

When Ile1182 was substituted with alanine, small changes were noted for DHP agonist and antagonist action.（*J Biol Chem. 1997 272:24952*）

訳 イソロイシン1182が，アラニンによって置換されたとき

★ replace（～を置換する）

他動詞．

❶ replaced by ～（～によって置換される）

For example, when Arg-325 is replaced by alanine or lysine, the resulting mutant enzymes possess no detectable asparagine synthetase activity.（*J Biol Chem. 1997 272:12384*）

訳 アルギニン-325が，アラニンかリジンによって置換されると

❷ replaced with ～（～によって置換される）

The highly conserved arginine at position 284 was replaced with alanine to construct UvrD-R284A.（*J Biol Chem. 1997 272:18614*）

訳 高度に保存された284番目のアルギニンが，アラニンによって置換された

6. 作製する／構築する 【create】

作製する	構築する	開発する
construct create generate synthesize	assemble build organize	develop

使い分け

◆ **construct**, **create**, **generate** は，「作製する」という意味に用いられる．

◆ **synthesize** は，「合成する」という意味で使われる．

◆ **assemble**, **build**, **organize** は「構築する」，**organize** は「組織化する」という意味に用いられる．

◆ **develop** は，「開発する」という意味で使われる．

共起・頻度分析 （数字：用例数）

直前の単語					直後の単語					
be動詞	we	to			a	by	from	into	to	in
1,353	603	0	constructed	2,672	440	224	203	3	126	146
1	27	362	construct(s)	2,343	180	24	13	52	80	232
601	254	0	created	1,547	237	360	37	0	45	112
0	0	645	create(s)	1,326	493	1	0	0	0	0
2,483	1,194	1	generated	8,238	604	1,732	555	0	110	584
1	21	2,153	generate(s)	4,373	797	2	1	0	1	11
429	131	0	synthesized	3,041	94	298	119	1	57	315
0	3	331	synthesize(s)	790	52	1	1	0	0	3

直前の単語					直後の単語					
be動詞	we	to			a	by	from	into	to	in
429	15	2	assembled	1,430	19	60	**94**	**180**	12	**132**
0	1	**289**	assemble (s)	1,074	63	12	10	**279**	33	66
126	24	0	built	320	34	15	**43**	9	18	23
0	10	**149**	build (s)	251	**77**	0	0	2	1	1
305	1	0	organized	852	1	48	1	**166**	16	**105**
0	0	**81**	organize	265	6	2	0	30	9	6
2,532	**1,044**	0	developed	9,685	**1,854**	184	80	68	**691**	**509**
0	**212**	**1,560**	develop	4,517	**740**	12	91	88	20	277

解説

◆ **construct**, **create**, **generate**, **synthesize** は，受動態で by を伴う用例が多い．

◆ **construct**, **create**, **generate**, **synthesize**, **assemble**, **organize**, **develop** は，受動態で in を伴う用例が多い．

◆ **assemble**, **organize** は，受動態で into を伴う用例が多い．

◆ **construct**, **generate**, **synthesize**, **assemble**, **build** は，受動態で from を伴う用例が多い．

◆ **develop** は，受動態で to 不定詞を伴う用例が多い．

◆ **construct**, **create**, **generate**, **develop** は，we を主語にし，a＋名詞を目的語にする用例も多い．

◆ **construct**, **create**, **develop** は過去形の用例が多く，**organize** は現在形の用例が多い．

★ construct（～を作製する／コンストラクト）

他動詞および名詞として用いられる．

❶ constructed a ～（～を作製した）

We constructed a series of deletion mutants of these four loops and tested their ability to form fully processed CaR as well as their ability to be activated by Ca^{2+}.（*J Biol Chem. 2001 276:32145*）

訳 われわれは一連の欠失変異体を作製した

❷ constructed by ～（～によって作製される）

A *mazG* deletion strain of *E. coli* was constructed by replacing the *mazG* gene with a kanamycin resistance gene.（*J Bacteriol. 2002 184:5323*）

訳 大腸菌の *mazG* 欠損株は，*mazG* 遺伝子をカナマイシン耐性遺伝子に置換することによって作製された

★ create（～を作製する）

他動詞.

❶ created by ～（～によって作製される）

A targeting vector was created by replacing a 3.2-kb segment of the gene encompassing the catalytic domain with a phosphoglycerokinase promoter-driven neomycin resistant (neor) gene cassette. (*Blood. 1998 92:168*)

訳 ターゲティングベクターは，…を置換することによって作製された

★ generate（～を作製する／～を産生する）

他動詞.

❶ generated by ～（～によって作製される）

A GADS-deficient mouse was generated by gene targeting, and the function of GADS in T cell development and activation was examined. (*Science. 2001 291:1987*)

訳 GADS欠損マウスが，ジーンターゲティングによって作製された

★ synthesize（～を合成する）

他動詞として用いられることが多いが，自動詞の用例もある.

❶ synthesized in ～（～において合成される）

Conjugated cholic acid and chenodeoxycholic acid were synthesized in the liver and secreted into bile but could not reenter the liver from portal blood and accumulated in serum. (*Gastroenterology. 2000 119:188*)

訳 抱合型のコール酸とケノデオキシコール酸は，肝臓において合成された

★ assemble（～を構築する／集合する）

自動詞として用いられることもあるが，他動詞の用例の方が多い.

❶ assembled into ～（～に構築される）

These structures were assembled into a three-dimensional construct by superimposing the overlapping sequences at the ends of each peptide. (*Biophys J. 2001 81:1029*)

訳 これらの構造は三次元のコンストラクトに構築された

build（～を構築する）

他動詞.

❶ built from ～（～から構築される）

Many proteins are built from structurally and functionally distinct domains.（*J Mol Biol. 2003 332:529*）

訳 多くのタンパク質は，構造的および機能的に異なるドメインから構築されている

★ organize（～を組織化する／～を構築する）

他動詞．

❶ organized into ～（～に組織化される）

We report that the three Min proteins are organized into extended membrane-associated coiled structures that wind around the cell between the two poles.（*Proc Natl Acad Sci USA. 2003 100:7865*）

訳 3 個の Min タンパク質は，延長した膜結合コイルド構造に組織化される

★★ develop（～を開発する／～を発症する／発生する）

「開発する／発症する」の意味では他動詞として，「発生する」の意味では自動詞として用いられることが多い．

❶ developed a ～（われわれは，～を開発した）

We have developed a novel method for quantitating protein phosphorylation by a variety of protein kinases.（*Anal Biochem. 2003 313:9*）

訳 われわれは，…を定量化するための新規の方法を開発した

❷ developed to ～（～するために開発される）

A new method was developed to identify and differentiate varicella-zoster virus（VZV）wild-type strains from the attenuated varicella Oka vaccine strain.（*J Clin Microbiol. 2000 38:3156*）

訳 新しい方法が，…を同定するために開発された

7. 導入する 【transfect】

導入する		
transfect	introduce	transform

使い分け

◆ **transfect** は「(細胞内に DNA などを) 移入する／導入する」場合に，**introduce** は「(遺伝子に変異などを) 導入する」場合に用いられる．

◆ **transform** は，「(遺伝子導入によって) 形質転換させる／癌化させる」という意味に用いられる．

共起・頻度分析

(数字：用例数)

直前の単語					直後の単語		
were	we	to			into	with	by
228	72	8	transfected	4,515	271	1,379	10
0	0	22	transfect (s)	35	0	1	2
284	188	0	introduced	2,023	661	8	162
0	239	144	introduce (s)	708	2	0	0
49	14	15	transformed	1,936	135	157	110
0	1	155	transform (s)	810	20	0	2

解説

◆ **transfect** は受動態で，with を伴う用例が非常に多い．また，into を伴う用例もかなりあるが，by を伴うものはほとんどない．

◆ **introduce** は受動態で，into を伴う用例が非常に多い．by を伴う用例もかなりあるが，with を伴うものはほとんどない．

◆ **transformed** は受動態で，with, into, by のいずれを伴う用例も多い．

★ transfect（～を移入する／～を導入する）

他動詞．

❶ **transfected with ～**（～を移入される）

Activation was observed in cells transfected with both CD14 and TLR4. (*J Immunol. 2001 166:4620*)

訳 活性化が，…を移入された細胞において観察された

★ introduce（～を導入する）

他動詞．

❶ **introduced into ～**（～に導入される）

In order to elucidate regions with the promoter required for activity, point mutations were introduced into the carQRS promoter between positions −151 and 6. (*J Bacteriol. 2004 186:7836*)

訳 点突然変異が，−151 番目と 6 番目の間の carQRS プロモータに導入された

⋆ transform（形質転換する／～を癌化させる）

他動詞の用例が多いが，自動詞としても用いられる．

❶ **transformed with ～**（～によって形質転換される）

To investigate the function of this conserved domain, d1blic mutant cells were transformed with constructs designed to express D1bLIC proteins with mutated P-loops.（*Mol Biol Cell. 2004 15:4382*）

訳 d1blic 変異細胞は，変異した P ループを持つ D1bLIC タンパク質を発現するように設計されたコンストラクトによって形質転換された

8. 測定する／定量する 【measure】

測定する	アッセイする	定量する
measure	assay	quantify quantitate

使い分け

◆ **measure** は「測定する」，**assay** は「アッセイする」，**quantify**, **quantitate** は「定量する」という意味の他動詞として用いられる．

共起・頻度分析 （数字：用例数）

直前の単語					直後の単語			
were/was	we	to			by	with	for	in
3,843	853	17	measured	9,926	2,512	427	229	1,289
4	49	1,403	measure	5,276	10	16	96	99
437	71	0	assayed	1,015	212	39	269	120
1	0	130	assay (s)	17,721	50	468	813	491
428	90	0	quantified	926	243	36	11	105
0	33	525	quantify/quantifies	783	1	0	1	1
108	21	0	quantitated	219	70	12	6	17
0	2	118	quantitate	174	0	0	0	0

解説

◆ **measure** は受動態で by または in を伴う用例が多く，また，with や for を伴うことも多い．

◆ **assay** は，受動態で for, by, in を伴う用例が多い．

◆ **quantify** は，受動態で by または in を伴う用例が多い．

◆ **quantitate** は受動態で by を伴うことが多く，また，能動態で to 不定詞の用例も多い．

◆ **assay** は，名詞の用例が多い．

★★ measure（～を測定する）

他動詞.

❶ measured by ～（～によって測定される）

The activity was measured by an RDH activity assay with recombinant RDH10 expressed in COS cells.（*Invest Ophthalmol Vis Sci. 2002 43:3365*）

訳 活性が，…を使ったRDH活性アッセイによって測定された

❷ measured in ～（～において測定される）

AVP levels were measured in patients with dilated cardiomyopathy（DCM）and CHF and in patients with large left-to-right intracardiac shunts.（*Circulation. 2004 109:2550*）

訳 AVPレベルが，…の患者において測定された

★ assay（～をアッセイする／アッセイ）

名詞の用例が多いが，他動詞としても用いられる.

❶ assayed for ～（～に関してアッセイされる）

Intact capsids were assayed for their ability to bind to heparin-agarose *in vitro*, and virions that packaged DNA were assayed for their ability to transduce normally permissive cell lines.（*J Virol. 2003 77:6995*）

訳 無傷のキャプシドが，…に結合するそれらの能力に関してアッセイされた

❷ assayed by ～（～によってアッセイされる）

Protein levels were assayed by immunoblot analysis, using the antibodies HMFG-2, 1G8, or OC125, which are specific to MUC1, -4 and -16, respectively.（*Invest Ophthalmol Vis Sci. 2004 45:114*）

訳 タンパク質レベルが，イムノブロット分析によってアッセイされた

★ quantify（～を定量する）

他動詞.

❶ quantified by ～（～によって定量される）

CEACAM6 expression was quantified by real-time polymerase chain reaction（PCR）and Western blot.（*Ann Surg. 2004 240:667*）

訳 CEACAM6発現は，リアルタイム・ポリメラーゼ連鎖反応法（PCR）とウエスタンブロットによって定量された

quantitate（～を定量する）

他動詞.

❶ to quantitate ～（～を定量するために）

An ELISA was used to quantitate serum levels of IGF-I and IGFBP-3 for both studies.（*Cancer Res. 2003 63:3991*）

訳 酵素結合免疫吸着検定法が，IGF-I と IGFBP-3 の血清レベルを定量するために使われた

9. 集める／回収する【collect】

集める	回収する
collect gather harvest	recover retrieve

使い分け
◆**collect**, **gather** は「集める」，**harvest** は「（細胞などを）集める」，**recover**, **retrieve** は「回収する」という意味の他動詞として用いられる.

頻度分析

	用例数		用例数
collect	2,277	❶ (be) collected from	460
gather	139	❶ (be) gathered from	18
harvest	675	❶ (be) harvested from	147
recover	1,981	❶ (be) recovered from	485
retrieve	285	❶ (be) retrieved from	64

解説
◆いずれの語も受動態で from を伴う用例が多い.

★ collect（～を集める）

他動詞.

❶ collected from ～（～から集められる）

Blood samples were collected from 1412 patients with severe angiographically defined CAD（stenosis >/=70%）.（*Circulation. 2000 102:1227*）

訳 血液サンプルが，…の1,412人の患者から集められた

gather（～を集める）

他動詞．

❶ gathered from ～（～から集められる）

Data on recent transplantations were gathered from the Scientific Registry of Transplant Recipients and directly from the transplantation programs.（*N Engl J Med. 2003 348:818*）

訳 最近の移植に関するデータが，Scientific Registry of Transplant Recipientsから集められた

harvest（～を集める／～を採取する）

他動詞．

❶ harvested from ～（～から集められる）

CD8+ T cells were harvested from Tcr transgenic OT-1 mice whose Tcr recognize an OVA peptide in the context of the class I major histocompatibility complex molecule Kb.（*Invest Ophthalmol Vis Sci. 2000 41:1803*）

訳 CD8+ T細胞が，Tcrトランスジェニック OT-1マウスから集められた

★ recover（～を回収する／～を回復する）

他動詞．受動態で「回収される」という意味で用いられることが多い．

❶ recovered from ～（～から回収される）

West Nile virus was recovered from the brain of a red-tailed hawk that died in Westchester County, N.Y., in February 2000.（*J Clin Microbiol. 2000 38:3110*）

訳 ウエストナイルウイルスは，ウエストチェスター郡で死んだアカオノスリの脳から回収された

retrieve（～を回収する）

他動詞．受動態の用例が多い．

❶ retrieved from ～（～から回収される）

In our clinical evaluation study, MCM2-positive cells were retrieved from 37 of 40 patients with symptomatic colorectal cancer, but from none of 25 healthy control individuals.（*Lancet. 2002 359:1917*）

訳 MCM2陽性細胞は，症候性の結腸直腸癌の患者40人中37人から回収された

10. 濃縮する 【enrich】

濃縮する	
enrich	concentrate

使い分け ◆ **enrich**, **concentrate** は，「濃縮する」という意味の他動詞として用いられる．

頻度分析

	用例数		用例数
enrich	2,106	❶（be）enriched in	666
concentrate	1,128	❶（be）concentrated in	296

★ enrich（～を濃縮する）

他動詞．

❶ **enriched in ～**（～において濃縮される）

B2R was highly enriched in this fraction, whereas B1R was not enriched.（*Biochemistry. 2002 41:14340*）

訳 B2R がこの分画において高度に濃縮された

★ concentrate（～を濃縮する）

他動詞．

❶ **concentrated in ～**（～において濃縮される）

Furthermore, NF-κB is concentrated in the nucleus of neurons in the brains of HD transgenic mice.（*J Neurosci. 2004 24:7999*）

訳 NF-κB は，HD トランスジェニックマウスの脳においてニューロンの核で濃縮される

11. 単離する／精製する 【isolate】

単離する	分離する	精製する
isolate	separate dissociate	purify

使い分け
◆ **isolate** は「単離する」, **separate**, **dissociate** は「分離する」, **purify** は「精製する」という意味の他動詞として用いられる.
◆ **dissociate** は,「解離する」という意味の自動詞としても使われる.

頻度分析

	用例数		用例数
isolate	15,904	❶ (be) isolated from	2,985
		❷ we isolated	564
separate	4,277	❶ (be) separated from	237
dissociate	1,091	❶ (be) dissociated from	116
purify	6,815	❶ (be) purified from	630

★★ isolate（～を単離する）

他動詞.

❶ isolated from ～（～から単離される）

An oncogenic variant of c-Jun was isolated from the acutely transforming retrovirus ASV17.（*Oncogene. 2000 19:3537*）

訳 c-Jun の発癌性の変異体が，急性に形質転換するレトロウイルス ASV17 から単離された

❷ we isolated ～（われわれは～を単離した）

We isolated a novel gene, BNI5, as a dosage suppressor of the cdc12-6 growth defect.（*Mol Cell Biol. 2002 22:6906*）

訳 われわれは，新規の遺伝子 BNI5 を…の用量抑制因子として単離した

★ separate（～を分離する／～を分ける／別個の）

他動詞および形容詞として用いられる.

❶ separated from ～（～から分離される）

Contaminating fibroblasts were separated from epithelial cells by differential trypsinization.（*Infect Immun. 2000 68:4200*）

訳 混入した線維芽細胞が，差動的なトリプシン処理によって上皮細胞から分離された

★ dissociate（〜を分離する／解離する）

自動詞・他動詞の両方の用例がある．

❶ **dissociated from 〜**（〜と分離される）

In this study, neurons were dissociated from brain slices prepared from prepubertal female GnRH-EGFP mice.（*J Neurosci. 2002 22: 2313*）

訳 ニューロンが脳のスライスから分離された

★ purify（〜を精製する）

他動詞．

❶ **purified from 〜**（〜から分離される）

βB_2-crystallin protein was purified from rat and bovine tissues by FPLC chromatography.（*Invest Ophthalmol Vis Sci. 2000 41:3056*）

訳 βB_2-クリスタリンタンパク質がラットとウシの組織から精製された

12. 調製する 【prepare】

調製する	処理する
prepare	process

使い分け ◆ **prepare** は「調製する」，**process** は「処理する／加工する」という意味の他動詞として用いられる．

頻度分析

	用例数		用例数
preapre	2,810	❶ (be) preapred from	705
process	24,352	❶ (be) processed by	132
		❷ (be) processed for	123

★ prepare（〜を調製する）

他動詞．

❶ **preapred from ～**（～から調製される）

MSCs were prepared from bone marrow aspirates obtained from the iliac crest or from the tibia/femur during joint surgery.（*Arthritis Rheum. 2002 46:704*）

訳 関葉系幹細胞は骨髄吸引液から調製された

★ process（～を処理する／～を加工する／～をプロセシングする／過程）

名詞として使われることが多いが，他動詞としても用いられる．

❶ **processed by ～**（～によってプロセシングされる）

We report here that p40 and p35 subunits are processed by disparate pathways.（*J Immunol. 2000 164:839*）

訳 p40とp35サブユニットが，異なる経路によってプロセシングされる

❷ **processed for ～**（～のために処理される）

The animals were killed at 2 or 4 weeks after injury, and the carotid arteries were harvested and processed for immunohistochemistry, scanning electron microscopy (SEM), and morphometric analysis of endothelialization and neointimal formation.（*Circulation. 2004 110:2039*）

訳 頸動脈が採取され，そして免疫組織化学のために処理された

13. 刺激する／処理する 【stimulate】

刺激する	処理する	添加する
stimulate prime	treat	add

使い分け ◆ **stimulate**, **prime** は「刺激する（ことによって活性化する）」，**treat** は「（薬剤などを使って）処理する／治療する」，**add** は「（薬剤などを）添加する」という意味の他動詞として用いられる．

共起・頻度分析

（数字：用例数）

直前の単語				直後の単語		
be動詞	to			by	with	to
963	8	stimulated	8,815	1,068	525	76
0	1,122	stimulate (s)	4,536	0	1	0

直前の単語				直後の単語		
be動詞	to			by	with	to
94	7	primed	869	**57**	**85**	5
11	**80**	prime (s)	513	1	1	1
1,932	1	treated	11,677	146	**4,600**	16
1	**782**	treat (s)	972	1	3	5
836	9	added	2,096	8	19	**779**
0	56	add (s)	550	0	0	**155**

解説

◆ **stimulate**, **prime** は，受動態で with または by を伴う用例が多い．
◆ **treat** は受動態で with を伴う用例が非常に多く，by を伴う用例は少ない．
◆ **add** は受動態で，to を伴う用例が多い．

★★ stimulate（～を刺激する）

他動詞．

❶ stimulated by ～（～によって刺激される）

All DExH/D proteins characterized to date hydrolyse nucleoside triphosphates and, in most cases, this activity is stimulated by the addition of RNA or DNA.（*Nature. 2000 403:447*）

訳 この活性は RNA あるいは DNA の添加によって刺激される

★ prime（～を刺激する）

他動詞．

❶ primed with ～（～によって刺激される）

T cells from wild-type mice were primed with staphylococcal enterotoxin B（SEB）*in vitro*, which induced an autoreactive proliferative response to syngeneic feeder cells.（*J Immunol. 2000 164:2994*）

訳 野生型マウス由来の T 細胞は，ブドウ球菌のエンテロトキシン B によって刺激された

★★ treat（～を処理する）

他動詞．

❶ treated with ～（～によって処理される）

But when cells were treated with cycloheximide（CHX）or emetine, expression levels were restored to those observed in primary and immortal（10）10 MEF cells.（*Oncogene. 2001 20:3306*）

訳 細胞がシクロヘキシミド（CHX）かエメチンで処理されたとき

★ add（～を添加する／～を加える）

他動詞の用例が多いが，自動詞としても用いられる．

❶ added to ～（～に添加される）

When IL-3 plus GM-CSF or IL-3 plus IL-5 were added to eosinophils cultured with NCI-H292 cells, MUC5AC mucin production increased; eosinophils or cytokines alone had no effect.（*J Immunol. 2001 167:5948*）

訳 IL-3 プラス GM-CSF あるいは IL-3 プラス IL-5 が，NCI-H292 細胞と培養された好酸球に添加されたとき

14. 暴露する 【expose】

暴露する	
expose	challenge

使い分け

◆ **expose** は受動態で，「（X 線や薬剤などに）暴露される」ときに用いられる．

◆ **challenge** は受動態で，「（細菌などによって）暴露される／（細菌などを）接種される」場合に使われる．

共起・頻度分析　（数字：用例数）

直前の単語			直後の単語	
were			to	with
514	exposed	4,117	2,400	1
0	expose (s)	200	0	0
164	challenged	916	8	416
0	challenge (s)	3,692	135	584

解説

◆ **expose** は，受動態で to を伴う用例が非常に多い．

◆ **challenge** は，受動態で with を伴う用例が非常に多い．

◆ **challenge** は，名詞の用例が多い．

★ expose（～を暴露する）

他動詞．

❶ exposed to ～（～に暴露される）

When cells were exposed to the above agents, only Complex caused an up-regulation of connexin43 protein (based on immunocytochemical and immunoblot analysis).（*Cancer Res. 2002 62:3544*）

訳 細胞が上記の薬剤に暴露されたとき

★ challenge（～を暴露する／～を接種する）

名詞の用例が多いが，他動詞としても用いられる．

❶ challenged with ～（～に暴露される）

Wild-type and CD14-deficient（CD14-D）mice were challenged with *Escherichia coli* LPS.（*Circulation. 2002 106:2608*）

訳 野生型およびCD14欠損（CD14-D）マウスが，大腸菌のリポ多糖に暴露された

15. 培養する／インキュベートする 【culture】

培養する	インキュベートする
culture cultivate	incubate

使い分け

◆ **culture** は「（細胞や菌などを）培養する」ときに，**cultivate** は「（細菌などを）培養する」ときに用いられることが多い．

◆ **incubate** は，「（細胞などの培養や酵素反応のために）インキュベートする」という意味で使われる．

共起・頻度分析

（数字：用例数）

直前の単語			直後の単語	
were			in	with
322	cultured	4,828	436	225
20	culture (s)	10,327	196	266
12	cultivated	125	22	4
0	cultivate (s)	8	0	0
337	incubated	1,017	184	608
0	incubate (s)	4	0	0

解説

◆ **culture**, **cultivate** は，受動態で in を伴う用例が多い．

◆ **culture**, **incubate** は，受動態で with を伴う用例が多い．

◆ **culture** は，名詞の用例の方が多い．

★ culture（〜を培養する／培養）

他動詞および名詞として用いられる．

❶ cultured in 〜（〜において培養される）

When the VA cells were cultured in the presence of ethanol, we observed a dramatic reduction in cell accumulation.（*Hepatology. 2002 35:1196*）

訳 VA細胞がエタノール存在下で培養されたとき

❷ cultured with 〜（〜とともに培養される）

Purified NK cells were cultured with 200 ng/m*l* IL-15 for 2 days in the presence or absence of 10–200 ng/m*l* PGE_2.（*J Immunol. 2001 166:885*）

訳 精製されたNK細胞は，200 ng/m*l*のIL-15と2日間培養された

cultivate（〜を培養する）

他動詞．

❶ cultivated in 〜（〜において培養される）

A collection of 70 strains were cultivated in liquid culture and extracted using the QIAamp DNA minikit.（*J Clin Microbiol. 2004 42:4175*）

訳 70菌株のコレクションは液体培地で培養された

★ incubate（〜をインキュベートする／〜を保温する／〜を培養する）

他動詞．

❶ incubated with 〜（〜とインキュベートされる）

Cultured NRK-52E cells were incubated with EGF and/or ammonia and the protein/DNA ratio was measured, as a marker of hypertrophy.（*J Am Soc Nephrol. 2000 11:1631*）

訳 培養されたNRK-52E細胞はEGFとインキュベートされた

❷ incubated in 〜（〜において培養される）

This complex dissociated when the cells were incubated in medium without Ca^{2+} or treated with a c-Yes inhibitor, CGP77675.（*Mol Biol Cell. 2002 13:1227*）

訳 細胞が…のない培地で培養されたとき

第2章

名詞編

名詞は，主に主語，目的語，補語として用いられる．名詞編では，機能，現象，概念などを表す抽象的な名詞を多く取り上げたが，これらの名詞の特徴として，可算名詞・不可算名詞の区別がわかりにくいという点があげられる．これは，論文執筆時には非常に大きな問題なので，本書でもこの点にできるだけ触れるようにした．同じ単語であっても使われる状況によって可算・不可算が変わることがあり，冠詞の付け方を含めて特に注意が必要である．また，名詞には，特定の前置詞やto不定詞，that節などとの結び付きが強いものがあるので，ここではそのような高頻出共起表現（連語表現）を多数抽出して，その出現頻度とともに示した．名詞を用いる際には参考にして，その用法に留意して論文を執筆しよう．

【I-A. 知識・仮説・目的】 知識・仮説・目的などを表す名詞

1. 知識／理解 【knowledge】

知識	理解	認識
knowledge	understanding	awareness

使い分け ◆ **knowledge** は「知識」，**understanding** は「理解」，**awareness** は「認識」という意味に用いられる．

頻度分析

	用例数		用例数
knowledge	1,931	❶ knowledge of	791
		❷ our knowledge	574
		❸ knowledge about	105
understanding	5,252	❶ understanding of	2,797
		❷ our understanding	959
awareness	197	❶ awareness of	94

解説 ◆いずれの語も後に of を伴う用例が非常に多い．

★ knowledge（知識）

❶ knowledge of ～（～の知識）

Now we fill this fundamental gap in our knowledge of the DNA double helix.（*J Mol Biol. 2004 342:775*）

訳 われわれは，DNA の二重らせんに関するわれわれの知識の基礎的なギャップを埋める

★ understanding（理解）

❶ understanding of ～（～の理解）

These findings contribute to our understanding of spore formation and function and will be useful in the detection, prevention, and early treatment of anthrax.（*J Bacteriol. 2004 186:164*）

訳 これらの知見は，胞子の形成と機能に関するわれわれの理解に貢献する

awareness（認識／自覚）

❶ **awareness of ～**（～の認識）

However, there is increasing awareness of its role as an extracellular messenger molecule.（*J Cell Biol. 2002 158:345*）

訳 細胞外メッセンジャー分子としての役割に対する認識が増えている

名詞

2. 予想 【prediction】

予想	
prediction	expectation

使い分け ◆**prediction** は「予想」，**expectation** は「予想／期待」という意味に用いられる．

頻度分析

	用例数		用例数
prediction	2,920	❶ prediction of	515
expectation	489	❶ expectations of	28

I

★ prediction（予想）

A

❶ **prediction of ～**（～の予想）

We have developed a new method for the prediction of peptide sequences that bind to a protein, given a three-dimensional structure of the protein in complex with a peptide.（*J Mol Biol. 2001 313:317*）

訳 われわれは，…するペプチド配列の予想のための新しい方法を開発した

expectation（予想／期待）

複数形の用例もかなり多い．

❶ **expectations of ～**（～の予想）

Interestingly, the African sample fits the expectations of an equilibrium model based on polymorphism and divergence levels and on frequency spectrum.（*Am J Hum Genet. 2001 69:831*）

訳 アフリカ人のサンプルは，…に基づく平衡モデルの予想に合う

3. 可能性／確率 【possibility】

可能性	確率／見込み
possibility **potential**	**probability** **likelihood** **chance**

使い分け

◆ **possibility** は「可能性／起こりうること」，**potential** は「可能性／潜在力」，**probability**，**likelihood** は「確率／見込み」，**chance** は「公算／見込み」という意味に用いられる．

◆ **possibility** と **probability** は，混同しやすいので注意が必要である．**possbility** は〈起こりうる事象そのもの〉を意味するのに対して，**probability** は〈あることが起こりうる確率〉を意味する．つまり，two **possibilities** という用例はあっても high **possibility** の用例はなく，逆に high **probability** という用例はあるが two **probabilities** の用例はない．

頻度分析

	用例数		用例数
possibility	2,943	❶ possibility that	1,547
		❷ possibility of	710
potential	1,6431	❶ potential of	1,540
		❷ potential for	1,027
		❸ potential to	970
probability	1,706	❶ probability of	829
		❷ probability that	96
likelihood	1,132	❶ likelihood of	436
		❷ maximum likelihood	244
		❸ likelihood ratio	111
chance	352	❶ chance of	91
		❷ by chance	87

★ possibility（可能性）

"possibility that" の用例が非常に多い．

❶ possibility that ～（～という可能性）

These results raise the possibility that tumors with COX2 methylation may be less sensitive to treatment using specific COX2 inhibitors. (*Cancer Res. 2000 60:4044*)

訳 これらの結果は，…という可能性を示唆する

★ potential（可能性／潜在力／潜在的な）

名詞および形容詞として用いられる．

❷ potential for ～（～に対する可能性）

Neutralization of the chemokine MIG/CXCL9 may have therapeutic potential for the treatment of chronic rejection after heart transplantation.（*Am J Pathol. 2002 161:1307*）

訳 ケモカインMIG/CXCL9の中和は，心臓移植後の慢性拒絶反応の処置に対する治療上の可能性を持つかもしれない

★ probability（確率／蓋然性／見込み）

“probability of” の用例が非常に多い．

❶ probability of ～（～の確率）

In women who present with 1 or more symptoms of UTI, the probability of infection is approximately 50%.（*JAMA. 2002 287:2701*）

訳 感染の確率はおよそ50％である

★ likelihood（見込み／確率／尤度）

❶ likelihood of ～（～の確率）

These results suggest that expression of RARα-PML increases the likelihood of chromosome 2 deletions in APL cells.（*Proc Natl Acad Sci USA. 2000 97:13306*）

訳 RARα-PMLの発現は第2染色体の欠損の確率を上げる

chance（公算／見込み／機会）

❶ chance of ～（～の公算）

Smokers had a 3.5-fold higher chance of developing lung cancer compared with nonsmokers.（*Radiology. 2000 217:257*）

訳 喫煙者は，非喫煙者と比較して肺癌を発症する公算が3.5倍高い

4. 仮説／概念 【hypothesis】

仮説／仮定	概念／考え	見解／考え
hypothesis assumption	concept notion	idea notion view

使い分け

◆**hypothesis**は「仮説」，**assumption**は「仮定」，**concept**は「概念」，**notion**は「概念／見解」，**idea**，**view**は「見解／考え」という意味に用いられる．

頻度分析

	用例数		用例数
hypothesis	7,504	❶ hypothesis that	4,010
		❷ hypothesis of	181
assumption	773	❶ assumption that	185
		❷ assumption of	85
concept	1,270	❶ concept that	361
		❷ concept of	310
notion	782	❶ notion that	553
		❷ notion of	78
idea	1,052	❶ idea that	625
		❷ idea of	63
view	1,682	❶ view of	536
		❷ view that	396

解説

◆いずれの語も後にthatを伴う用例が多い．

★ hypothesis（仮説）

❶ hypothesis that ～（～という仮説）

These results support the hypothesis that IL-6 may play an important role in the control of micrometastatic disease in breast cancer. (*Cancer Res. 2003 63:8051*)

訳 これらの結果は，…という仮説を支持する

assumption（仮定／推定）

❶ assumption that ～（～という仮定）

Our approach is based on the assumption that many pathways exhibit two properties: their genes exhibit a similar gene expression profile, and the protein products of the genes often interact.（*Bioinformatics. 2003 19:i264*）

訳 われわれのアプローチは，…という仮定に基づいている

★concept（概念）

❶ concept that ～（～という概念）

These data support the concept that the life-span of neutrophil in the air spaces is modulated during acute inflammation.（*Crit Care Med. 2000 28:1*）

訳 これらのデータは，…という概念を支持する

notion（概念／考え）

❶ notion that ～（～という概念）

These results support the notion that the peripheral expansion of the $CD4^{+}CD25^{+}$ T cells is controlled in part by costimulation.（*J Immunol. 2004 173:2428*）

訳 これらの結果は，…という概念を支持する

★idea（考え／見解）

❶ idea that ～（～という考え）

Our results support the idea that epistatic variance may be more common in natural populations than was once suspected.（*Heredity. 2001 86:144*）

訳 われわれの結果は，…という考えを支持する

★view（見解／考え）

❷ view that ～（～という見解）

These findings support the view that CLC/CLF is a target-derived factor required for the survival of specific pools of motoneurons.（*J Neurosci. 2003 23:8854*）

訳 これらの知見は，…という見解を支持する

5. 計画 【design】

計画		
design	program	schedule

使い分け ◆ **design** は「計画／設計」，**program** は「プログラム／計画」，**schedule** は「スケジュール／予定」という意味に用いられる．

頻度分析

	用例数		用例数
design	3,512	❶ design of	1,000
program	3,450	❶ program of	227
		❷ program that	113
schedule	402	❶ schedule of	54

★ design（計画／設計／設計する）

名詞および動詞として用いられる．

❶ design of ～（～の設計／～の計画）

These findings have implications for the design of immunotherapeutic strategies and for testing candidate HIV vaccines.（*J Virol. 2004 78:4463*）

訳 これらの知見は，免疫療法的戦略の設計のための意味を持つ

★ program（プログラム／計画／計画する）

名詞および動詞として用いられる．

❶ program of ～（～のプログラム）

In response to environmental stress, cells induce a program of gene expression designed to remedy cellular damage or, alternatively, induce apoptosis.（*Mol Cell Biol. 2004 24:1365*）

訳 細胞は，細胞の傷害を治すように設計された遺伝子発現のプログラムを誘導する

schedule（スケジュール／予定／予定する）

名詞および動詞として用いられる．

❶ schedule of ～（～のスケジュール）

Finally, the development of DTH was dependent on the schedule of administration of the vaccine, specifically, the timing of an induction dose administered at the beginning of the treatment program. (*J Clin Oncol. 2004 22:403*)

訳 DTHの発生は，ワクチンの投与のスケジュールに依存していた

6. 目的／試み 【purpose】

目的	試み	努力
purpose aim goal objective end	attempt	effort

使い分け ◆ **purpose**, **aim**, **goal**, **objective**, **end** は「目的」，**attempt** は「試み」，**effort** は「努力」という意味に用いられる．

頻度分析

	用例数		用例数
purpose	1,871	❶ the purpose of	1,304
aim	1,971	❶ the aim of	780
goal	1,655	❶ the goal of	571
objective	1,447	❶ the objective of	541
end	9,737	❶ To this end,	192
attempt	1,255	❶ in an attempt to	327
effort	1,939	❶ in an effort to	408

解説 ◆ **purpose**, **aim**, **goal**, **objective** は，後にofを伴う用例が非常に多い．
◆ **attempt**, **effort** は，後にto不定詞を伴う用例が多い．

★ purpose（目的）

❶ the purpose of ～（～の目的）

The purpose of this study was to determine whether HP1 functions as a transcriptional repressor in the absence of chromosome rearrangements. (*Proc Natl Acad Sci USA. 2001 98:11423*)

訳 この研究の目的は，HP1が転写の抑制因子として機能するかどうかを決定することであった

名詞

I

A

★ aim（目的／目指す）

名詞および動詞として用いられる．

❶ the aim of ～（～の目的）

The aim of this study was to investigate whether pancreatic ductal epithelial cells could be differentiated into insulin-secreting cells by exposing them to GLP-1.（*Diabetes. 2001 50:785*）

訳 この研究の目的は，膵管上皮細胞がインスリン分泌細胞に分化できるかどうかを精査することであった

★ goal（目的）

❶ the goal of ～（～の目的）

The goal of this study was to determine the role of mTOR in type I collagen regulation.（*J Biol Chem. 2004 279:23166*）

訳 この研究の目的は，…におけるmTORの役割を決定することであった

★ objective（目的／客観的な）

名詞の用例が多いが，形容詞としても用いられる．

❶ the objective of ～（～の目的）

The objective of this study was to examine the effect of leukocyte activation on PSGL-1 expression and PSGL-1-mediated leukocyte adhesion to P-selectin.（*J Immunol. 2000 165:2764*）

訳 この研究の目的は，PSGL-1発現に対する白血球活性化の効果を調べることであった

★★ end（目的／末端／終わる）

名詞の用例が多いが，動詞としても用いられる．

❶ To this end, ～（この目的のために，～）

To this end, we used RNA interference to specifically down-regulate DNMT1 protein expression in NCI-H1299 lung cancer and HCC1954 breast cancer cells.（*Cancer Res. 2004 64:3137*）

訳 この目的のために，われわれはRNA干渉を使った

★ attempt（試み）

名詞の用例が多いが，動詞としても用いられる．

❷ in an attempt to ～（～する試みにおいて／～しようとして）

In an attempt to identify genes involved in the terminal esterification stage of bacteriochlorophyll biosynthesis, a previously uncharacterized 5-kb region of this cluster was sequenced.（*J Bacteriol. 2000 182:3175*）

訳 …に関与する遺伝子を同定する試みにおいて

★ effort（努力／試み）

❶ in an effort to ～（～する努力において／～しようとして）

In an effort to understand how this occurs, we have investigated whether these factors cause disregulation of cholesterol ester metabolism in J774.2 macrophages.（*J Biol Chem. 2002 277:42557*）

訳 どのようにこれが起こるのかを理解する努力において

7. 研究／検査 【study】

研究／調査	検査	分析
study investigation research work survey search dissection	examination test	analysis

使い分け

◆ **study**，**investigation**，**research**，**work** は「研究／調査」，**survey** は「調査」，**search** は「探索」，**dissection** は「精査」，**examination** は「検査」，**test** は「テスト／検査」，**analysis** は「分析」という意味に用いられる．

頻度分析

	用例数		用例数
study	68,829	❶ studies of	3,837
		❷ studies in	1,389
		❸ studies on	808
investigation	2,754	❶ investigation of	776
		❷ further investigation	315
		❸ investigation into	76
research	3,356	❶ of research	206
		❷ research on	191
		❸ research in	131
work	5,073	❶ this work	1,580
survey	1,304	❶ survey of	258
search	2,202	❶ search for	646
		❷ search of	179
dissection	449	❶ dissection of	212
examination	3,001	❶ examination of	1,277
test	10,582	❶ test of	248
analysis	42,055	❶ analysis of	14,298

解説

◆ **investigation**, **dissection**, **examination**, **analysis** は，後に of を伴う用例が非常に多い．

◆ **research** は後に on または in を，**search** は後に for を伴うことが多い．

★★ study（研究／研究する）

名詞および動詞として用いられる．

❶ **studies of ～**（～の研究）

These results were consistent with previous studies of human airway tissues.（*Am J Respir Crit Care Med. 2001 164:1059*）

訳 これらの結果はヒトの気道組織の以前の研究と一致した

★ investigation（研究／調査）

❶ **investigation of ～**（～の研究）

Collectively these findings identify the BNP gene as a potential model for the investigation of TR-dependent gene regulation in the heart.（*J Biol Chem. 2003 278:15073*）

訳 これらの知見は，心臓における TR 依存性遺伝子発現調節の研究のための潜在的なモデルとして BNP 遺伝子を同定した

★ research（研究／研究する）

名詞の用例が多いが，動詞としても用いられる．

❷ **research on ～**（～に関する研究）

A central unresolved problem in research on Alzheimer disease is the nature of the molecular entity causing dementia.（*Nat Neurosci. 2005 8:79*）

訳 アルツハイマー病に関する研究における中心的な未解決の問題は，痴呆を起こす分子実体の性質である

★ work（研究／働く）

名詞の用例が多いが，動詞としても用いられる．

❶ **this work**（この研究）

In this work, we have examined the interaction between carboxyl-terminal residues within secretin and the prototypic secretin receptor.（*J Biol Chem. 2000 275:26032*）

訳 この研究において，われわれは…の間の相互作用を調べた

★ survey（調査／調査する）

名詞の用例が多いが，動詞としても用いられる．

❶ survey of ～（～の調査）

In a survey of human and mouse tissues, expression was highest in testis.（*Nucleic Acids Res. 2003 31:6117*）

訳 ヒトとマウスの組織の調査において，精巣における発現が最高であった

★ search（探索）

名詞の用例が多いが，動詞としても用いられる．

❶ search for ～（～の探索）

In the search for individual transcription factor binding sites, multiple alignments markedly increase the signal-to-noise ratio compared to pairwise alignments.（*Genome Res. 2004 14:313*）

訳 個々の転写因子結合部位の探索において，マルチプルアラインメントがＳＮ比を顕著に上昇させる

dissection（精査）

❶ dissection of ～（～の精査）

To obtain insight into the mechanisms of RHD3 regulation, we conducted a molecular genetic dissection of RHD3 gene expression and function.（*Plant Physiol. 2002 129:638*）

訳 われわれは，RHD3遺伝子の発現と機能の分子遺伝学的精査を行った

★ examination（検査／検討）

❶ examination of ～（～の検査）

Histological examination of postischemic cerebral microvessels revealed a strong upregulation of E-and P-selectin expression.（*Circ Res. 2002 91:907*）

訳 虚血後の脳微小血管の組織学的検査は，E-およびP-セレクチン発現の強力な上方制御を明らかにした

★ test（テスト／テストする）

動詞の用例が多いが，名詞としても用いられる．

❶ test of ～（～のテスト）

As a test of the model, we propose mutations that should reverse these outcomes.（*J Mol Biol. 2003 332:953*）

訳 モデルのテストとして

☆★ analysis（分析）

❶ **analysis of ～**（～の分析）

These data suggest that these novel recombinant proteins are useful for the analysis of antibody responses to Msg.（*J Infect Dis. 2002 186:644*）

訳 これらの新規の組換えタンパク質は，Msgに対する抗体反応の分析に役立つ

名詞

8. 同定／発見／事実 【identification】

同定／検出	発見／知見	観察／事実	特徴づけ
identification detection	discovery finding	observation fact	characterization

使い分け

◆ **identification** は「同定」，**detection** は「検出」，**discovery** は「発見」，**finding** は「発見／知見」，**observation** は「観察」，**fact** は「事実」，**characterization** は「特徴づけ」という意味に用いられる．

I

頻度分析

	用例数		用例数
identification	5,310	❶ identification of	3,963
detection	5,981	❶ detection of	2,625
discovery	1,681	❶ discovery of	763
finding	17,208	❶ findings suggest	2,790
		❷ findings indicate	1,390
		❸ finding that	737
		❹ findings in	388
observation	6,925	❶ observations suggest	796
		❷ observation that	692
		❸ observation of	438
fact	1,780	❶ the fact that	1,047
		❷ in fact	548
characterization	3,406	❶ characterization of	2,874

B

解説

◆ **identification**, **detection**, **discovery**, **characterization** は，後に of を伴う用例が非常に多い．

◆ **finding**, **observation**, **fact** は，後に that を伴う用例が多い．

★ identification（同定）

❶ identification of ～（～の同定）

Genetic and biochemical studies have led to the identification of the Stat3-Interacting Protein StIP1.（*Proc Natl Acad Sci USA. 2000 97:10120*）

訳 遺伝的および生化学的研究は，…の同定につながった

★ detection（検出）

❶ detection of ～（～の検出）

To this end, we developed a new sensitive method for the detection of factor IXa based on its affinity to antithrombin III.（*Biochemistry. 2004 43:11883*）

訳 われわれは，…の検出のための新しい高感度の方法を開発した

★ discovery（発見）

❶ discovery of ～（～の発見）

Here, we report the discovery of a new mechanism by which glucagon suppresses insulin secretion.（*Diabetes. 2000 49:1681*）

訳 われわれは，グルカゴンがインスリン分泌を抑制する新しい機構の発見を報告する

★★ finding（知見／発見）

複数形の用例が多い．

❶ findings suggest ～（知見は～を示唆する）

These findings suggest that macrophage-derived MMP-9 and mesenchymal cell MMP-2 are both required and work in concert to produce AAA.（*J Clin Invest. 2002 110:625*）

訳 これらの知見は，…ということを示唆する

❸ finding that ～（～という発見）

This conclusion is supported by the finding that deleting the SGS1 helicase also suppressed heteroduplex rejection.（*Proc Natl Acad Sci USA. 2004 101:9315*）

訳 この結論は，…という発見によって支持される

★ observation（観察）

❷ observation that ～（～という観察）

This hypothesis is supported by the observation that the loss of PEX19 results in degradation of PMPs and/or mislocalization of PMPs to the mitochondrion. (*J Cell Biol. 2000 148:931*)

訳 この仮説は，…という観察によって支持される

★ fact（事実）

❶ the fact that ～（～という事実）

Despite the fact that mTERF binds DNA as a monomer, the presence in its sequence of three leucine-zipper motifs suggested the possibility of mTERF establishing intermolecular interactions with proteins of the same or different type. (*J Biol Chem. 2004 279:15670*)

訳 mTERF が単量体として DNA に結合するという事実にもかかわらず

★ characterization（特徴づけ）

❶ characterization of ～（～の特徴づけ）

We report the characterization of ScPex8p, which is essential for peroxisomal biogenesis in *Saccharomyces cerevisiae*. (*J Biol Chem. 2000 275:3593*)

訳 われわれは，ScPex8p の特徴づけを報告する

9. 評価／比較 【assessment】

評価	評価／推定	比較
assessment evaluation	estimation	comparison

使い分け

◆ **assessment**，**evaluation** は「評価」，**estimation** は「評価／推定」，**comparison** は「比較」という意味に用いられる．

頻度分析

	用例数		用例数
assessment	2,313	❶ assessment of	1,099
evaluation	2,304	❶ evaluation of	1,137
estimation	464	❶ estimation of	242
comparison	6,891	❶ comparison of	2,176
		❷ comparison with	937
		❸ comparison to	537

解説 ◆いずれの語も後にofを伴う用例が非常に多い.

★ assessment（評価）

❶ assessment of ～（～の評価）

This animal model may be used for the assessment of new IPVs and of combination vaccines containing an IPV component.（*J Infect Dis. 2004 190:1404*）

訳 この動物モデルは，新しいIPVの評価のために使われるかもしれない

★ evaluation（評価）

❶ evaluation of ～（～の評価）

T2-weighted MR imaging was useful in the evaluation of cysts and lymph nodes.（*Radiology. 2004 230:637*）

訳 T2強調MR画像は，嚢胞とリンパ節の評価に役立った

estimation（評価／推定）

❶ estimation of ～（～の評価／～の推定）

The method is applied in the estimation of growth rates of a human population based on 37 SNP loci.（*Genetics. 2000 154:931*）

訳 その方法は，…の成長率の評価に適用される

★ comparison（比較）

後にof，withまたはtoを伴う用例が多い.

❶ comparison of ～（～の比較）

Based on a comparison of the R subunit sequences, we predict that the linker regions are the likely cause of these large differences in shape among the isoforms.（*J Mol Biol. 2004 37:1183*）

訳 Rサブユニット配列の比較に基づいて，われわれは…ということを予想する

10. 決定 【determination】

決定	
determination	decision

使い分け
◆ **determionation** は「決定」，**decision** は「決定／決断」という意味に用いられる．

頻度分析

	用例数		用例数
determination	2,189	❶ determination of	1,210
decision	1,223	❶ decision making	232

解説
◆ **determionation** は，後に of を伴う用例が非常に多い．

★ determination（決定）

❶ determination of ～（～の決定）

A new method for the determination of enantiomeric compositions of a variety of drugs including propranolol, atenolol, and ibuprofen has been developed.（*Anal Biochem. 2004 325:206*）

訳 プロプラノロール，アテノロールおよびイブプロフェンを含む種々の薬剤の鏡像異性的成分を決定する新しい方法が開発された

★ decision（決定／決断）

❶ decision making（判断／意思決定）

This prognostic model may help clinicians and patients in clinical decision making, as well as investigators in research planning.（*J Clin Oncol. 2005 23:175*）

訳 この予後のモデルは，臨床的判断の際に臨床家と患者を助けるかもしれない

名詞

I

B

11. 証明／証拠 【evidence】

証明	証拠
demonstration proof	evidence

使い分け
◆ **demonstration** は「実証」，**proof** は「証明」，**evidence** は「証拠」という意味に用いられる．
◆ **evidence** を積み重ねたものを **proof** という．

頻度分析

	用例数		用例数
demonstration	959	❶ demonstration of	529
proof	304	❶ proof of	212
evidence	16,415	❶ evidence that	4,909
		❷ evidence for	3,554
		❸ evidence of	3,061

解説
◆ **demonstration**, **proof** は，of を後に伴う用例が非常に多い．
◆ **evidence** は，of だけでなく that や for を後に伴う用例も多い．

★ demonstration（実証／証明／実演）

❶ demonstration of ～（～の実証）

This is the first demonstration of a prognostic significance for STAT proteins in a malignancy.（*Blood. 2002 99:252*）

訳 これは，…の最初の実証である

proof（証明／校正刷り）

❶ proof of ～（～の証明）

This study provides proof of principle for a targeting strategy that would be generally useful for many gene therapy applications.（*J Virol. 2001 75:8016*）

訳 この研究は，ターゲティング戦略の原理の証明を提供する

★★ evidence（証拠）

❶ evidence that ～（～という証拠）

We provide evidence that endogenous COUP-TF activity represses

the COL7A1 promoter.（*J Biol Chem. 2004 279:23759*）
訳 われわれは，…という証拠を提供する

❷ **evidence for ～**（～の証拠）

These results provide evidence for a new function of BMP4.（*Development. 2000 127:1431*）
訳 これらの結果は，BMP4の新しい機能の証拠を提供する

❸ **evidence of ～**（～の証拠）

There was no evidence of interaction between radiotherapy and tamoxifen.（*Lancet. 2003 362:95*）
訳 放射線治療とタモキシフェンの間に相互作用の証拠はなかった

12. 報告／結論／説明／示唆 【report】

報告／記述	結論	説明	示唆
report description	conclusion	explanation interpretation	suggestion

使い分け

◆ **report** は「報告」，**description** は「記述」，**conclusion** は「結論」，**explanation** は「説明」，**interpretation** は「解釈／説明」，**suggestion** は「示唆」という意味に用いられる．

頻度分析

	用例数		用例数
report	16,159	❶ this report	2,164
description	759	❶ description of	497
conclusion	2,740	❶ in conclusion	1,436
		❷ conclusion that	385
explanation	1,181	❶ explanation for	685
		❷ explanation of	112
interpretation	1,049	❶ interpretation of	547
suggestion	389	❶ suggestion that	157

解説

◆ **description**，**interpretation** は後に of を，**conclusion**，**suggestion** は後に that を，**explanation** は後に for を伴う用例が多い．

名詞

I

B

★ report（報告／報告する）

動詞および名詞として用いられる．

❶ this report（この報告）

In this report, we show that NIK kinase activity is specifically increased in cells stimulated by two EphRs, EphB1 and EphB2.（*Mol Cell Biol. 2000 20:1537*）

訳 この報告において，われわれは…ということを示す

description（記述）

❶ description of ～（～の記述）

This is the first description of a biological mechanism and functional significance of PNI.（*Cancer Res. 2004 64:6082*）

訳 これは，PNIの生物学的機構と機能的重要性に関する最初の記述である

★ conclusion（結論）

❷ conclusion that ～（～という結論）

Overall, these data support the conclusion that MCDF is a partial agonist at the ER.（*Cancer Res. 2004 64:2889*）

訳 これらのデータは，…という結論を支持する

★ explanation（説明）

❶ explanation for ～（～に対する説明）

These findings may provide an explanation for some of the racial differences in colon cancer incidence.（*Am J Epidemiol. 2003 158:951*）

訳 これらの知見は，人種差のいくつかに対する説明を提供するかもしれない

★ interpretation（解釈／説明）

❶ interpretation of ～（～の解釈）

These findings may have implications for the interpretation of BNP levels in the assessment of patients with heart and lung disease.（*Circulation. 2004 109:2872*）

訳 これらの知見は，心臓と肺の疾患を持つ患者の評価におけるBNPレベルの解釈にとっての意味を持つかもしれない

suggestion（示唆）

❶ suggestion that ～（～という示唆）

This fact leads to the suggestion that primordial proteins most closely related to the PST family were the evolutionary precursors of all members of the MOP superfamily.（*Eur J Biochem. 2003 270:799*）

訳 これらの事実は，…という示唆につながる

【Ⅱ-A. 発生・増加】 発生・増加などを表す名詞

1. 存在 【presence】

存在	
presence	existence

使い分け ◆**presence** は「存在」，**existence** は「存在／実存」という意味に用いられる．

頻度分析

	用例数		用例数
presence	18,824	❶ presence of	16,936
existence	1,742	❶ existence of	1,639

解説 ◆どちらも後に of を伴う用例が圧倒的に多い．

★★ presence（存在）

“in the presence of” の形での用例が非常に多い．

❶ presence of ～（～の存在）

Formation of TRAP-AT complexes occurs only in the presence of tryptophan.（*J Mol Biol. 2004 338:669*）

訳 TRAP-AT 複合体の形成は，トリプトファンの存在下においてのみ起こる

★ existence（存在／実存）

❶ existence of ～（～の存在）

In this review, we present evidence for the existence of three highly conserved regions（CRs）shared by MDM2 proteins and MDMX proteins of different species.（*Gene. 2000 242:15*）

訳 われわれは，3 つの高度に保存された領域の存在の証拠を提示する

2. 発生／発現 【appearance】

名詞

発生／出現	発現
appearance occurrence emergence outbreak	expression

使い分け ◆**appearance**, **occurrence**, **emergence** は「発生／出現」, **outbreak** は「急激な発生」, **expression** は「発現」という意味に用いられる.

頻度分析

	用例数		用例数
appearance	1,576	❶ appearance of	1,113
occurrence	1,239	❶ occurrence of	873
		❷ occurrence in	69
emergence	640	❶ emergence of	489
outbreak	740	❶ outbreak of	103
expression	80,412	❶ expression of	29,956
		❷ gene expression	11,386
		❸ expression in	7,905

II

解説 ◆いずれの語も後に of を伴う用例が多い.

A

★ appearance（出現／発生）

❶ **appearance of ～** （～の出現）

Western blot analysis indicated that MAEBL expression was better correlated with the appearance of the canonical ORF1 transcript. (*J Mol Biol. 2004 343:589*)

訳 MAEBL 発現は，…の出現とよりよく相関した

★ occurrence（発生／出現）

❶ **occurrence of ～** （～の発生）

There are sex differences in the occurrence of microsatellite instability (MSI) in colon tumors. (*Cancer Res. 2001 61:126*)

訳 大腸腫瘍におけるマイクロサテライト不安定性（MSI）の発生には性差がある

emergence（発生／出現）

❶ emergence of ～（～の発生）

Passage of the virus-infected cells under these conditions led to the emergence of a viral variant that was able to replicate efficiently in this culture system.（*J Virol. 2000 74:10882*）

訳 このような条件下でのウイルス感染細胞の継代は，ウイルスの変異体の発生につながった

outbreak（急激な発生）

❶ outbreak of ～（～の急激な発生）

In March 1998, an outbreak of acute gastroenteritis occurred among students at a Texas university.（*J Infect Dis. 2000 181:1467*）

訳 急性胃腸炎の急激な発生がテキサス大学で学生の間に起こった

⁑ expression（発現）

❶ expression of ～（～の発現）

In considering the potential involvement of other MRP family isoforms, a 3-fold increase in the expression of MRP5 was observed in MCF7/VP cells.（*Cancer Res. 2001 61:5461*）

訳 MRP5の発現の3倍の上昇がMCF7/VP細胞において観察された

❸ expression in ～（～における発現）

Here oligonucleotide microarrays have been used to compare the patterns of gene expression in preadipocytes and adipocytes *in vitro* and *in vivo*.（*J Biol Chem. 2001 276:34167*）

訳 オリゴヌクレオチドマイクロアレイは，前脂肪細胞と脂肪細胞における遺伝子発現のパターンを比較するために使われてきた

3. 由来／原因 【origin】

由来	原因
origin source	cause

使い分け ◆ **origin** は「起源」，**source** は「出所／ソース」，**cause** は「原因」という意味に用いられる．

頻度分析

	用例数		用例数
origin	3,794	❶ origin of	932
		❷ origin for	73
		❸ origin in	62
source	3,427	❶ source of	1,644
		❷ source for	165
		❸ source in	58
cause	12,442	❶ cause of	2,017
		❷ cause for	70

解説 ◆いずれの語も後に of を伴う用例が多い．

★ origin（起源／発生／始まり）

❶ **origin of ～**（～の起源／～の発生）

Stem cells are crucial for normal development and homeostasis, and their misbehavior may be related to the origin of cancer. (*Development. 2004 131:337*)

訳 それらの不正な挙動は癌の起源に関連するかもしれない

★ source（ソース／源）

❶ **source of ～**（～のソース）

Umbilical cord blood has been increasingly used as a source of hematopoietic stem cells. (*Blood. 2002 99:364*)

訳 臍帯血は，造血性幹細胞のソースとしてますます使われている

★ cause（原因／〜を引き起こす）

名詞および動詞として用いられる．

❶ cause of 〜（〜の原因）

Ovarian cancer is a major cause of cancer death in women.（*Cancer Res. 2002 62:2923*）

訳 卵巣癌は女性における癌による死亡の主な原因である

4．産生 【production】

産生	収率／収量
production generation	yield

使い分け ◆ **production**，**generation** は「産生」，**yield** は「収率／収量」という意味に用いられる．

頻度分析

	用例数		用例数
production	13,458	❶ production of	4,548
		❷ production in	1,138
		❸ production by	825
generation	5,344	❶ generation of	2,560
		❷ generation in	184
		❸ generation by	101
yield	3,333	❶ yield of	297

解説 ◆いずれの語も後に of を伴う用例が多い．

★★ production（産生）

❶ production of 〜（〜の産生）

The activation by HVT resulted in the production of pp38 protein.（*J Virol. 2000 74:10176*）

訳 HVT による活性化は pp38 タンパク質の産生という結果になった

★ generation（産生）

❶ **generation of ～**（～の産生）

These bidomain effects play an important role in the generation of the whole-heart magnetocardiogram and cannot be ignored.（*Biophys J. 2004 87:4326*）

訳 これらのバイドメイン効果は，…の産生において重要な役割を果たす

★ yield（収率／収量）

❶ **yield of ～**（～の収率）

This leads to the decrease in the yield of monomers at lower pressures.（*J Biol Chem. 2001 276:6253*）

訳 これは単量体の収率の減少につながる

5．増加／上昇 【increase】

増加／上昇	上方制御
increase elevation augmentation	up-regulation

使い分け ◆ **increase**，**elevation**，**augmentation** は「増加／上昇」，**up-regulation** は「上方制御」という意味に用いられる．

頻度分析

	用例数		用例数
increase	30,090	❶ increase in	13,623
		❷ increase of	873
		❸ increase with	158
elevation	1,836	❶ elevation of	670
		❷ elevation in	252
augmentation	363	❶ augmentation of	203
up-regulation	1,971	❶ up-regulation of	1,476
		❷ up-regulation in	81

解説
◆ **increase**，**elevation** は，後に in を伴う用例が多い．
◆ **elevation**，**augmentation**，**up-regulation** は，後に of を伴う用例が非常に多い．

名詞 II A

★★ increase（増加／上昇）

名詞および動詞として用いられる．

❶ increase in ～（～の増加／～の上昇）

This effect is due to an increase in the amount of gag-pol mRNA.（*J Virol. 2000 74:4839*）

訳 この効果は gag-pol メッセンジャー RNA の量の増加のせいである

★ elevation（上昇）

❶ elevation of ～（～の上昇）

Furthermore, the elevation of intracellular Ca^{2+} induced NDRG-1/Cap43 mRNA in HIF-1α-deficient cells.（*Mol Cell Biol. 2002 22:1734*）

訳 細胞内 Ca^{2+} の上昇は NDRG-1/Cap43 メッセンジャー RNA を誘導した

❷ elevation in ～（～の上昇）

An elevation in the levels of fatty acid transporter gene expression was also observed.（*J Biol Chem. 2000 275:27117*）

訳 脂肪酸トランスポーター遺伝子発現のレベルの上昇もまた観察された

augmentation（上昇／増大）

❶ augmentation of ～（～の増大）

The induced levels of hMLH1 and hPMS1 are important for the augmentation of p53 phosphorylation by ATM in response to DNA damage.（*Mol Cell Biol. 2004 24:6430*）

訳 誘導されたレベルの hMLH1 と hPMS1 は，ATM による p53 リン酸化の増大にとって重要である

★ up-regulation（上方制御）

❶ up-regulation of ～（～の上方制御）

Mutation of p53 may play a role in the up-regulation of bcl-2 expression in some B cell lymphomas.（*Oncogene. 2001 20:240*）

訳 p53 の変異は，bcl-2 発現の上方制御において役割を果たすかもしれない

6. 増強　【enhancement】

増強	活性化
enhancement potentiation	activation

使い分け ◆ **enhancement**, **potentiation** は「増強」, **activation** は「活性化」という意味に用いられる.

頻度分析

	用例数		用例数
enhancement	2,572	❶ enhancement of	1,303
		❷ enhancement in	185
potentiation	997	❶ potentiation of	341
activation	43,186	❶ activation of	17,699
		❷ activation in	1,911
		❸ activation by	1,890

解説 ◆いずれの語も後に of を伴う用例が非常に多い.

★ enhancement（増強）

❶ **enhancement of ～**（～の増強）

An adaptor protein SH2-Bβ is involved in the enhancement of the tyrosine kinase activity of Jak2 following ligand/receptor interaction.（*Oncogene. 2002 21:7137*）

訳 アダプタータンパク質 SH2-Bβは，Jak2 のチロシンキナーゼ活性の増強に関与する

★ potentiation（増強）

❶ **potentiation of ～**（～の増強）

Secretion of MDA-7 was not required for the potentiation of TNF-induced NF-κB activation.（*J Immunol. 2004 173:4368*）

訳 MDA-7 の分泌は，TNF に誘導される NF-κB 活性化の増強に必要とされなかった

★★ activation（活性化）

❶ activation of ～（～の活性化）

Human Rad17（hRad17）is centrally involved in the activation of cell-cycle checkpoints by genotoxic agents or replication stress.（*EMBO J. 2004 23:4660*）

訳 ヒトのRad17（hRad17）は，遺伝毒性剤による細胞周期チェックポイントの活性化に中心的に関与する

7. 促進／誘導 【promotion】

促進	誘導
promotion **acceleration** **facilitation**	**induction**

使い分け ◆**promotion**，**acceleration**，**facilitation**は「促進」，**induction**は「誘導」という意味に用いられる．

頻度分析

	用例数		用例数
promotion	480	❶ promotion of	250
acceleration	423	❶ acceleration of	193
facilitation	575	❶ facilitation of	155
induction	11,949	❶ induction of	7,562
		❷ induction by	502
		❸ induction in	396

解説 ◆いずれの語も後にofを伴う用例が非常に多い．

promotion（促進）

❶ promotion of ～（～の促進）

The loss of cell cycle control is believed to be an important mechanism in the promotion of carcinogenesis.（*Cancer Res. 2004 64:1997*）

訳 細胞周期調節の喪失は，発癌の促進における重要な機構であると信じられている

acceleration（促進）

❶ **acceleration of ～**（～の促進）

Our results show that while ectopic overexpression of p75 c-Myb results in the acceleration of cell death, similar overexpression of p89 c-Myb results in the protection of cells from apoptotic death.（*Mol Cell Biol. 2003 23:6631*）

訳 p75 c-Myb の異所性の過剰発現は細胞死の促進という結果になる

facilitation（促進／亢進）

❶ **facilitation of ～**（～の促進）

This suggests that mental stress may lead to sudden death through the facilitation of lethal ventricular arrhythmias.（*Circulation. 2000 101:158*）

訳 精神的なストレスは，致死的な心室性不整脈の促進による突然死につながるかもしれない

★★ induction（誘導）

❶ **induction of ～**（～の誘導）

The present studies showed that Fas-FasL signaling plays a major role in the induction of apoptosis in lymphocytes after exposure to IAV.（*J Virol. 2001 75:59210*）

訳 Fas-FasL シグナル伝達は，リンパ球のアポトーシスの誘導において主要な役割を果たす

8. 進行／進歩 【progression】

進行／進歩		
progression	progress	advance

使い分け ◆いずれの語も「進行／進歩」という意味に用いられる．

頻度分析

	用例数		用例数
progression	6,230	❶ progression of	1,499
progress	1,167	❶ progress in	246
advances	1,228	❶ advances in	555

名詞

II

A

解説 ◆ **progression** は後に of を，**progress**，**advance** は後に in を伴う用例が多い．

★ progression（進行／進歩）

❶ progression of ～（～の進行）

This study supports the concept that coronary endothelial dysfunction may play a role in the progression of coronary atherosclerosis. (*Circulation. 2000 101:948*)

訳 冠状動脈内皮の機能障害は，冠状動脈のアテローム硬化症の進行において役割を果たすかもしれない

★ progress（進歩／進行）

名詞の用例が多いが，動詞としても用いられる．

❶ progress in ～（～における進歩）

In spite of considerable recent progress in identifying multiple roles of MMPs in disease, our understanding of MMP function in cancer is far from complete. (*Oncogene. 2000 19:6642*)

訳 MMP の複数の役割を同定することにおけるかなりの最近の進歩にもかかわらず

★ advance（進歩／進行）

動詞の用例が多いが，名詞としても用いられる．名詞としては複数形の用例が多い．

❶ advances in ～（～における進歩）

Despite recent advances in reconstruction techniques, total gastrectomy is still accompanied by various complications. (*Transplantation. 2003 76:61*)

訳 再構築の技術における最近の進歩にもかかわらず

9. 増殖 【growth】

増殖	複製
growth proliferation	replication

使い分け

◆ **growth**, **proliferation** は「増殖」, **replication** は「複製」という意味で用いられる.

◆ **growth**, **proliferation** は細胞や細菌の増殖について用いられ, **replication** は DNA やウイルスの複製に対して使われる.

頻度分析

	用例数		用例数
growth	30,235	❶ growth of	2,432
		❷ cell growth	2,257
		❸ growth in	1,451
proliferation	9,967	❶ cell proliferation	3,273
		❷ proliferation of	1,434
		❸ proliferation in	781
replication	10,790	❶ DNA replication	2,001
		❷ replication in	1,020
		❸ replication of	859

解説 ◆いずれの語も後に of または in を伴う用例が多い.

★ growth（増殖／成長）

❶ growth of ～（～の増殖）

In addition, 17-AAG significantly inhibited the growth of a glioma xenograft in nude mice.（*Cancer Res. 2001 61:4010*）

訳 17-AAG はグリア細胞腫異種移植片の増殖を有意に抑制した

★ proliferation（増殖）

❷ proliferation of ～（～の増殖）

Here we investigate whether CLT inhibits the proliferation of lung mesenchymal cells.（*Transplantation. 2000 70:1263*）

訳 われわれは, CLT が肺の間葉系細胞の増殖を抑制するかどうかを精査する

★ replication（複製）

❷ replication in ～（～の複製）

The DnaA protein is essential for initiation of DNA replication in a wide variety of bacterial and plasmid replicons.（*J Mol Biol. 2001 310:69*）

訳 DnaA タンパク質は, さまざまな細菌およびプラスミドのレプリコンの DNA 複製の開始に必須である

❸ **replication of ～**（～の複製）

We investigated the role of PG in the replication of HIV-1 in primary macrophages.（*J Biol Chem. 2002 277:16913*）

訳 われわれは，HIV-1の複製におけるPGの役割を精査した

10. 拡大／伸展 【expansion】

拡大／伸展	
expansion	extension

使い分け ◆**expansion**は「拡大／増大」，**extension**は「伸展／拡大」という意味に用いられる．

頻度分析

	用例数		用例数
expansion	3,190	❶ expansion of	1,209
extension	2,256	❶ extension of	515

解説 ◆どちらも後にofを伴う用例が多い．

★ expansion（拡大／増大）

❶ **expansion of ～**（～の拡大）

Huntington's disease is caused by the expansion of CAG repeats coding for a polyglutamine tract in the huntingtin protein.（*Proc Natl Acad Sci USA. 2002 99:727*）

訳 ハンチントン病はCAGリピートの拡大によって引き起こされる

★ extension（伸展／拡大）

❶ **extension of ～**（～の伸展）

Cell spreading and motility require the extension of the plasma membrane in association with the assembly of actin.（*J Cell Biol. 2000 148:127*）

訳 細胞延展と運動性は原形質膜の伸展を必要とする

11. 低下／減少 【decrease】

低下／減少	下方制御
decrease reduction fall loss decline attenuation	down-regulation

使い分け

◆ **decrease**, **reduction**, **fall** は「低下／減少」, **loss** は「減少／喪失」, **decline**, **attenuation** は「低下／減衰」, **down-regulation** は「下方制御」という意味に用いられる.

頻度分析

	用例数		用例数
decrease	10,848	❶ decrease in	5,455
		❷ a decrease	2,086
		❸ decrease of	410
reduction	11,541	❶ reduction in	5,040
		❷ reduction of	2,720
		❸ a reduction	1,498
fall	699	❶ fall in	200
loss	14,662	❶ loss of	9,854
		❷ loss in	713
decline	2,032	❶ decline in	890
		❷ decline of	115
attenuation	1,059	❶ attenuation of	487
		❷ attenuation in	59
down-regulation	1,911	❶ down-regulation of	1,408
		❷ down-regulation in	56

解説

◆ **decrease**, **reduction**, **fall**, **decline** は後に in を, **loss**, **attenuation**, **down-regulation** は後に of を伴う用例が非常に多い.

★ decrease（低下／減少／低下する／減少させる）

名詞および動詞として用いられる．

❶ decrease in ～（～の低下／～の減少）

The activation is associated with a decrease in K_m and an increase in V_{max}, suggesting positive allosteric activation.（*Biochemistry. 2001 40:15318*）

訳 活性化は K_m の低下と V_{max} の上昇に伴われた

★★ reduction（低下／減少／還元）

❶ reduction in ～（～の低下／～の減少）

Improvement in the clinical symptoms of RA was associated with a reduction in serum VEGF levels.（*Arthritis Rheum. 2001 44:2055*）

訳 RAの臨床症状の改善は血清VEGFレベルの低下に伴われた

fall（低下／低下する）

名詞および動詞として用いられる．

❶ fall in ～（～の低下）

Removal of doxycycline caused a rapid fall in levels of mutant K-Ras RNA and concomitant apoptotic regression of both the early proliferative lesions and the tumors.（*Genes Dev. 2001 15:3249*）

訳 ドキシサイクリンの除去は，変異K-Ras RNAレベルの急速な低下を引き起こした

★★ loss（低下／喪失／損失）

❶ loss of ～（～の低下／～の喪失）

Amino acid replacement of any one of these three loop residues results in a significant loss of catalytic activity.（*Biochemistry. 2004 43:4447*）

訳 これら3つのループ残基のどの1つのアミノ酸置換も触媒活性の有意な低下という結果になる

★ decline（低下／減衰）

動詞としても用いられるが，名詞の用例の方が多い．

❶ decline in ～（～の低下）

Alzheimer's disease is a neurodegenerative disorder associated with a decline in cognitive abilities.（*Mol Pharmacol. 2004 66:538*）

訳 アルツハイマー病は認識能の低下を伴う神経変性疾患である

★ attenuation（低下／減衰）

❶ **attenuation of ～**（～の低下）

Coexpression of IL-15 resulted in the attenuation of virulence of vaccinia virus, and mice inoculated with 10^5 plaque-forming units or less resolved the infection successfully.（*Proc Natl Acad Sci USA. 2001 98:5146*）

訳 IL-15の共発現は，ワクシニアウイルスの病原性の低下という結果になった

★ down-regulation（下方制御）

❶ **down-regulation of ～**（～の下方制御）

Because Mrp3 is capable of transporting toxic bile acids, up-regulation of Mrp3 may compensate for the down-regulation of Mrp2 in obstructive cholestasis.（*Hepatology. 2001 33:783*）

訳 Mrp3の上方制御はMrp2の下方制御を代償するかもしれない

12. 抑制 【inhibition】

抑制	干渉
inhibition suppression repression depression	interference

使い分け

◆ **inhibition**, **suppression**, **repression**, **depression** は「抑制」，**interference** は「干渉」という意味に用いられる．

頻度分析

	用例数		用例数
inhibition	19,184	❶ inhibition of	11,294
		❷ inhibition by	1,012
suppression	3,921	❶ suppression of	2,292
repression	3,423	❶ repression of	1,234
		❷ repression by	258
		❸ repression in	140

depression	2,554	❶ depression of	162
		❷ depression in	157
interference	1,906	❶ interference with	263
		❷ interference in	86
		❸ interference of	72

解説

◆ **inhibition**, **suppression**, **repression** は，後に of を伴う用例が非常に多い．

◆ **interference** は，後に with を伴う用例が多い．

★★ inhibition（抑制）

“inhibition of”の用例が非常に多い．

❶ inhibition of ～（～の抑制）

Further, transient transfection of activated ErbB2 resulted in the inhibition of Muc1 transcriptional activation in luciferase reporter assays.（*Oncogene. 2004 23:697*）

訳 活性化された ErbB2 の一過性のトランスフェクションは，Muc1 転写活性化の抑制という結果になった

★ suppression（抑制）

“suppression of”の用例が非常に多い．

❶ suppression of ～（～の抑制）

Maspin is a unique serpin involved in the suppression of tumor growth and metastasis.（*Oncogene. 2000 19:6053*）

訳 Maspin は，腫瘍の増殖と転移の抑制に関与するユニークなセルピンである

★ repression（抑制）

❶ repression of ～（～の抑制）

RB pathway activation resulted in the repression of Plk1 promoter activity, and this action was dependent on the SWI/SNF chromatin remodeling complex.（*J Biol Chem. 2004 279:29278*）

訳 RB 経路の活性化は Plk1 プロモータ活性の抑制という結果になった

★ depression（抑制／抑鬱）

❶ depression of ～（～の抑制）

At >20 Hz, the depression of IPSCs in the FB pathway was greater

than in the FF pathway.（*J Comp Neurol. 2004 475:361*）
訳 FB 経路における IPSC の抑制は FF 経路より大きかった

★ interference（干渉）

❶ interference with 〜（〜との干渉）

Here we demonstrate that measles virus can inhibit cytokine responses by direct interference with host STAT protein-dependent signaling systems.（*J Virol. 2003 77:7635*）
訳 麻疹ウイルスは，宿主の STAT タンパク質依存性シグナル伝達系との直接の干渉によってサイトカイン応答を抑制できる

名詞

13. 阻止／遮断 【block】

阻止／遮断	非存在
block blockade	absence

使い分け ◆ **block** は「阻止／遮断」，**blockade** は「遮断」，**absence** は「非存在」という意味に用いられる．

Ⅱ

頻度分析

	用例数		用例数
block	6,800	❶ block in	299
		❷ block of	287
blockade	2,114	❶ blockade of	988
absence	10,828	❶ absence of	10,072

B

解説
◆ **block** は，後に in を伴う用例が多い．
◆ **blockade**，**absence** は，後に of を伴う用例が非常に多い．

★ block（阻止／遮断／〜を阻止する）

動詞および名詞として用いられる．

❶ block in 〜（〜の阻止／〜の遮断）

Dephosphorylation of the RS domain led to a block in U2 snRNP binding to the substrate.（*Mol Cell Biol. 2002 22:5337*）
訳 RS ドメインの脱リン酸化は，U2 低分子リボ核タンパク質の基質への

結合の阻止につながった

★ blockade（遮断）

❶ blockade of ～（～の遮断）

Blockade of FGF signaling downregulates the expression of members of the RAR signaling pathway, RARα, RALDH2 and CYP26.（*Development. 2004 131:2653*）

訳 FGF シグナル伝達の遮断は，RAR シグナル伝達経路のファミリーである RARα，RALDH2 および CYP26 の発現を下方制御する

★★ absence（非存在／存在しないこと）

“in the absence of” の形での用例が非常に多い．

❶ absence of ～（～の非存在）

CYP2E1 is a loosely coupled enzyme; formation of reactive oxygen species occurs even in the absence of added substrate.（*J Biol Chem. 2000 275:15563*）

訳 活性酸素種の形成は，たとえ加えられた基質が存在しなくても起こる

14. 破壊／喪失／切断 【disruption】

破壊	喪失／欠損	切断	除去
disruption **destruction** **ablation**	**loss** **deletion** **defect** **deficiency**	**cleavage** **truncation**	**elimination** **removal**

使い分け

◆ **disruption**, **destruction**, **ablation** は「破壊」，**loss** は「喪失」，**deletion** は「欠失」，**defect**, **deficiency** は「欠損」，**cleavage**, **truncation** は「切断」，**elimination**, **removal** は「除去」という意味に用いられる．

頻度分析

	用例数		用例数
disruption	3,426	❶ disruption of	2,374
		❷ disruption in	120
destruction	1,060	❶ destruction of	364
		❷ destruction in	81

ablation	1,070	❶ ablation of	377
loss	14,662	❶ loss of	9,854
		❷ loss in	713
deletion	7,477	❶ deletion of	2,987
		❷ deletion in	329
defect	8,963	❶ defects in	2,320
		❷ defects of	251
deficiency	3,532	❶ deficiency in	479
		❷ deficiency of	385
cleavage	7,215	❶ cleavage of	1,742
		❷ cleavage site	512
		❸ proteolytic cleavage	254
truncation	1,004	❶ truncation of	247
elimination	1,259	❶ elimination of	713
removal	2,625	❶ removal of	1,984

解説

◆ **disruption**, **destruction**, **loss**, **deletion**, **elimination**, **removal** は，後に of を伴う用例が非常に多い．

◆ **defect**, **deficiency** は，後に in を伴う用例が多い．

★ disruption（破壊／分裂）

"disruption of" の用例が非常に多い．

❶ **disruption of ～**（～の破壊）

The reduction in MZ B cells was not due to the disruption of splenic architecture, disregulated terminal differentiation, nor to increased apoptosis within the MZ B cell compartment. (*J Immunol. 2003 171:2783*)

訳 MZ B 細胞の減少は脾臓構造の破壊のせいではなかった

★ destruction（破壊）

❶ **destruction of ～**（～の破壊）

CD8$^+$ T cells are essential for the destruction of the insulin-producing pancreatic β cells characterizing this disease. (*J Immunol. 2004 173:2530*)

訳 CD8$^+$ T 細胞は，インスリン産生膵臓 β 細胞の破壊にとって必須である

名詞

II

B

★ ablation（破壊）

❶ ablation of ～（～の破壊）

We report that genetic ablation of C5a receptor expression completely protects mice from arthritis.（*J Exp Med. 2002 196:1461*）

訳 C5a受容体発現の遺伝的な破壊はマウスを関節炎から完全に保護した

★★ loss（喪失／損失／低下）

❶ loss of ～（～の喪失／～の低下）

Deletions in the highly conserved C terminus result in a complete loss of activity.（*J Biol Chem. 2000 275:10983*）

訳 高度に保存されたC末端の欠失は活性の完全な喪失という結果になる

★ deletion（欠失／欠損）

❶ deletion of ～（～の欠失）

This leads to the deletion of 35 nucleotides and results in a frameshift with a premature stop codon at amino acid position 498.（*Blood. 2002 99:3063*）

訳 これは35塩基の欠失につながる

★ defect（欠損）

複数形の用例が多い.

❶ defects in ～（～の欠損）

Mutants with defects in several TRAPP subunits are temperature-sensitive in their ability to displace GDP from Ypt1p.（*J Cell Biol. 2000 151:289*）

訳 いくつかのTRAPPサブユニットの欠損をもつ変異体は温度感受性である

★ deficiency（欠損／欠損症）

❶ deficiency in ～（～の欠損）

It has been postulated that this impairment relates to a deficiency in mesocortical DA function.（*J Neurosci. 2002 22:3708*）

訳 この障害は中脳皮質ドーパミン機能の欠損に関連する

★ cleavage（切断）

❶ cleavage of ～（～の切断）

CA is generated by proteolytic cleavage of the Gag precursor polyprotein during viral maturation.（*Nat Struct Biol. 2002 9:537*）

訳 CAは，Gag前駆体ポリタンパク質のタンパク質分解性の切断によって産生される

★ truncation（切断）

❶ truncation of ～（～の切断）

N-terminal truncation of Vav3 activated its transforming potential, as measured by focus-formation assays.（*Mol Cell Biol. 2000 20:9212*）

訳 Vav3のN末端の切断はその形質転換能を活性化した

★ elimination（除去）

❶ elimination of ～（～の除去）

Catechol-*O*-methyltransferase（COMT）is a key enzyme in the elimination of dopamine in the prefrontal cortex of the human brain.（*Am J Hum Genet. 2004 75:807*）

訳 カテコール*O*メチル基転移酵素（COMT）は，ドパミンの除去の際の鍵になる酵素である

★ removal（除去）

❶ removal of ～（～の除去）

It has long been known that Nef, Env, and Vpu participate in the removal of the viral receptor from the cell surface.（*J Virol. 2002 76:4125*）

訳 Nef，EnvおよびVpuは，細胞表面からのウイルス受容体の除去に関与する

【Ⅱ-C. 変化・移動・影響】 変化・移動・影響などを表す名詞

15. 変化／置換 【change】

変化	修飾	変換／置換
change alteration	modification	conversion substitution replacement

使い分け

◆ **change**, **alteration** は「変化」, **modification** は「修飾」, **conversion** は「変換」, **substitution**, **replacement** は「置換」という意味に用いられる.

頻度分析

	用例数		用例数
change	29,160	❶ changes in	11,326
		❷ conformational change	1,196
		❸ changes of	496
alteration	4,363	❶ alterations in	1,887
		❷ alteration of	530
modification	4,759	❶ modification of	1,263
		❷ modifications in	125
		❸ modification by	124
conversion	2,516	❶ conversion of	1,198
		❷ conversion to	271
		❸ conversion from	64
substitution	6,088	❶ substitution of	1,065
		❷ substitutions in	428
		❸ substitutions at	356
replacement	2,512	❶ replacement of	929
		❷ replacement therapy	246
		❸ replacement with	80

解説

◆ **change**, **alteration** は，後に in を伴う用例が非常に多い.

◆ **modification**, **conversion**, **substitution**, **replacement** は，後に of を伴う用例が非常に多い.

★★ change（変化）

名詞の用例が多いが，動詞としても用いられる．名詞としては複数形の用例が多く，特に“changes in”の用例が非常に多い．

❶ changes in ～（～の変化）

In addition, alterations in p53 transcription levels occur in response to changes in the cell cycle.（*Gene. 2001 274:129*）

訳 p53 転写レベルの変化は細胞周期の変化に応答して起こる

★ alteration（変化）

❶ alterations in ～（～の変化）

Infection leads to alterations in cell morphology and impairment of cell replication but not increased cell death.（*J Neurosci. 2004 24:4875*）

訳 感染は細胞形態の変化と細胞複製の障害につながる

★ modification（修飾／修正）

❶ modification of ～（～の修飾）

We show that this silencing involves changes in the modification of histones in FLC chromatin.（*Nature. 2004 427:159*）

訳 このサイレンシングは，FLC クロマチンにおけるヒストンの修飾の変化を含む

★ conversion（変換）

❶ conversion of ～（～の変換）

This enzyme catalyzes the conversion of pyruvate and ATP to phosphoenolpyruvate（PEP）, AMP, and phosphate and is thought to function in gluconeogenesis.（*J Bacteriol. 2001 183:709*）

訳 この酵素は，ピルビン酸と ATP のホスホエノールピルビン酸（PEP），AMP とリン酸への変換を触媒する

★ substitution（置換）

❶ substitution of ～（～の置換）

We report on the unexpected structural changes caused by substitution of acidic amino acids in the Q(B) binding pocket of the bacterial photosynthetic reaction center by alanines.（*Biochemistry. 2002 41:5998*）

訳 われわれは，酸性アミノ酸の置換によって引き起こされる予想外の構造的変化について報告する

★ replacement（置換）

❶ replacement of ～（～の置換）

Adhesion was restored by replacement of TSP2 and by inhibitors of MMP2 activity.（*J Biol Chem. 2001 276:8403*）

訳 接着はTSP2の置換とMMP2活性の抑制によって回復された

16. 移動／移行／伝達 【transfer】

移動	遊走	移行／侵入	伝達
transfer movement	migration	shift transition translocation entry	transmission transduction

使い分け ◆ **transfer**, **movement**は「移動」, **migration**は「遊走」, **shift**, **transition**, **translocation**は「移行」, **entry**は「侵入／移行」, **transmission**, **transduction**は「伝達」という意味に用いられる.

頻度分析

	用例数		用例数
transfer	8,823	❶ transfer of	1,639
		❷ transfer to	463
		❸ transfer from	395
movement	3,887	❶ movement of	795
		❷ movement in	118
migration	5,362	❶ cell migration	1,326
		❷ migration of	843
		❸ migration in	314
shift	5,155	❶ shift in	667
		❷ shift of	256
transition	3,862	❶ transition state	1,144
		❷ transition from	541
		❸ transition of	271
translocation	5,048	❶ translocation of	1,523
		❷ nuclear translocation	770
		❸ translocation to	192

entry	3,771	❶ entry into	635
		❷ entry of	436
		❸ entry in	123
transmission	3,862	❶ transmission of	686
		❷ transmission in	276
		❸ transmission at	100
transduction	4,633	❶ signal transduction	3,170
		❷ transduction of	459
		❸ transduction in	213

解説 ◆ **shift** は後に in を，**transition** は後に from を，**entry** は後に into を伴う用例が多い．

★ transfer（移動）

名詞の用例が多いが，動詞としても用いられる．

❶ **transfer of ～**（～の移動）

This enzyme catalyzes the transfer of a methyl group from 5-methyltetrahydrofolate to homocysteine, generating tetrahydrofolate and methionine.（*Mol Cell Biol. 2001 21:1058*）

訳 この酵素は，5メチルテトラヒドロ葉酸からホモシステインへのメチル基の移動を触媒する

★ movement（移動）

❶ **movement of ～**（～の移動）

Electron microscopy studies showed that CD29 was involved in the movement of neutrophils from the interstitium into alveoli.（*J Immunol. 2001 166:3484*）

訳 CD29は好中球の間質から肺胞への移動に関与した

★ migration（遊走）

❷ **migration of ～**（～の遊走）

Loss of DCC function results in the migration of many LHRH neurons to inappropriate destinations.（*J Neurosci. 2001 21:911*）

訳 DCC機能の喪失は，多くのLHRHニューロンの不適当な行き先への遊走という結果になる

★ shift（移行／変化）

動詞としても用いられるが，名詞の用例の方が多い．

❶ shift in ～（～の移行）

These effects were found to correlate with a shift in Bax/Bcl-2 ratio more towards apoptosis.（*Oncogene. 2002 21:3727*）

訳 これらの効果はBax/Bcl-2比の移行と相関することが見つけられた

★ transition（移行／遷移）

❷ transition from ～（～から移行）

We identified 700 genes that were differentially expressed during the transition from primary growth to secondary growth.（*Plant Physiol. 2004 135:1069*）

訳 われわれは，一次成長から二次成長への移行の間に差動的に発現した700遺伝子を同定した

★ translocation（移行）

❶ translocation of ～（～の移行）

Inhibitors of the nuclear translocation of angiogenin abolish the angiogenic activities of these factors.（*Oncogene. 2005 24:445*）

訳 アンジオゲニンの核移行の抑制は，これらの因子の血管形成活性を消滅させる

★ entry（侵入／移行）

❶ entry into ～（～への侵入）

HIV uses either of the chemokine receptors CCR5 and CXCR4 for entry into CD4-positive cells.（*Lancet. 2004 363:2040*）

訳 HIVは，CD4陽性細胞への侵入のためにケモカイン受容体CCR5とCXCR4のどちらかを使う

★ transmission（伝達／伝染）

❶ transmission of ～（～の伝達）

To examine the functions of electrical synapses in the transmission of signals from rod photoreceptors to ganglion cells, we generated connexin36 knockout mice.（*Neuron. 2002 36:703*）

訳 桿体光受容体から神経節細胞へのシグナルの伝達における電気的シナプスの機能を調べるために

★ transduction（伝達）

❷ transduction of 〜（〜の伝達）

JNK and p38 protein kinases are involved in the signal transduction of apoptotic stimulus.（*Anal Biochem. 2003 317:147*）

訳 JNK と p38 タンパク質キナーゼは，アポトーシス性の刺激のシグナル伝達に関与する

名詞

17. 取り込み【uptake】

取り込み	
uptake	**incorporation**

使い分け ◆ **uptake**, **incorporation** は「取り込み」という意味に用いられる．

頻度分析

	用例数		用例数
uptake	5,586	❶ uptake of	1,142
		❷ uptake in	469
		❸ glucose uptake	442
		❹ uptake by	281
incorporation	2,305	❶ incorporation of	1,081
		❷ incorporation into	336
		❸ thymidine incorporation	156

解説 ◆どちらも後に of を伴う用例が多い．

II

C

★ uptake（取り込み）

❶ uptake of 〜（〜の取り込み）

Helicobacter pylori cells are naturally competent for the uptake of both plasmid and chromosomal DNA.（*Mol Microbiol. 2000 37:1052*）

訳 ピロリ菌細胞は，自然にプラスミドおよび染色体 DNA の両方を取り込む能力がある

★ incorporation（取り込み）

❶ **incorporation of ～**（～の取り込み）

The incorporation of 6-thioguanine（S6G）into DNA is an essential step in the cytotoxic activity of thiopurines.（*Nucleic Acids Res. 2003 31:1331*）

訳 6チオグアニン（S6G）のDNAへの取り込みは，チオプリンの細胞傷害活性における必須のステップである

18. 影響 【influence】

影響		
influence	impact	effect

使い分け ◆**influence**は「影響」，**impact**は「影響／衝撃」，**effect**は「影響／効果」という意味に用いられる．

頻度分析

	用例数		用例数
influence	6,316	❶ influence of	1,231
		❷ influence on	501
impact	3,140	❶ impact of	1,277
		❷ impact on	840
effect	52,164	❶ effect of	9,570
		❷ effect on	7,336
		❸ no effect	3,546

解説 ◆いずれの語も後にofまたはonを伴う用例が多い．

★ influence（影響／影響する）

動詞としても用いられるが，名詞の用例の方が多い．

❶ **influence of ～**（～の影響）

Furthermore, we examined the influence of TSA treatment on enhancer activity in macrophage cells and lysozyme non-expressing cells, including multipotent macrophage precursors.（*Nucleic Acids Res. 2001 29:4551*）

訳 われわれはエンハンサー活性に対するTSA処理の影響を調べた

★ impact（影響／衝撃）

❶ impact of ～（～の影響／～の衝撃）

To assess the impact of activation of calpain on CaR protein in caveolar fractions, we analyzed the effects of m-calpain on the CaR.（*J Biol Chem. 2003 278:31167*）

訳 CaRタンパク質に対するカルパインの活性化の影響を査定するために

★★ effect（影響／効果）

❷ effect on ～（～に対する影響／～に対する効果）

However, CCI-779 inhibited expression of c-myc in CCI-sensitive PTEN-null myeloma cells but had no effect on expression in CCI-resistant cells.（*Cancer Res. 2002 62:5027*）

訳 CCI-779はCCI感受性のPTEN欠損骨髄腫細胞においてc-mycの発現を抑制したが，CCI抵抗性細胞での発現には影響しなかった

【Ⅱ-D．障害・疾患】 障害・疾患などを表す名詞

19．障害／疾患　【disorder】

疾患／疾病	障害／損傷	閉塞
disease illness disorder	impairment deficit disturbance barrier dysfunction damage	occlusion

使い分け

◆ **disease**, **illness** は「疾患」，**disorder** は「疾患／障害」，**impairment** は「障害／損傷」，**deficit** は「障害／欠損」，**disturbance**, **barrier** は「障害」，**dysfunction** は「機能障害」，**damage** は「損傷」，**occlusion** は「閉塞」という意味に用いられる．

頻度分析

	用例数		用例数
disease	30,817	❶ disease in	1,258
		❷ disease progression	641
		❸ disease activity	273
illness	1,350	❶ of illness	248
disorder	6,904	❶ disorder characterized by	259
		❷ disorder in	176
		❸ disorder of	166
impairment	1,722	❶ impairment of	437
		❷ impairment in	277
		❸ cognitive impairment	158
deficit	1,766	❶ deficits in	431
disturbance	439	❶ disturbances in	68
barrier	2,797	❶ barrier to	304
		❷ barrier for	93
dysfunction	2,788	❶ dysfunction in	331
damage	6,692	❶ DNA damage	2,729
		❷ damage in	446
		❸ damage to	440

occlusion	969	❶ artery occlusion	173
		❷ occlusion of	103
		❸ occlusion in	41

解説 ◆ **disease**, **disorder**, **deficit**, **disturbance**, **dysfunction**, **damage** は後に in を，**barrier** は後に to を伴う用例が多い．

名詞

★★ disease（疾患）

❶ **disease in ～**（～における疾患）

We examined whether a known TAP1 polymorphism in the ATPase domain altered the severity of disease in patients with RRP.（*J Infect Dis. 2004 189:871*）

訳 われわれは，ATPase ドメインの既知の TAP1 多型性が RRP 患者の疾患重症度を変えるかどうかを調べた

★ illness（疾病／疾患）

❶ **～ of illness**（疾病の～）

In stratified care, initial treatment is based on measurement of the severity of illness or other factors.（*JAMA. 2000 284:2599*）

訳 初期治療は疾病の重症度の計測に基づいている

II

★ disorder（疾患／障害）

動詞としても用いられるが，名詞の用例が多い．

❶ **disorder characterized by ～**（～によって特徴づけられる疾患）

Spinal muscular atrophy（SMA）is an autosomal recessive disorder characterized by a loss of α motoneurons in the spinal cord.（*J Cell Biol. 2003 162:919*）

訳 脊髄性筋萎縮症（SMA）は，脊髄における α 運動ニューロンの欠損によって特徴づけられる常染色体劣性遺伝疾患である

D

★ impairment（障害／損傷）

❶ **impairment of ～**（～の障害）

We thus postulated that the impairment of nucleolar function might stabilize p53 by preventing its degradation.（*EMBO J. 2003 22:6068*）

訳 核小体機能の障害は，その分解を抑制することによって p53 を安定化するかもしれない

★ deficit（障害／欠損）

❶ deficits in ～（～の障害）

Patients affected by schizophrenia show deficits in both visual perception and working memory.（*Am J Psychiatry. 2000 157:781*）

訳 統合失調症に冒された患者は視知覚と作業記憶の両方の障害を示す

disturbance（障害）

❶ disturbances in ～（～の障害）

We hypothesize that disturbances in neuroendocrine and cardiac autonomic activity（CAA）contribute to development of MS.（*Circulation. 2002 106:2659*）

訳 神経内分泌と心臓自律神経活動（CAA）の障害はMSの発生に寄与する

★ barrier（障害）

動詞としても用いられるが，名詞の用例の方が多い．

❶ barrier to ～（～に対する障害）

Cellular rejection may not constitute a direct major barrier to xenotransplantation.（*Transplantation. 2004 78:1569*）

訳 細胞性の拒絶反応は，異種移植に対する直接の主な障害ではないかもしれない

★ dysfunction（機能障害）

動詞としても用いられるが，名詞の用例の方が多い．

❶ dysfunction in ～（～における機能障害）

We have shown that allopurinol improves endothelial dysfunction in chronic heart failure.（*Circulation. 2002 106:221*）

訳 アロプリノールは慢性心不全における内皮細胞機能障害を改善する

★ damage（損傷／障害）

動詞としても用いられるが，名詞の用例の方が多い．

❷ damage in ～（～における損傷）

Transcription-coupled repair（TCR）is essential for the rapid, preferential removal of DNA damage in active genes.（*Proc Natl Acad Sci USA. 2002 99:4239*）

訳 転写に共役した修復（TCR）は，活動的な遺伝子における急速で優先的なDNA損傷の除去にとって必須である

occlusion（閉塞）

❷ **occlusion of ～**（～の閉塞）

We demonstrate that COX-2-deficient mice have a significant reduction in the brain injury produced by occlusion of the middle cerebral artery.（*Proc Natl Acad Sci USA. 2001 98:1294*）

訳 COX-2 欠損マウスでは，中大脳動脈の閉塞によって起こる脳障害の有意な減少がある

20. 発症／感染 【development】

発症／発病	感染
development **onset** **morbidity**	**infection**

使い分け ◆ **development** は「発症／発生」，**onset** は「発症／開始」，**morbidity** は「発病／罹患率」，**infection** は「感染」という意味に用いられる．

頻度分析

	用例数		用例数
development	23,227	❶ development of	9,394
		❷ development in	1,047
onset	4,209	❶ onset of	2,122
morbidity	1,253	❶ morbidity and mortality	547
infection	21,724	❶ infection of	1,606
		❷ infection with	1,566
		❸ infection in	1,307

解説 ◆ **development**，**onset** は，後に of を伴う用例が非常に多い．

★★ development（発症／発生／開発）

❶ **development of ～**（～の発症）

Human prostate tumors have elevated levels of 15-lipoxygenase-1 (15-LOX-1) and data suggest that 15-LOX-1 may play a role in the development of prostate cancer.（*J Biol Chem. 2002 277:40549*）

名詞

II

D

訳 15-LOX-1 は，前立腺癌の発症において役割を果たすかもしれない

★ onset（発症／開始）

❶ onset of ～（～の発症）

Histologic examination of affected joints was performed approximately 20 days after the onset of arthritis.（*Arthritis Rheum. 2000 43:2668*）

訳 関節炎の発症後およそ 20 日目に患部の関節の組織学的検査が行われた

★ morbidity（発病／罹患率／病的状態）

“morbidity and mortality” の用例が多い．

❶ morbidity and mortality（発病と死亡）

Coronary heart disease（CHD）is the leading cause of morbidity and mortality in persons with type 2 diabetes mellitus（T2DM）.（*Circulation. 2004 110:2817*）

訳 冠動脈心疾患（CHD）は 2 型糖尿病の人における発病および死亡の主な原因である

★★ infection（感染）

❷ infection with ～（～による感染）

Mice lacking LXRs are highly susceptible to infection with the intracellular bacteria *Listeria monocytogenes*（LM）.（*Cell. 2004 119:299*）

訳 LXR を欠損しているマウスは，細胞内細菌リステリアによる感染に高度に感受性が高い

21. 症状／徴候 【symptom】

症状	指標／徴候
symptom manifestation sign	indication

使い分け ◆ **symptom**, **manifestation**, **sign** は「症状／徴候」, **indication** は「指標／徴候」という意味に用いられる．

頻度分析

	用例数		用例数
symptom	3,722	❶ symptoms of	471
		❷ of symptoms	235
		❸ symptoms in	199
manifestation	680	❶ manifestations of	215
		❷ clinical manifestations	137
sign	1,300	❶ signs of	441
		❷ clinical signs	135
indication	547	❶ indication of	115

解説 ◆いずれの語も後にofを伴う用例が多い.

★ symptom（症状／徴候）

複数形の用例が多い.

❶ symptoms of ～（～の症状）

In patients with symptoms of heart failure, elevations in B-type natriuretic peptide（BNP）accurately identify ventricular dysfunction.（*Circulation. 2003 108:2987*）

訳 心不全の症状を持つ患者において，B型ナトリウム利尿ペプチド（BNP）の上昇は心室の機能障害を正確に同定する

manifestation（症状／徴候）

複数形の用例が多い.

❶ manifestations of ～（～の症状）

The variety of clinical manifestations of human cytomegalovirus infection probably results from both viral and host factors.（*J Infect Dis. 2001 183:218*）

訳 ヒトサイトメガロウイルス感染の臨床症状の多様性は，おそらくウイルスと宿主の両方の因子に起因する

★ sign（症状／徴候）

複数形の用例が多い.

❶ signs of ～（～の徴候／～の症状）

No signs of HCV infection were detected in the 8 months following the injection.（*J Virol. 2000 74:2046*）

訳 HCV 感染の徴候は認められなかった

indication（指標／徴候）

❶ indication of ～（～の指標／～の徴候）

Second, the ratio of ADP-Glc to UDP-Glc was used as an indication of the intracellular location of the AGPase activity in a wide range of starch-synthesizing organs.（*Plant Physiol. 2001 125:818*）

訳 ADP-Glc 対 UDP-Glc の比は，AGPase 活性の細胞内局在の指標として使われた

【Ⅲ-A. 関連・異同・識別】 関連・異同・識別を表す名詞

1. 関与 【involvement】

関与	
involvement	participation

使い分け ◆ **involvement**, **participation** は「関与」という意味に用いられる．

頻度分析

	用例数		用例数
involvement	2,613	❶ involvement of	1,699
		❷ involvement in	441
participation	566	❶ participation of	259
		❷ participation in	166

解説 ◆どちらも後に of または in を伴う用例が多い．

★ involvement（関与）

❶ involvement of ～（～の関与）

In this work we provide evidence for the involvement of the N terminus of Cdc25 in the regulation of its activity.（*Genetics. 2000 154:1473*）

訳 われわれは，活性の調節への Cdc25 の N 末端の関与についての証拠を提供する

❷ involvement in ～（～への関与）

Although the physiological role of apolipoprotein A-II is unclear, evidence for its involvement in free fatty acid metabolism in mice has recently been obtained.（*Biochemistry. 2002 41:11681*）

訳 マウスの遊離脂肪酸代謝へのそれの関与の証拠が最近得られた

participation（関与／参加）

❶ participation of ～（～の関与）

Induction of apoptotic cell death generally requires the participation of cysteine proteases belonging to the caspase family.（*J Virol. 2000 74:7470*）

訳 アポトーシス性細胞死の誘導は，一般にシステインプロテアーゼの関与を必要とする

2. 関連 【relation】

関連	相関
relation relationship association connection link	correlation

使い分け
◆ **relation** は「関連」，**relationship** は「関連性」，**association**，**connection** は「関連／結合」，**link** は「リンク／関連」，**correlation** は「相関」という意味に用いられる．

頻度分析

	用例数		用例数
relation	1,954	❶ relation to	636
		❷ in relation to	557
		❸ relation between	500
		❹ relation of	123
relationship	6,630	❶ relationship between	2,659
		❷ relationship of	414
		❸ relationship to	305
association	12,280	❶ association of	2,600
		❷ association with	2,425
		❸ association between	1,839
		❹ in association with	581
connection	1,425	❶ connection between	214
		❷ connections with	82
		❸ connections of	78
link	3,625	❶ link between	1,171
correlation	4,918	❶ correlation between	1,593
		❷ correlation with	354
		❸ correlation of	302

解説 ◆いずれの語も後にbetweenを伴う用例が多い．

★ relation（関連／関係）

"in relation to" "relation between" の用例が多い．

❷ in relation to ～（～に関して）

The results are discussed in relation to the molecular mechanisms that support latent inhibition of cued fear conditioning.（*Behav Neurosci. 2004 118:1444*）

訳 結果が，…する分子機構に関して議論される

❸ relation between ～（～の間の関連）

We examined the relation between blood pressure and urinary sodium as a marker of dietary intake.（*Am J Clin Nutr. 2004 80:1397*）

訳 われわれは血圧と尿中ナトリウムの間の関連を調べた

★ relationship（関連性）

❶ relationship between ～（～の間の関連性）

In this study we investigated the relationship between cdk5 activity and regulation of the mitogen-activated protein (MAP) kinase pathway.（*J Biol Chem. 2002 277:528*）

訳 われわれは，cdk5活性とマイトジェン活性化プロテイン（MAP）キナーゼ経路の調節との間の関連性を精査した

★★ association（関連／結合）

❸ association between ～（～の間の関連）

The authors therefore examined the association between BMI and cardiovascular risk factors in a very lean population in China.（*Am J Epidemiol. 2000 151:88*）

訳 それゆえ，著者らはBMIと心臓血管病の危険因子の間の関連を調べた

❹ in association with ～（～と関連して）

A literature search indicated that 28 of the remaining 97 genes have been reported in association with pancreatic cancer, validating this approach.（*Am J Pathol. 2002 160:1239*）

訳 残りの97のうちの28遺伝子は，膵臓癌と関連して報告されている

★ connection（関連／結合）

複数形の用例の方が多い．

❶ connection between ～（～の間の関連）

In addition, antisense morpholino oligonucleotide-mediated loss of FGF-8 expression *in vivo* substantially reduced the phenotypic effects in EphA4Y928F expressing embryos, suggesting a connection between Eph and FGF signaling.（*Mol Biol Cell. 2004 15:1647*）

訳 Eph と FGF シグナル伝達の間の関連を示唆している

★ link（リンク／連結／関連／連結する）

動詞の用例が多いが，名詞としても用いられる．

❶ link between ～（～の間のリンク）

Thus, the Arc signal transduction system provides a link between the electron transport chain and gene expression.（*Science. 2001 292:2314*）

訳 Arc シグナル伝達系が，電子伝達系と遺伝子発現との間のリンクを提供する

★ correlation（相関）

❶ correlation between ～（～の間の相関）

In cell lines, we examined the correlation between methylation status and mRNA expression by reverse transcription-PCR.（*Cancer Res. 2002 62:3382*）

訳 われわれはメチル化状態と mRNA 発現の間の相関を調べた

3. 一致／類似 【agreement】

一致	類似性	相同性
agreement concordance	similarity	homology

使い分け

◆ **agreement**, **concordance** は「一致」，**similarity** は「類似性」，**homology** は「相同性」という意味に用いられる．

頻度分析

	用例数		用例数
agreement	1,992	❶ agreement with	1,105
		❷ agreement between	239
concordance	260	❶ concordance between	68
similarity	3,676	❶ similarity to	945
		❷ sequence similarity	707
		❸ similarity between	246
homology	4,139	❶ homology to	1,034
		❷ sequence homology	541
		❸ homology with	467

解説 ◆**agreement**, **concordance** は後に between を，**similarity**, **homology** は後に to を伴う用例が多い.

★ agreement（一致）

❷ **agreement between ～**（～の間の一致）

There was good agreement between MRI and histological findings, with a value of κ=0.69（0.53 to 0.85）.（*Circulation. 2001 104:2051*）

訳 磁気共鳴画像法と組織学的知見がよく一致した

concordance（一致）

❶ **concordance between ～**（～の間の一致）

There was excellent concordance between the standard MSP assay and the real-time assay（91%, P<0.0001）.（*Am J Pathol. 2002 161:629*）

訳 標準的な MSP アッセイとリアルタイムアッセイの間によい一致があった

★ similarity（類似性）

❶ **similarity to ～**（～への類似性）

ComA, however, has no significant sequence similarity to any known enolase.（*J Biol Chem. 2003 278:45858*）

訳 しかし，ComA はどの既知のエノラーゼにも有意な配列の類似性を持たない

★ homology（ホモロジー／相同性）

❶ **homology to ～**（～に対するホモロジー）

The vaccinia virus（VV）B8R gene encodes a secreted protein with homology to the γ interferon（IFN-γ）receptor.（*J Virol. 2001 75:11*）

訳 ワクシニアウイルス（VV）B8R遺伝子は，γインターフェロン（IFN-γ）受容体に対するホモロジーを持つ分泌タンパク質をコードする

4. 違い／識別 【difference】

違い	変異	識別／区別
difference	variation	discrimination distinction

使い分け
◆ **difference** は「違い」，**variation** は「変異」，**discrimination**，**distinction** は「識別／区別」という意味に用いられる．

頻度分析

	用例数		用例数
difference	16,302	❶ differences in	5,655
		❷ differences between	1,767
		❸ significant differences	1,185
variation	5,462	❶ variation in	1,635
		❷ genetic variation	426
		❸ variation of	342
		❹ of variation	320
discrimination	1,070	❶ discrimination of	161
		❷ discrimination between	117
distinction	362	❶ distinction between	126

解説
◆ **difference**, **variation** は，後に in を伴う用例が多い．
◆ **discrimination**，**distinction** は，後に of または between を伴う用例が多い．

★ difference（違い）

❶ **differences in ～**（～の違い）

At baseline, there were no significant differences in blood pressure between groups.（*Circulation. 2001 104:839*）

訳 グループ間に血圧の有意な違いはなかった

★ variation（変異／変動）

❶ variation in 〜（〜の変異）

We determined the influence of genetic variation in MCP-1 on HIV-1 pathogenesis in large cohorts of HIV-1-infected adults and children. (*Proc Natl Acad Sci USA. 2002 99:13795*)

訳 われわれは，HIV-1の病因に対するMCP-1の遺伝的変異の影響を決定した

★ discrimination（識別／区別）

❶ discrimination of 〜（〜の識別）

They rated their confidence in the discrimination of hemangiomas from malignant tumors. (*Radiology. 2001 219:699*)

訳 彼らは，血管腫の悪性腫瘍との識別の際のそれらの信頼度を評価した

distinction（区別／差異）

❶ distinction between 〜（〜の間の区別）

The distinction between the two mechanisms inducing coagulation in the xenograft provides an opportunity for specific intervention. (*Transplantation. 2000 69:475*)

訳 異種移植における凝固を含む2つの機構の間の区別は特異的な介入の機会を与える

【Ⅲ-B. 性質】 性質などを表す名詞

5. 維持／保持 【maintenance】

維持	保持
maintenance	retention

使い分け
◆**maintenance**は「維持」，**retention**は「保持／維持」という意味に用いられる．

頻度分析

	用例数		用例数
maintenance	3,233	❶ maintenance of	2,028
retention	1,704	❶ retention of	693

解説
◆どちらも後にofを伴う用例が非常に多い．

★ maintenance（維持）

❶ maintenance of ～（～の維持）

Membrane skeletons play an important role in the maintenance of cell shape and integrity in many cell types.（*J Biol Chem. 2002 277:41240*）

訳 膜骨格は細胞の形と統合性の維持に重要な役割を果たす

★ retention（保持／維持）

❶ retention of ～（～の保持）

Binding of p300 is associated with the retention of p53 in the nucleus, which results in the accumulation of p53 in an acetylase-independent manner.（*J Biol Chem. 2001 276:45928*）

訳 p300の結合は核におけるp53の保持と関連している

6. 耐性／寛容 【resistance】

耐性／抵抗	寛容／耐性
resistance	tolerance

使い分け ◆**resistance** は「耐性／抵抗」，**tolerance** は「寛容／耐性」という意味に用いられる．

頻度分析

	用例数		用例数
resistance	9,959	❶ resistance to	2,693
		❷ resistance in	678
		❸ resistance of	475
tolerance	3,353	❶ tolerance to	544
		❷ glucose tolerance	388
		❸ tolerance in	273

解説 ◆どちらも後に to を伴う用例が多い．

★★ resistance（耐性／抵抗）

❶ **resistance to ～**（～に対する耐性）

These data illustrate that a single molecule can confer resistance to humoral and cellular immune attack.（*Transplantation. 2003 75:542*）

訳 単一の分子が体液性および細胞性の免疫攻撃に対する耐性を与えうる

★ tolerance（寛容／耐性）

❶ **tolerance to ～**（～に対する寛容）

Alternatively, induction of immune tolerance to tumor antigens may enable cancer progression.（*J Exp Med. 2004 200:1581*）

訳 腫瘍抗原に対する免疫寛容の誘導は癌の進行を可能にするかもしれない

名詞

III

B

7. 保存 【conservation】

保存	
conservation	preservation

使い分け ◆**conservation** は「(進化的な) 保存」を意味し，**preservation** は「(物質的あるいは機能的な) 保存」を意味する．

頻度分析

	用例数		用例数
conservation	1,441	❶ conservation of	602
preservation	528	❶ preservation of	278

解説 ◆どちらも後に of を伴う用例が非常に多い．

★ conservation（保存）

❶ conservation of ～（～の保存）

The proposed model provides a plausible explanation for the conservation of this acidic residue among the LDL-A modules.（*J Biol Chem. 2004 279:16629*）

訳 提唱されたモデルは，この酸性の残基の保存に対する妥当な説明を提供する

preservation（保存）

❶ preservation of ～（～の保存）

Thus, VEGFR-1 agonists may have therapeutic potential for preservation of organ function in certain liver disorders.（*Blood. 2001 98:3534*）

訳 VEGFR-1 アゴニストは，器官機能の保存のための治療上の可能性を持つかもしれない

8. 貯蔵 【pool】

貯蔵		
pool	storage	store

使い分け
◆いずれの語も「貯蔵」という意味に用いられる．
◆ **pool** は「(小胞の) 貯蔵」，**storage** は「(情報の) 貯蔵」，**store** は「(Ca^{2+}の) 貯蔵」などに用いられる．

頻度分析

	用例数		用例数
pool	2,425	❶ pool of	609
storage	1,261	❶ storage of	115
store	1,410	❶ stores in	60

⋆ pool（プール／貯蔵／プールする）

動詞としても用いられるが，名詞の用例の方が多い．

❶ **pool of ～**（～のプール）

During synaptic activity, however, synapsin was detected in the pool of vesicles proximal to the active zone. (*J Cell Biol. 2003 161:737*)

訳 シナプシンは小胞のプールの中に検出された

⋆ storage（貯蔵）

❶ **storage of ～**（～の貯蔵）

Working memory involves transient storage of information and the ability to manipulate that information for short-range planning and prediction. (*Nat Neurosci. 2003 6:66*)

訳 作業記憶は情報の一時的な貯蔵を含む

⋆ store（貯蔵／貯蔵する）

動詞としても用いられるが，名詞の用例の方が多い．

❶ **stores in ～**（～における貯蔵）

In HEK 293 cells stably expressing type 1 parathyroid (PTH) receptors, PTH stimulated release of intracellular Ca^{2+} stores in only 27% of cells, whereas 96% of cells responded to carbachol. (*J Biol*

Chem. 2000 275:1807）
訳 PTHは，わずか27%の細胞においてのみ細胞内Ca^{2+}貯蔵の放出を刺激した

9. 蓄積／沈着 【accumulation】

蓄積	沈着	沈着物
accumulation	**deposition**	**deposit**

使い分け ◆**accumulation**は「蓄積」，**deposition**は「沈着」，**deposit**は「沈着物」という意味に用いられる．

頻度分析

	用例数		用例数
accumulation	6,971	❶ accumulation of	3,732
		❷ accumulation in	702
		❸ nuclear accumulation	315
deposition	1,652	❶ deposition of	451
		❷ deposition in	192
deposit	607	❶ deposits in	87
		❷ amyloid deposits	73
		❸ deposits of	71

解説 ◆いずれの語も後にofまたはinを伴う用例が多い．

★ accumulation（蓄積）

❶ accumulation of ～（～の蓄積）

Zmpste24 deficiency results in the accumulation of prelamin A within cells, a complete loss of mature lamin A, and misshapen nuclear envelopes.（*Proc Natl Acad Sci USA. 2004 101:18111*）
訳 Zmpste24欠損症は細胞内でのプレラミンAの蓄積という結果になる

★ deposition（沈着）

❶ deposition of ～（～の沈着）

Alzheimer's disease（AD）is characterized by the deposition of senile plaques (SPs) and neurofibrillary tangles (NFTs) in vulnerable brain regions. (*Am J Pathol. 2000 156:15*)

訳 アルツハイマー病 (AD) は老人斑の沈着によって特徴づけられる

名詞

deposit (沈着物／沈着する)

動詞および名詞として用いられる.

❶ **deposits in ～** (～における沈着物)

ABri was found to be the main component of amyloid deposits in FBD brains. (*J Mol Biol. 2001 310:157*)

訳 ABri は，FBD の脳におけるアミロイド沈着物の主な構成成分であることが見つけられた

10. 組成 【composition】

組成	構成成分
composition	component constituent

使い分け ◆ **composition** は「組成」, **component**, **constituent** は「構成成分」という意味に用いられる.

III

B

頻度分析

	用例数		用例数
composition	2,826	❶ composition of	869
		❷ composition in	89
component	14,078	❶ component of	3,866
		❷ component in	290
constituent	659	❶ constituent of	151

★ composition (組成)

❶ **composition of ～** (～の組成)

These alterations in the composition of irx4 secondary cell walls had a dramatic effect on the morphology and architecture of the walls, which expand to fill most of the cell, and also on the physical properties of irx4 stems. (*Plant J. 2001 26:205*)

訳 これらの irx4 の二次細胞壁の組成の変化は形態に対して劇的な効果があった

⁑ component（構成成分）

❶ component of ～（～の構成成分）

The strand transferase RAD51 is a component of the homologous recombination repair pathway.（*Nucleic Acids Res. 2001 29:1534*）

訳 鎖転移酵素 RAD51 は相同組換え修復経路の構成成分である

constituent（構成成分）

❶ constituent of ～（～の構成成分）

Amyloid β protein（AβP）is the major constituent of senile plaques associated with Alzheimer's disease（AD）.（*FASEB J. 2001 15:2433*）

訳 アミロイドβタンパク質（AβP）はアルツハイマー病に関連する老人斑の主な成分である

11. 構造 【structure】

構造	構築
structure conformation	architecture

使い分け ◆ **structure** は「構造」，**conformation** は「高次構造／構造」，**architecture** は「構築／構造」という意味に用いられる．

頻度分析

	用例数		用例数
structure	33,888	❶ structure of	7,096
		❷ crystal structure	2,520
		❸ secondary structure	1,141
		❹ structure in	714
conformation	4,197	❶ conformation of	964
		❷ conformation in	244
		❸ closed conformation	148
architecture	1,191	❶ architecture of	308

⁑ structure（構造）

❶ **structure of ～**（～の構造）

Analyses of CD spectra reveal changes in the structure of ΔE18M (loss of β-sheet, gain of unordered structure).（*Biochemistry. 2002 41:10038*）

訳 CD スペクトルの分析は ΔE18M の構造の変化を明らかにする

* conformation（高次構造／構造）

❶ **conformation of ～**（～の高次構造）

The binding of ligands to the TS molecule leads to dramatic changes in the conformation of the enzyme, particularly within the C-terminal domain.（*Biochemistry. 2004 43:1972*）

訳 TS 分子へのリガンドの結合は，その酵素の高次構造の劇的な変化につながる

* architecture（構築／構造）

❶ **architecture of ～**（～の構築／～の構造）

A common feature in the architecture of neuronal networks is a high degree of seemingly redundant synaptic connectivity.（*J Neurosci. 2004 24:5230*）

訳 ニューロンネットワークの構築における共通の特徴は，一見重複のように思われる高度のシナプスの結合性である

12. 要求／必要　【requirement】

要求	必要
requirement **demand**	**need** **necessity**

使い分け ◆ **requirement** は「要求／必要」，**demand** は「要求／需要」，**need** は「必要／要求」，**necessity** は「必要性／必要」という意味に用いられる．

頻度分析

	用例数		用例数
requirement	3,758	❶ requirement for	1,651
		❷ requirement of	325
demand	525	❶ demand for	96
need	3,108	❶ need for	1,251
		❷ need to	930
necessity	170	❶ necessity of	81

解説 ◆ **requirement**, **demand**, **need** は後に for を，**necessity** は後に of を伴う用例が多い．

★ requirement（要求／必要）

❶ requirement for ～（～に対する要求）

In addition, overexpression of Rad26 partially bypasses the requirement for Rad7 in GGR, specifically in the repair of non-transcribed sequences.（*Mol Microbiol. 2004 52:1653*）

訳 Rad26 の過剰発現は Rad7 に対する要求を部分的に回避する

demand（要求／需要／要求する）

名詞および動詞として用いられる．

❶ demand for ～（～に対する要求）

De novo pyrimidine biosynthesis is activated in proliferating cells in response to an increased demand for nucleotides needed for DNA synthesis.（*J Biol Chem. 2003 278:3403*）

訳 DNA 合成に必要とされるヌクレオチドに対する要求の増加に応答して，新規のピリミジン生合成は増殖する細胞において活性化される

★ need（必要／要求／必要とする）

名詞および動詞として用いられる．助動詞として用いられることは少ない．

❶ need for ～（～の必要）

Identical HLA matching has enabled these individuals to be transplanted without the need for immunosuppressive medication.（*Gastroenterology. 2002 123:1341*）

訳 同一の HLA 適合は，免疫抑制の薬物療法の必要なしにこれらの個人が移植を受けることを可能にした

necessity（必要性／必要）

❶ **necessity of ～**（～の必要性）

These findings also underscore the necessity of screening current and future drugs at h5-HT2B receptors for agonist actions before their use in humans.（*Mol Pharmacol. 2003 63:1223*）

訳 これらの知見はまた，現在および未来の薬物をスクリーニングすることの必要性を強調する

13. 制限／限界 【restriction】

制限	限界
restriction	limitation

使い分け ◆ **restriction** は「制限」，**limitation** は「限界／制限」という意味に用いられる．

頻度分析

	用例数		用例数
restriction	1,973	❶ restriction of	202
limitation	1,243	❶ limitations of	224

★ restriction（制限）

❶ **restriction of ～**（～の制限）

Restriction of growth of PstDC3000 in coi1-20 leaves is partially dependent on NPR1 and fully dependent on SA, indicating that SA-mediated defenses are required for restriction of PstDC3000 growth in coi1-20 plants.（*Plant J. 2001 26:509*）

訳 SAに仲介される防御がPstDC3000の増殖の制限のために必要とされる

★ limitation（限界／制限）

❶ **limitations of ～**（～の限界）

This overcomes the limitations of *in vitro* methods for generating large constructs based on restriction digestion, ligation, and transformation of DNA into *Escherichia coli* cells.（*Genomics. 2000 70:165*）

訳 これは，大きなコンストラクトを作製するための試験管内の方法の限界を克服する

14. 領域／部位　【region】

領域	部位	局在化	位置
region area	site locus location	localization	position

使い分け

◆ **region**, **area** は「領域／部位」, **site** は「部位」, **locus** は「部位／座位」, **location** は「部位／局在」, **localization** は「局在／局在化」, **position** は「位置／番目」という意味に用いられる.

◆ **region**, **area** は, **site** より広い場所を意味する.

◆ **locus** は, 染色体上の位置に用いられることが多い.

頻度分析

	用例数		用例数
region	35,926	❶ region of	6,703
		❷ region in	796
		❸ region that	544
area	8,421	❶ areas of	953
		❷ in areas	299
		❸ area under	289
site	47,221	❶ binding site	5,123
		❷ active site	4,444
		❸ site of	3,322
		❹ site in	1,373
		❺ site for	1,153
locus（複数形：**loci**）	8,974	❶ locus in	313
		❷ locus on	227
		❸ locus of	223
location	4,116	❶ location of	1,114
		❷ location in	137
localization	7,183	❶ localization of	2,895
		❷ nuclear localization	1,105
		❸ subcellular localization	639
		❹ localization to	303
position	8,913	❶ at position	1,179
		❷ position of	1,121
		❸ position in	271

共起・頻度分析

（数字：用例数）

直前の単語		
at	in	
5	405	the region 1,419
3	126	the area 498
316	23	the site 1,562
22	37	the locus 380
15	30	the location 817
1,179	111	position 5,817

解説

◆いずれの語も場所を表すために用いられるが，the **region**, the **area** の前には in が使われることが多く，the **site** の前には at が用いられることが多い.

◆at **position** は，アミノ酸の番号などを示すときに使われる.

★★ region（領域／部位）

❶ region of ～（～の領域）

Here we describe the identification of a nuclear localization signal (NLS) in the N-terminal region of Smad 3, the major Smad protein involved in TGF-β signaling. (*Proc Natl Acad Sci USA. 2000 97:7853*)

訳 われわれは，Smad 3 の N 末端領域にある核局在シグナルの同定について述べる

★ area（領域／部位）

❶ areas of ～（～の領域）

Induced MIG, IP-10, and I-TAC protein expression was localized in areas of inflammation at 2 to 3 days and was temporally associated with increased levels of CXCR3$^+$ lymphocytes in bronchoalveolar lavage fluid. (*Infect Immun. 2005 73:485*)

訳 誘導された MIG，IP-10 および I-TAC タンパク質発現は，炎症の領域に局在した

★★ site（部位）

❸ site of ～（～の部位）

The results suggest that the relative amount of TNF-α at the site of infection determines whether the cytokine is protective or destructive. (*Infect Immun. 2000 68:6954*)

訳 感染の部位における TNF-α の相対量はサイトカインが保護的であるか

名詞

III

B

破壊的であるかを決定する

★ locus（部位／座位）

複数形は loci.

❶ locus in ～（～における部位）

Mutations of this locus in humans have been identified as the cause of the craniofacial disorder Saethre-Chotzen syndrome.（*Development. 2002 129:2761*）

訳 ヒトのこの部位の変異は，頭蓋顔面疾患 Saethre-Chotzen 症候群の原因として同定されている

★ location（部位／局在）

❶ location of ～（～の部位／～の局在）

The aim of this study was to determine the location of the TNFR1 receptor（p55）in the medulla using immunocytochemical methods.（*Brain Res. 2004 1004:156*）

訳 この研究の目的は，髄質における TNFR1 受容体（p55）の局在を決定することだった

★ localization（局在／局在化）

❶ localization of ～（～の局在）

In addition, the interaction with DDB1 has been implicated in the nuclear localization of HBx.（*J Virol. 2001 75:10383*）

訳 DDB1 との相互作用は HBx の核局在に関与している

★ position（位置／番目／位置する）

動詞として用いられるもことがあるが，名詞の用例の方が多い.

❶ at position ～（～番目の…）

These studies suggest that the residue at position 166 is involved in the interaction between the Na^+ and sugar transport pathways.（*Biochemistry. 2001 40:11897*）

訳 166 番目の残基は，Na^+ と糖輸送経路の間の相互作用に関与する

15. 状態 【state】

名詞

状態	条件	状況
state status	condition	situation

使い分け

◆ **state**, **status** は「状態」, **condition** は「条件／状態」, **situation** は「状況」という意味に用いられる.

頻度分析

	用例数		用例数
state	18,661	❶ steady state	2,642
		❷ state of	1,869
		❸ state in	350
status	4,058	❶ status of	836
		❷ health status	178
		❸ status in	177
condition	12,620	❶ under conditions	905
		❷ conditions of	674
		❸ conditions in	468
situation	689	❶ situation in	73

III

B

⁑ state（状態）

❷ state of ～（～の状態）

We also show that chromosome condensation is independent of the state of DNA replication in the early embryo.（*Nat Cell Biol. 2000 2:609*）

訳 染色体凝縮はDNA複製の状態とは無関係である

* status（状態）

❶ status of ～（～の状態）

Here we report studies on the status of members of E2F family in cycling HEp-2 and HeLa cells and quiescent serum-starved, contact-inhibited human lung fibroblasts.（*J Virol. 2000 74:7842*）

訳 われわれは，E2Fファミリーメンバーの状態に関する研究を報告する

★★ condition（条件／状態／条件づける）

動詞として用いられることもあるが，名詞の用例の方が多い．名詞としては複数形で使われることが非常に多い．

❶ under conditions ～（～の条件）

This IRES causes an increase in translation under conditions of amino acid starvation.（*J Biol Chem. 2001 276:12285*）

訳 アミノ酸飢餓の条件下では，このIRESは翻訳の増加を引き起こす

situation（状況）

❶ situation in ～（～における状況）

In contrast to the situation in plants, only one of the two isoforms（IDI1）is highly conserved, ubiquitously expressed and most likely responsible for housekeeping isomerase activity.（*J Mol Evol. 2003 57:282*）

訳 植物における状況とは対照的に，2つのアイソフォームのうちの1つ（IDI1）だけが高度に保存されている

16. 重要性 【importance】

重要性	意味／含意
importance significance	implication

使い分け ◆ **importance**，**significance**は「重要性」，**implication**は「意味／含意／密接な関係／重要性」という意味に用いられる．

頻度分析

	用例数		用例数
importance	4,217	❶ importance of	3,086
		❷ importance in	346
		❸ importance for	171
significance	2,407	❶ significance of	1,392
		❷ significance in	103
		❸ significance for	86

implication	3,025	❶ implications for	1,744
		❷ important implications	652
		❸ implications of	628

解説 ◆ **importance**, **significance** は後に of を，**implication** は後に for を伴う用例が非常に多い．

★ importance（重要性）

❶ **importance of ～**（～の重要性）

These results highlight the importance of clinical trials of ICDs in patients with low ejection fractions who have had myocardial infarction.（*Ann Intern Med. 2001 135:870*）

訳 これらの結果は ICDs の臨床治験の重要性を強調する

★ significance（重要性／有意性）

❶ **significance of ～**（～の重要性）

V/Q scans are useful to assess the functional significance of PV stenosis.（*Circulation. 2003 108:3102*）

訳 V/Q スキャンは PV 狭窄症の機能的重要性を評価するのに役に立つ

★ implication（意味／含意／密接な関係／重要性）

複数形で使われることが多く，特に“implications for”の用例が非常に多い．

❶ **implications for ～**（～に対する意味／～にとっての意味）

These findings have important implications for HIV vaccine design.（*J Infect Dis. 2003 187:1053*）

訳 これらの知見は HIV ワクチン設計にとっての重要な意味を持つ

17. 特徴 【feature】

特徴／形質	特性／プロファイル
feature characteristic character	profile

使い分け ◆**feature**，**characteristic** は「特徴」，**character** は「形質」，**profile** は「特性／プロファイル」という意味に用いられる．

頻度分析

	用例数		用例数
feature	7,292	❶ features of	2,102
		❷ structural features	500
		❸ clinical features	213
		❹ features in	192
characteristic	6,989	❶ characteristics of	1,636
character	674	❶ character of	179
profile	5,271	❶ profile of	712

解説 ◆いずれの語も後に of を伴う用例が多い．

★ feature（特徴／特集する）

動詞としても用いられるが，名詞の用例の方が多い．名詞としては複数形の用例が多い．

❶ **features of ～**（～の特徴）

The unique structural features of JDTic will make this compound highly useful in further characterization of the κ receptor.（*J Med Chem. 2003 46:3127*）

訳 JDTic の独特な構造的特徴はこの化合物を高度に有用にする

★ characteristic（特徴／特徴的な）

名詞および形容詞として用いられる．名詞としては複数形の用例が多い．

❶ **characteristics of ～**（～の特徴）

Oxidative stress is one of the characteristics of diabetes and is thought to be responsible for many of the pathophysiological changes caused by the disease.（*J Biol Chem. 2004 279:25172*）

訳 酸化ストレスは糖尿病の特徴の１つである

character（形質／特徴）

❶ **character of ～**（～の形質）

Retinoids have been shown to change the character of pancreatic ductal cancer cells to a less malignant phenotype.（*Gastroenterology. 2002 123:1331*）

訳 レチノイドは，膵管癌細胞の形質をより悪性度の低い表現型に変えることが示されてきた

★ profile（特性／プロファイル）

❶ **profile of ～**（～の特性）

To detect altered gene expression associated with mouse lung tumor progression, we compared the gene expression profile of lung adenocarcinomas with that of lung adenomas and normal lungs. (*Oncogene. 2002 21:5814*)

訳 われわれは，肺腺癌の遺伝子発現プロファイルを肺腺腫のそれと比較した

18. 効果／効力／効率 【effect】

効果	効力	効率
effect	efficacy potency	efficiency

使い分け

◆ **effect** は「効果」，**efficacy** は「有効性／効力」，**potency** は「効力」，**efficiency** は「効率」という意味に用いられる．

頻度分析

	用例数		用例数
effect	52,164	❶ effects of	12,690
		❷ effect on	7,336
		❸ no effect	3,546
efficacy	3,623	❶ efficacy of	1,640
		❷ efficacy in	270
		❸ therapeutic efficacy	128
potency	1,723	❶ potency of	399
		❷ potency in	75
		❸ of potency	69
efficiency	4,069	❶ efficiency of	1,379
		❷ catalytic efficiency	273
		❸ efficiency in	149

解説 ◆いずれの語も後にofを伴う用例が多い．

★★ effect（効果／影響）

❶ effects of ～（～の効果／～の影響）

Here, we examined the effects of GTPs on aortic smooth muscle cell (SMC) proliferation.（*FASEB J. 2003 17:702*）

訳 われわれは，大動脈平滑筋細胞（SMC）増殖に対するGTPの効果を調べた

★ efficacy（有効性／効力）

❶ efficacy of ～（～の有効性）

Nonhuman-primate models are needed to evaluate the efficacy of candidate vaccines.（*J Infect Dis. 2004 189:1013*）

訳 非ヒトの霊長類モデルは，候補ワクチンの有効性を評価するために必要とされる

★ potency（効力）

❶ potency of ～（～の効力）

Our goal was to determine whether the use of DC as direct antigen-presenting cells would improve the potency of 3H1 as vaccine.（*Cancer Res. 2003 63:2844*）

訳 われわれの目的は，直接の抗原提示細胞としてのDCの使用がワクチンとしての3H1の効力を改善するかどうかを決定することであった

★ efficiency（効率）

❶ efficiency of ～（～の効率）

The reduction in viral titer is due to a decrease in the efficiency of viral genomic RNA encapsidation.（*J Virol. 2000 74:9937*）

訳 ウイルス力価の低下は，ウイルスゲノムRNAキャプシド形成効率の減少のせいである

19. 能力／潜在力 【ability】

能力	威力	潜在力
ability capacity capability competence	power	potential

使い分け ◆**ability**, **capacity**, **capability**, **competence** は「能力」, **power** は「威力」, **potential** は「潜在力」という意味に用いられる.

頻度分析

	用例数		用例数
ability	14,297	❶ ability to	6,892
		❷ ability of	6,216
capacity	4,061	❶ capacity to	971
		❷ capacity of	959
		❸ capacity for	331
capability	863	❶ capability of	146
competence	511	❶ competence of	71
power	1,754	❶ power of	401
potential	16,431	❶ potential of	1,540
		❷ potential for	1,027
		❸ potential to	970

解説 ◆**ability**, **capacity** は，後に to を伴う用例が多い.

★★ ability（能力）

❶ **ability to ～**（～する能力）

This region is characterized by its ability to bind to different RNA species, including double-stranded RNA (dsRNA), a known potent inducer of IFNs. (*J Virol. 2003 77:13257*)

訳 この領域は，異なる RNA 種に結合する能力によって特徴づけられる

★ capacity（能力）

❶ **capacity to ～**（～する能力）

Substantial differences were found among strains in the capacity to induce IL-12 and TNF-α production in the DC. (*J Immunol. 2002 168:171*)

訳 実質的な違いが，IL-12およびTNF-α産生を誘導する能力において株間で見られた

capability（能力）

❶ capability of ～（～の能力）

Immunomodulatory activity was associated with the capability of the peptides to modulate heme oxygenase (HO) activity. (*J Biol Chem. 2000 275:17051*)

訳 免疫調節性の活性は，ペプチドのヘムオキシゲナーゼ（HO）活性を調節する能力に関連した

competence（能力）

❶ competence of ～（～の能力）

Overexpression of this mutant form of Smad2 can prolong the competence of endogenous mesodermal genes to respond to activin signalling. (*Nat Cell Biol. 2002 4:519*)

訳 この変異型のSmad2の過剰発現は，内在性の中胚葉性遺伝子の能力を延長できる

★ power（威力／能力）

❶ power of ～（～の威力）

These data illustrate the power of comparative genomic analysis for the study of human disease and identifies a novel BBS gene. (*Am J Hum Genet. 2004 75:475*)

訳 これらのデータは，ヒトの疾患の研究のための比較ゲノム分析の威力を例証する

★ potential（潜在力／可能性／潜在的な）

形容詞および名詞として用いられる．

❶ potential of ～（～の潜在力／～の可能性）

These data demonstrate the potential of species-specific PCR for the identification of *B. gladioli*. (*J Clin Microbiol. 2000 38:282*)

訳 これらのデータは，*B. gladioli*の同定のための種特異的なPCRの可能性を実証する

20．応答／反応 【response】

応答	反応
response	reaction

使い分け ◆ **response** は「応答／反応」，**reaction** は「（化学）反応」という意味に用いられる．

頻度分析

	用例数		用例数
response	51,817	❶ response to	14,129
		❷ in response to	8,464
		❸ immune response	2,205
		❹ response of	1,519
		❺ response in	1,291
reaction	14,642	❶ reaction of	931
		❷ reaction with	390
		❸ reaction in	326

⁑ response（応答／反応）

❷ in response to ～（～に応答して）

c-Jun N-terminal kinase（JNK）regulates gene expression in response to various extracellular stimuli.（*J Biol Chem. 2000 275:22868*）

訳 c-Jun N-terminal kinase（JNK）は，さまざまな細胞外刺激に応答して遺伝子発現を調節する

⁑ reaction（反応）

❶ reaction of ～（～の反応）

Peroxynitrite is a strong oxidizing agent that is formed in the reaction of nitric oxide and superoxide anion.（*Biochemistry. 2002 41:7508*）

訳 過酸化亜硝酸は，一酸化窒素とスーパーオキサイドアニオンの反応において形成される強力な酸化剤である

21. 認識／認知 【recognition】

認識／認知	認知
recognition	perception cognition

使い分け ◆ **recognition** は「（ものによる）認識」，**perception**, **cognition** は「（人や動物による）認知」という意味に用いられることが多い．

頻度分析

	用例数		用例数
recognition	5,828	❶ recognition of	1,369
		❷ recognition by	413
		❸ substrate recognition	237
perception	663	❶ perception of	160
cognition	216	❶ cognition in	25

★ recognition（認識／認知）

❶ recognition of ～（～の認識）

The type-II receptor（BMPRII）is required for recognition of all BMPs, and targeted deletion of BMPRII in mice results in fetal lethality before gastrulation.（*J Bacteriol. 2004 186:164*）

訳 2型受容体（BMPRII）はすべてのBMPの認識に必要とされる

perception（知覚／認知）

❶ perception of ～（～の認知）

phyC mutants flowered early when grown in short-day photoperiods, indicating that phyC plays a role in the perception of daylength.（*Plant Cell. 2003 15:1962*）

訳 phyCは日長の認知において役割を果たす

cognition（認知／認識）

❶ cognition in ～（～における認知）

Amyloid β（Aβ）protein immunotherapy lowers cerebral Aβ and improves cognition in mouse models of Alzheimer's disease（AD）.（*Am J Pathol. 2004 165:283*）

訳 アミロイドβ（Aβ）タンパク質免疫療法は大脳のAβを低下させ，そしてアルツハイマー病のマウスモデルにおける認知を改善する

22. 結合／接着 【binding】

結合	結合／関連	相互作用	接着／付着
binding bond bonding	association connection	interaction	adhesion attachment

使い分け

◆ **binding**, **bond**, **bonding**, **association**, **connection** は，「結合」という意味に用いられる．

◆ **bond**, **bonding** は，水素結合などの意味で用いられることが多い．

◆ **interaction** は「相互作用」という意味だが，結合を表す場合も多い．

◆ **adhesion** は「接着」，**attachment** は「付着／結合」という意味で使われる．

◆ **association**, **connection** は，「関連」という意味でも使われる．

頻度分析

	用例数		用例数
binding	65,386	❶ binding of	7,373
bond	6,207	❶ hydrogen bond	779
		❷ bond between	206
		❸ bond of	156
		❹ bond to	153
bonding	1,297	❶ hydrogen bonding	854
		❷ bonding between	77
		❸ bonding to	64
association	12,280	❶ association of	2,600
		❷ association with	2,425
connection	1,425	❶ synaptic connections	101
interaction	32,673	❶ interaction with	3,462
		❷ interaction between	3,119
		❸ interaction of	2,964

adhesion	7,272	❶ cell adhesion	1,911
		❷ adhesion molecule	1,022
		❸ adhesion to	666
		❹ adhesion of	462
attachment	2,274	❶ attachment of	475
		❷ attachment to	289
		❸ attachment protein	143
		❹ covalent attachment	122

★★ binding（結合）

動名詞として用いられることも多いが，名詞の用例も多い．

❶ binding of ～（～の結合）

Whereas the human cyclin T1 is required for the binding of Tat to TAR from HIV, it is unknown how Tat from EIAV interacts with its TAR.（*J Virol. 2000 74:892*）

訳 ヒトのサイクリンT1は，TatのTARへの結合に必要とされる

★ bond（結合／結合する）

動詞としても用いられるが，名詞の用例の方が多い．

❷ bond between ～（～の間の結合）

The low affinity of BmCYP-1 for CsA arises from incomplete preorganization of the binding site so that the formation of a hydrogen bond between His132 of BmCYP-1 and N-methylleucine 9 of CsA is associated with a shift in the backbone of approximately 1Å in this region.（*Biochemistry. 2000 39:592*）

訳 BmCYP-1のヒスチジン132とCsAのN-メチルロイシン9の間の水素結合の形成は，バックボーンのおよそ1Åのシフトに関連する

★ bonding（結合）

現在分詞として用いられることも多いが，名詞の用例も多い．

❷ bonding between ～（～の間の結合）

Thus the data confirm the presence of hydrogen bonding between the heme a formyl group and the R52 side chain, as suggested from crystallographic data.（*Biochemistry. 2000 39:2989*）

訳 データは，ヘムaホルミル基とR52側鎖の間の水素結合の存在を確認する

★★ association（結合／関連）

後に of または with を伴う用例が多い.

❶ association of 〜（〜の結合）

We show the association of endogenous FIP200 with FAK, which is decreased upon integrin-mediated cell adhesion concomitant with FAK activation.（*Mol Biol Cell. 2002 13:3178*）

訳 われわれは内在性 FIP200 の FAK との結合を示す

❷ association with 〜（〜との結合）

In addition, Claspin is phosphorylated in response to replication stress, and this phosphorylation appears to be required for its association with Chk1.（*J Biol Chem. 2003 278:30057*）

訳 このリン酸化は Chk1 とそれの結合に必要とされるように思われる

★ connection（結合／関連）

❶ synaptic connections（シナプス結合）

Precision of synaptic connections within neural circuits is essential for the accurate processing of sensory information.（*Dev Biol. 2004 269:26*）

訳 神経回路内におけるシナプス結合の精度は感覚情報の正確な処理にとって必須である

★★ interaction（相互作用）

❶ interaction with 〜（〜との相互作用）

We further determined that the carboxyl region of Dpb11 is required for its interaction with Ddc1.（*Genetics. 2002 160:1295*）

訳 Dpb11 のカルボキシル領域は，Ddc1 とそれの相互作用に必要とされる

❷ interaction between 〜（〜の間の相互作用）

We have found that the interaction between Skp1p and Sgt1p is critical for the assembly of CBF3 complexes.（*Mol Biol Cell. 2004 15:3366*）

訳 Skp1p と Sgt1p の間の相互作用は，CBF3 複合体の構築に決定的に重要である

★ adhesion（接着）

❸ adhesion to 〜（〜への接着）

Cell adhesion to extracellular matrix is an important physiological stimulus for organization of the actin-based cytoskeleton.（*J Cell Biol. 2000 150:807*）

訳 細胞外基質への細胞接着は，アクチンを基礎にした細胞骨格の組織化

のための重要な生理学的刺激である

★ attachment（結合／付着）

❶ attachment of ～（～の結合／～の付着）

This process is initiated by the covalent attachment of UMP to the terminal protein VPg, yielding VPgpU and VPgpUpU.（*J Virol. 2000 74:10359*）

訳 この過程は，末端タンパク質 VPg への UMP の共有結合によって開始される

23. 機能／作用 【function】

機能／作用		
function	action	operation

使い分け ◆**function** は「機能」，**action** は「作用」，**operation** は「作動／操作」という意味に用いられる．

頻度分析

	用例数		用例数
function	46,607	❶ function of	7,647
action	8,550	❶ action of	2,213
operation	791	❶ operation of	143

★★ function（機能／機能する）

名詞の用例が多いが，動詞としても用いられる．

❶ function of ～（～の機能）

The mammalian ING1 gene encodes a tumor suppressor required for the function of p53.（*Mol Cell Biol. 2002 22:5047*）

訳 哺乳類の ING1 遺伝子は，p53 の機能に必要とされる癌抑制因子をコードする

★ action（作用／作動）

❶ action of ～（～の作用）

The mechanism of action of HuR is not well understood.（*Nucleic*

Acids Res. 2000 28:2695)

訳 HuRの作用の機構はよく理解されてはいない

operation（作動／操作／手術）

❶ operation of ～（～の作動）

An intimate interaction between the PM and the ER/SR is essential for the operation of this calcium signalling pathway.（*Nat Cell Biol. 2002 4:379*）

訳 PMとER/SRの間の緊密な相互作用は，カルシウムシグナル伝達経路の作動にとって必須である

24. 調節 【regulation】

調節	補正
regulation **control** **modulation**	**adjustment**

使い分け ◆ **regulation**, **control**, **modulation** は「調節」, **adjustment** は「補正／調整」という意味に用いられる.

頻度分析

	用例数		用例数
regulation	18,735	❶ regulation of	12,378
		❷ regulation by	670
		❸ regulation in	627
control	30,319	❶ control of	3,991
modulation	3,100	❶ modulation of	2,075
adjustment	1,438	❶ after adjustment for	652

解説 ◆ **regulation**, **modulation** は後に of を, **adjustment** は後に for を伴う用例が非常に多い.

★★ regulation（調節）

❶ regulation of ～（～の調節）

NKT cells play important roles in the regulation of diverse immune

responses.（*J Immunol. 2003 171:2960*）
訳 NKT細胞は多様な免疫応答の調節において重要な役割を果たす

★★ control（調節／コントロール／調節する）

名詞の用例が多いが，動詞としても用いられる．

❶ control of ～（～の調節）

Wnt pathways are involved in the control of gene expression, cell behavior, cell adhesion, and cell polarity.（*Science. 2002 296:1644*）
訳 Wnt経路は遺伝子発現の調節に関与する

★ modulation（調節）

❶ modulation of ～（～の調節）

Thus, this enzyme may play a critical role in the modulation of nociceptor activity and plasticity of primary sensory neurons.（*J Comp Neurol. 2002 448:102*）
訳 この酵素は，侵害受容器活性の調節において決定的に重要な役割を果たすかもしれない

★ adjustment（補正／調整）

❶ after adjustment for ～（～に対する補正の後）

Polymorphism in the IL-6 gene was associated with the severity of appendicitis, even after adjustment for duration of symptoms.（*Ann Surg. 2004 240:269*）
訳 症状の期間に対する補正の後でさえ，IL-6遺伝子の多型性は虫垂炎の重症度と関連があった

25. 形成／集合 【formation】

形成	構築／集合	凝集
formation	assembly	aggregation

使い分け ◆ **formation**は「形成」，**assembly**は「構築／集合」，**aggregation**は「凝集／集合」という意味に用いられる．

頻度分析

	用例数		用例数
formation	20,894	❶ formation of	8,671
		❷ formation in	1,404
		❸ complex formation	990
assembly	7,318	❶ assembly of	2,070
		❷ assembly in	230
		❸ self-assembly	297
aggregation	2,285	❶ aggregation of	411
		❷ platelet aggregation	335
		❸ aggregation in	90

解説 ◆いずれの語も後にofを伴う用例が多い．

★★ formation（形成）

❶ **formation of ～**（～の形成）

Ligation of short DNA fragments results in the formation of linear and circular multimers of various lengths.（*Biophys J. 2000 79:2692*）

訳 短いDNA断片のライゲーションは，さまざまな長さの直鎖状および環状の多量体の形成という結果になる

★ assembly（構築／集合）

❶ **assembly of ～**（～の構築）

From these results we suggest a simple model for the assembly of functional initiation complexes in the *Xenopus* system.（*Genes Dev. 2000 14:1528*）

訳 われわれは，機能的開始複合体の構築に対する単純なモデルを示唆する

★ aggregation（凝集／集合）

❶ **aggregation of ～**（～の凝集）

Aβ 16-20e also inhibits the aggregation of the Aβ 1-40 peptide and disassembles preformed Aβ 1-40 fibrils.（*Biochemistry. 2003 42:475*）

訳 Aβ16-20eはまたAβ1-40ペプチドの凝集を抑制する

名詞

III

C

26. 機構／役割 【mechanism】

機構	基盤	役割
mechanism machinery mode	basis	role

使い分け
◆ **mechanism** は「機構／機序」，**machinery** は「機構／装置」，**mode** は「様式／機構」，**basis** は「基盤／基礎」，**role** は「役割」という意味に用いられる．

頻度分析

	用例数		用例数
mechanism	65,386	❶ mechanism of	5,470
		❷ mechanism for	3,040
		❸ mechanism by which	2,162
machinery	1,660	❶ machinery in	94
		❷ machinery that	87
		❸ machinery of	84
mode	1,660	❶ mode of	1,058
basis	8,327	❶ basis of	3,872
		❷ basis for	2,336
		❸ molecular basis	1,066
role	38,179	❶ role in	16,667
		❷ role of	12,346
		❸ role for	6,367

解説
◆いずれの語も後に of を伴う用例が多く，また，**machinery**，**role** は後に in を伴う用例も多い．

⁑ mechanism（機構／機序）

“mechanism of”“mechanism for”“mechanism by which”の用例が多い．

❸ **mechanism by which ～**（それによって～である機構）

In this study, we investigated the mechanism by which Ipc1 regulates melanin production. (*J Biol Chem. 2004 279:21144*)

訳 われわれは Ipc1 がメラニン産生を調節する機構を精査した

★ machinery（機構／装置）

❶ machinery in ～（～における機構／～の機構）

D-type cyclins（cyclins D1, D2, and D3）are key components of cell cycle machinery in mammalian cells.（*Proc Natl Acad Sci USA. 2001 98:194*）

訳 D型サイクリン（サイクリンD1，D2およびD3）は，哺乳類細胞の細胞周期機構の重要な構成成分である

★ mode（様式／機構）

❶ mode of ～（～の様式）

However, the mode of action of Bcl-2 proteins remains unclear.（*Mol Cell Biol. 2000 20:5680*）

訳 Bcl-2タンパク質の作用の様式は不明なままである

★ basis（基礎／基盤）

❶ basis of ～（～の基盤）

Understanding how these proteins interact with DNA is central to understanding the molecular basis of transposition.（*Science. 2000 289:77*）

訳 どのようにこれらのタンパク質がDNAと相互作用するかを理解することは，転位の分子基盤を理解することの中心を成す

★★ role（役割）

❶ role in ～（～における役割）

The ATM/p53-dependent DNA damage response pathway plays an important role in the progression of lymphoid tumors.（*Blood. 2004 103:291*）

訳 ATM/p53依存性のDNA損傷応答経路は，リンパ系腫瘍の増悪において重要な役割を果たす

❷ role of ～（～の役割）

For this study, we investigated the role of cyclooxygenase induction in the replication and growth of PRV.（*J Virol. 2004 78:12964*）

訳 われわれは，PRVの複製と増殖におけるシクロオキシゲナーゼ誘導の役割を精査した

【Ⅲ-D. 方法】 研究の方法などを表す名詞

27. 選択 【selection】

選択	選択肢
selection choice	option

使い分け
◆ **selection**, **choice** は「選択」, **option** は「選択肢／選択」という意味に用いられる.

頻度分析

	用例数		用例数
selection	5,205	❶ selection of	854
		❷ selection for	291
		❸ selection in	249
choice	1,052	❶ choice of	296
		❷ of choice	148
		❸ choice for	102
option	514	❶ option for	112

★ selection（選択）

❶ selection of ～（～の選択）

Differential regulation of viral replication may be important in the selection of specific viral variants as a result of an antiviral immune response.（*J Virol. 2001 75:8937*）

訳 ウイルス複製の差動的な調節は，特異的なウイルス変異株の選択において重要であるかもしれない

★ choice（選択）

❶ choice of ～（～の選択）

The computational speed of this method depends strongly on the choice of cutoff distance used to define interactions as measured by the density of entries of the constant linking/contact matrix.（*Biophys J. 2002 83:1620*）

訳 この方法の計算速度はカットオフ距離の選択に強く依存する

option（選択肢／選択）

"option for" の用例が多い.

❶ **option for ～**（～のための選択肢）

RF ablation is a cost-effective treatment option for patients with CRC liver metastases.（*Radiology. 2004 233:729*）

訳 RF 切除は，CRC 肝転移の患者のための対費用効果の高い治療の選択肢である

名詞

28. 使用／利用 【use】

使用	利用
use	**usage** **utilization**

使い分け ◆ **use** は「使用」，**usage**, **utilization** は「利用」という意味に用いられる.

頻度分析

	用例数		用例数
use	16,851	❶ use of	7,803
		❷ use in	865
usage	593	❶ usage of	130
utilization	1,142	❶ utilization of	387

解説 ◆いずれの語も後に of を伴う用例が多い.

III

D

★★ use（使用）

動詞の用例が多いが，名詞としても用いられる.

❶ **use of ～**（～の使用）

Conversely, $CD8^+$ T cells induce GVL effects primarily through the use of perforin and minimally through FasL mechanisms.（*Blood. 2000 96:1047*）

訳 $CD8^+$ T 細胞は，主にパーフォリンの使用によって GVL 効果を誘導する

usage（利用）

❶ usage of ～（～の利用）

A small amount of mRNA results from the usage of cryptic splice sites within exon 16.（*Blood. 2000 96:1113*）

訳 少量のメッセンジャーRNAが，エキソン16内の潜在性のスプライス部位の利用から生ずる

★ utilization（利用）

❶ utilization of ～（～の利用）

The differences between the ACTZ and CSTZ animals may be due to metabolic differences in the utilization of glucose.（*Brain Res. 2000 872:29*）

訳 ACTZ動物とCSTZ動物の間の違いは，グルコースの利用における代謝の違いのせいかもしれない

29. 置換／交換 【substitution】

置換	交換
substitution **replacement**	**exchange**

使い分け

◆ **substitution**, **replacement** は「置換」, **exchange** は「交換」という意味に用いられる.

頻度分析	用例数		用例数
substitution	6,088	❶ substitution of	1,065
		❷ substitutions in	428
		❸ substitutions at	356
replacement	2,512	❶ replacement of	929
exchange	4,369	❶ exchange of	353
		❷ exchange factor	313
		❸ exchange between	129

★ substitution（置換）

❶ substitution of ～（～の置換）

This variation produces a non-conservative substitution of arginine for glutamine at amino acid position 5345（Gln5345Arg).（*Hum Mol Genet. 2003 12:3315*）

訳 この変異は，アルギニンのグルタミンとの非保存的な置換を生じる

★ replacement（置換）

❶ replacement of ～（～の置換）

Crystal structures showed that the replacement of Gln 143 with Ala made no significant change in the overall structure of the mutant enzyme.（*Biochemistry. 2000 39:7131*）

訳 グルタミン 143 のアラニンによる置換は，変異酵素全体の構造の有意な変化をもたらさなかった

★ exchange（交換）

動詞としても用いられるが，名詞の用例の方が多い．

❸ exchange between ～（～の間の交換）

Primary T-cell responses in lymph nodes（LNs）require contact-dependent information exchange between T cells and dendritic cells（DCs).（*Nature. 2004 427:154*）

訳 リンパ節（LNs）における一次 T 細胞応答は，T 細胞と樹状細胞の間の接触依存性の情報交換を必要とする

30. 測定／定量 【measurement】

測定／アッセイ	定量
measurement	quantification
assay	quantitation

使い分け ◆ **measurement** は「測定」，**assay** は「アッセイ」，**quantification, quantitation** は「定量化／定量」という意味に用いられる．

名詞

III

D

頻度分析

	用例数		用例数
measurement	6,718	❶ measurements of	1,353
		❷ measurements in	185
		❸ measurements on	131
assay	17,721	❶ assay for	547
quantification	618	❶ quantification of	398
quantitation	443	❶ quantitation of	291

解説 ◆ **measurement**, **quantification**, **quantitation** は後に of を，**assay** は後に for を伴う用例が多い．

★ measurement（測定）

❶ measurements of ～（～の測定）

In some studies, direct measurements of secreted cytokine levels were performed.（*Crit Care Med. 2001 29:S8*）

訳 分泌されたサイトカインレベルの直接測定が行われた

★★ assay（アッセイ／アッセイする）

動詞としても用いられるが，名詞の用例の方が多い．

❶ assay for ～（～のアッセイ）

Here we describe an assay for analysing fat storage and mobilization in living *Caenorhabditis elegans*.（*Nature. 2003 421:268*）

訳 われわれは，脂肪の貯蔵と動員を分析するためのアッセイについて述べる

quantification（定量化／定量）

❶ quantification of ～（～の定量）

The method was applied to the quantification of anandamide, PEA, and OEA in plasma prepared from rat blood collected either by cardiac puncture or by decapitation.（*Anal Biochem. 2000 280:87*）

訳 その方法は，血清中のアナンダミド，PEA および OEA の定量のために適用された

quantitation（定量化／定量）

❶ quantitation of ～（～の定量）

An assay for quantitation of plasmid copy numbers in bacterial cell cultures has been developed and validated.（*Anal Biochem. 2000 284:70*）

訳 培養細菌細胞におけるプラスミドのコピー数の定量のためのアッセイが開発され，そして検証された

名詞

31．単離／精製 【isolation】

単離	分離	精製
isolation	separation segregation	purification

使い分け

◆ **isolation**（単離）は，タンパク質，遺伝子あるいは細胞などを単一にすることを意味する．

◆ **separation**, **segregation** は，「分離」という意味に用いられる．

◆ **purification** は，「（タンパク質などを）精製する」ときに用いられる．

頻度分析

	用例数		用例数
isolation	1,701	❶ isolation of	694
separation	2,501	❶ separation of	640
segregation	1,003	❶ segregation of	274
purification	1,023	❶ purification of	348

解説 ◆いずれの語も後に of を伴う用例が多い．

III

★ isolation（単離）

D

❶ isolation of ～（～の単離）

Here we report the isolation of two related but distinct cDNA clones, hOsa1 and hOsa2, that encode the largest subunits of human SWI/SNF.（*J Biol Chem. 2002 277:41674*）

訳 われわれは，2つの関連するが異なる cDNA クローンの単離を報告する

★ separation（分離）

❶ separation of ～（～の分離）

Interestingly, inhibition of dephosphorylation by okadaic acid results in the separation of the actin bundles from the plasma membrane.（*J Cell Biol. 2000 148:87*）

訳 オカダ酸による脱リン酸化の抑制は，原形質膜からのアクチン束の分離という結果になる

★ segregation（分離）

❶ segregation of ～（～の分離）

The centromere plays a critical role in the segregation of chromosomes during mitosis.（*J Cell Biol. 2001 153:1199*）

訳 中心体は，有糸分裂の間の染色体の分離において決定的に重要な役割を果たす

★ purification（精製）

❶ purification of ～（～の精製）

Here we report the purification of the Dmc1 protein from the budding yeast *Saccharomyces cerevisiae* and present basic characterization of its biochemical activity.（*J Biol Chem. 2001 276:41906*）

訳 われわれは，…からのDmc1タンパク質の精製を報告する

32. 処理 【treatment】

処理	添加	刺激
treatment	**addition**	**stimulation**

使い分け

◆ **treatment** は「(薬剤などによる) 処理」，**addition** は「(薬剤などの) 添加」，**stimulation** は「(電気などによる) 刺激」という意味に用いられる．

頻度分析

	用例数		用例数
treatment	27,112	❶ treatment of	6,339
		❷ treatment with	4,240
		❸ of treatment	1,125
addition	18,959	❶ in addition	13,370
		❷ addition of	4,538

stimulation	9,854	❶ stimulation of	3,404
		❷ stimulation with	660
		❸ stimulation by	323

解説 ◆いずれの語も後にofを伴う用例が多いが，**treatment**, **stimulation**は後にwithを伴う用例も多い．

名詞

⁑ treatment（処理／治療）

❷ treatment with ～（～による処理）

DNA synthesis induced by TNF-α was inhibited by treatment with transforming growth factor-β.（*Am J Pathol. 2003 163:465*）

訳 TNF-αによって誘導されたDNA合成は，トランスフォーミング成長因子βによる処理によって抑制された

⁑ addition（添加）

"in addition（そのうえ／加えて）" の用例が非常に多い．

❷ addition of ～（～の添加）

This uncoating is reversed by the addition of calcium.（*J Biol Chem. 2001 276:34148*）

訳 この脱コートは，カルシウムの添加によって逆戻りさせられる

⁑ stimulation（刺激）

❶ stimulation of ～（～の刺激）

Excitatory postsynaptic currents（EPSCs）evoked by stimulation of reticulospinal axons were recorded in ventral horn neurons.（*J Physiol. 2001 532:323*）

訳 網様体脊髄の軸索の刺激によって誘起される興奮性シナプス後電流（EPSCs）が前角ニューロンで記録された

III

D

33. 培養／インキュベーション 【culture】

培養	インキュベーション
culture	incubation

使い分け
◆ **culture** は，「（細胞などの）培養」という意味に用いられる．
◆ **incubation**（インキュベーション／培養）は，細胞だけでなくタンパク質などにも用いられ，しばらくの間一定の条件下に置くことを意味する．

頻度分析

	用例数		用例数
culture	5,993	❶ in culture	1,268
		❷ cell culture	844
		❸ tissue culture	445
incubation	1,837	❶ incubation of	638
		❷ incubation with	418
		❸ incubation in	102

★ culture（培養／培養する）

動詞としても用いられるが，名詞の用例の方が多い．

❶ **～ in culture**（培養中の～）

The activity of Qsulf1 in cells in culture results in the release of Noggin from the cell surface and a restoration of BMP responsiveness to the cells.（*J Biol Chem. 2004 279:5604*）

訳 培養中の細胞における Qsulf1 の活性は，細胞表面からの Noggin の放出という結果になる

★ incubation（インキュベーション／培養）

"incubation of" "incubation with" の用例が多い．

❶ **incubation of ～**（～のインキュベーション）

Prolonged incubation of eosinophils with IL-5 induced significant eosinophil-derived neurotoxin release.（*J Immunol. 2000 165:2198*）

訳 IL-5 と好酸球の延長したインキュベーションは，好酸球由来神経毒の有意な放出を誘導した

第3章

形容詞編

形容詞には，名詞を修飾する用法と動詞の後で補語となる用法との主に２つの用法がある．名詞を修飾する場合は通常は名詞の前に置かれるが，後ろに置かれる場合もある．名詞や動詞の後では，特定の前置詞とともに使われることがよくある．修飾する名詞および後にくる前置詞との組み合わせについて，個々の単語の用法をよく調べることが重要である．

【Ⅰ-A. 程度】 程度を表す形容詞

1. 著しい 【marked】

有意な	著しい	劇的な	注目すべき
significant	marked striking prominent intense	dramatic drastic	remarkable notable noteworthy

使い分け

◆ **significant** は「（統計的に）有意な」という意味で用いられることが多いが，「著しく」という意味で使われることもある．

◆ **marked**, **striking**, **prominent**, **intense** は「著しい」，**dramatic** は「劇的な」，**drastic** は「激烈な」，**remarkable**, **notable**, **noteworthy** は「注目すべき」という意味に用いられる．

頻度分析

	用例数		用例数
significant	20,650	❶ significant differences	1,185
		❷ significant increase in	950
		❸ significant reduction	743
marked	3,108	❶ a marked increase in	254
		❷ marked reduction	242
		❸ marked decrease	145
striking	1,237	❶ striking differences	73
		❷ striking contrast	70
		❸ striking increase	41
prominent	1,336	❶ prominent in	167
		❷ prominent role	135
		❸ prominent feature	58
intense	750	❶ of intense	105
		❷ intense interest	32
dramatic	1,750	❶ a dramatic increase in	189
		❷ dramatic reduction	137
		❸ dramatic changes	89
drastic	128	❶ drastic reduction	28
remarkable	754	❶ remarkable ability to	21

notable	297	❶ most notable	51
noteworthy	81	❶ it is noteworthy that	34

☆☆ significant〔(統計的に) 有意な／著しい／重要な〕

❶ **significant differences**（有意な違い）

There were no significant differences in the levels of seven additional cytokines evaluated.（*Infect Immun. 2001 69:7178*）

訳 …のレベルの有意な違いはなかった

❷ **significant increase in ～**（～の有意な増加）

Overexpression of noggin resulted in a significant increase in the number of neurons in the trigeminal and dorsal root ganglia.（*Development. 2004 131:1175*）

訳 nogginの過剰発現は，…におけるニューロンの数の有意な増加という結果をもたらした

☆ marked（顕著な／著しい）

❶ **a marked increase in ～**（～の顕著な上昇）

Exposure of HCs to IFN-α resulted in a marked increase in tumor necrosis factor-α（TNF-α）secretion.（*Blood. 2002 100:647*）

訳 有毛細胞のIFN-αへの曝露は，腫瘍壊死因子-α（TNF-α）分泌の顕著な上昇という結果になった

☆ striking（著しい／顕著な）

❶ **striking differences**（著しい違い）

Analysis of BDNF(NT3/NT3) mice showed striking differences in the ability of NT3 to promote survival, short-range innervation and synaptogenesis in different sensory systems.（*Development. 2003 130:1479*）

訳 BDNF(NT3/NT3) マウスの分析は，…を促進するNT3の能力の著しい違いを示した

☆ prominent（顕著な／著しい）

❶ **prominent in ～**（～において顕著な）

The decrease was most prominent in zone 4（r = −0.76, P = 0.007).（*Invest Ophthalmol Vis Sci. 2002 43:3312*）

訳 減少はゾーン4においてもっとも顕著であった

形容詞

I

A

★ intense（大きな／著しい）

❷ intense interest（大きな関心）

The development of an effective vaccine against *Mycobacterium tuberculosis* is a research area of intense interest.（*J Immunol. 2001 166:6227*）

訳 結核菌に対する効果的なワクチンの開発は，大きな関心のある研究領域である

★ dramatic（劇的な）

❶ a dramatic increase in ～（～の劇的な上昇）

Loss of Brn3a leads to a dramatic increase in apoptosis and severe loss of neurons in sensory ganglia.（*J Cell Biol. 2004 167:257*）

訳 Brn3aの欠損はアポトーシスの劇的な上昇につながる

drastic（激烈な）

❶ drastic reduction（激烈な減少）

A drastic reduction in BRCA1 gene expression is a characteristic feature of aggressive sporadic breast carcinoma.（*Mol Cell Biol. 2003 23:2225*）

訳 BRCA1遺伝子発現の激烈な減少は，活動的で散発性の乳癌に特有の特徴である

★ remarkable（注目すべき／著しい）

❶ remarkable ability to ～（～する注目すべき能力）

We further found that while growing cells are rapidly killed by iron starvation, stationary-phase cells show a remarkable ability to survive iron depletion.（*Infect Immun. 2003 71:6510*）

訳 定常期細胞は鉄欠乏を生き残る注目すべき能力を示す

notable（注目すべき／著しい）

❶ most notable ～（もっとも注目すべき～）

The most notable difference between dogs and primates is seen in the fraction of parent drug excreted unchanged in the urine, 50% in the dog and < 1% in the primate.（*Cancer Res. 2000 60:4433*）

訳 イヌと霊長類の間のもっとも注目すべき違いが，…の画分に見られる

noteworthy（注目すべき／顕著な）

❶ **it is noteworthy that ～**（～ということは注目に値する）

It is noteworthy that this SNP has been linked to attention deficit hyperactivity disorder and has been associated with schizophrenic patients that do not respond to treatment with clozapine.（*Mol Pharmacol. 2004 66:1293*）

訳 このSNPが…に関連してきたということは注目に値する

2. 強力な 【strong】

強力な	
strong powerful	robust potent

使い分け ◆修飾する名詞が，それぞれ異なることに注意が必要である．

頻度分析

	用例数		用例数
strong	5,839	❶ strong evidence	360
		❷ strong correlation	192
		❸ strong association	111
		❹ strong expression	102
powerful	1,424	❶ powerful tool	234
robust	1,758	❶ robust to	70
		❷ robust in	50
		❸ robust expression	33
		❹ robust increase	33
potent	5,405	❶ more potent	597
		❷ most potent	411
		❸ potent inhibitor	375

★★ strong（強力な／強い）

❶ **strong evidence**（強力な証拠）

Taken together, these results provide strong evidence that EBNA2 plays an important role in regulating Cp activity. (*J Virol. 2000 74:11115*)

訳 これらの結果は，…という強力な証拠を提供する

★ powerful（強力な／力強い）

❶ powerful tool（強力な手段）

An inducible promoter system provides a powerful tool for studying the genetic basis for virulence. (*Infect Immun. 2001 69:7851*)

訳 誘導性のプロモータシステムは，…のための強力な手段を提供する

★ robust（強力な／強固な）

❸ robust expression（強力な発現）

Thus, the robust expression of early inflammatory cytokines in *C. neoformans* 52D-infected mice promoted the development of protective immunity even in the absence of MIP-1α (*Infect Immun. 2001 69:6256*)

訳 *C. neoformans* 52D に感染したマウスにおける初期の炎症性サイトカインの強力な発現は，保護的な免疫の発達を促進した

★★ potent（強力な／効力のある）

前に more や most を伴う用例が多い．

❶ more potent（より強力な）

However, consistent with its DAT activity, piperidine 14 was found to be about 2.5-fold more potent than cocaine in enhancing stereotypic movements. (*J Med Chem. 2000 43:1215*)

訳 ピペリジン 14 は，コカインより約 2.5 倍強力であることが見つけられた

3. 大きな 【large】

大きな	大量の
large	massive
huge	
great	

使い分け ◆ **large**は「大きな」，**huge**は「巨大な／大きな」，**great**は「大きな／偉大な」，**massive**は「大量の／大きな」という意味に用いられる．

頻度分析

	用例数		用例数
large	13,662	❶ large scale	938
		❷ a large number of	697
huge	67	❶ huge amounts of	5
great	1,023	❶ of great interest	50
massive	514	❶ massive apoptosis	19

形容詞

★★ large（大きな）

❶ large scale（大規模）

Large-scale gene expression profiling was performed on embryo-derived stem cell lines to identify molecular signatures of pluripotency and lineage specificity.（*Genome Res. 2002 12:1921*）

訳 大規模な遺伝子発現プロファイリングが，胚由来の幹細胞株に対して実行された

I

huge（巨大な／大きな）

❶ huge amounts of ～（大量の～）

Although huge amounts of genomic data are at hand, current experimental protein interaction assays must overcome technical problems to scale-up for high-throughput analysis.（*Bioinformatics. 2003 19:125*）

訳 大量のゲノムデータが手元にあるけれども

A

★ great（大きな／非常な／偉大な）

❶ of great interest（非常に興味ある）

Thus, it is of great interest to understand the mechanisms underlying insulin receptor binding kinetics.（*J Theor Biol. 2000 205:355*）

訳 …の根底にある機構を理解することには非常に興味がある

★ massive（大量の／大きな）

❶ massive apoptosis（大量のアポトーシス）

However, Cited2 was required for the survival of neuroepithelial cells and its absence led to massive apoptosis in dorsal neuroectoderm around the FB-MB boundary and in a restricted transverse domain in the hindbrain.（*Hum Mol Genet. 2002 11:283*）

訳 それの欠如は，背側の神経外胚葉の大量のアポトーシスにつながった

4. 主な 【major】

主な	
major	**dominant**
main	**predominant**

使い分け

◆ **major**, **main**, **dominant**, **predominant** は，「主な／主要な」という意味に用いられる．

頻度分析

	用例数		用例数
major	14,045	❶ major histocompatibility	927
		❷ major role	702
		❸ major depression	362
		❹ major cause	309
main	1,810	❶ the main cause	29
dominant	14,045	❶ dominant negative	3,430
		❷ autosomal dominant	601
		❸ dominant role	101
predominant	1,205	❶ predominant role	46
		❷ predominant form	44
		❸ predominant in	40

解説

◆ **major**, **dominant**, **predominant** は role の前，**main** は cause の前などで用いられる．

☆ major（主な／主要な）

❷ **major role**（主な役割）

DNA damage caused by UV radiation is thought to play a major role in carcinogenesis induction.（*Cancer Res. 2000 60:5612*）

訳 UV照射によって引き起こされるDNA損傷は，発癌誘導において主な役割を果たすと考えられる

★ main（主な）

❶ **the main cause**（主な原因）

Infection with *Helicobacter pylori* is the main cause of peptic-ulcer disease.（*Lancet. 2000 355:1665*）

訳 ヘリコバクターピロリの感染は消化性潰瘍疾患の主な原因である

★★ dominant（主な／主要な／支配的な）

"dominant negative" の用例が非常に多い．

❸ **dominant role**（主な役割）

These data support a model in which glycolysis plays a dominant role in glucose-stimulated insulin secretion.（*J Biol Chem. 2002 277:30914*）

訳 解糖は，グルコースに刺激されるインスリン分泌の際に主な役割を果たす

★ predominant（主な／優性の）

❶ **predominant role**（主な役割）

Thus PKA plays a predominant role in the cGMP-induced phosphorylation of VASP and platelet inhibition in human platelets.（*Blood. 2003 101:4423*）

訳 PKAは，cGMPに誘導されるVASPのリン酸化の際に主な役割を果たす

5. 完全な 【complete】

完全な		
complete	full	perfect

使い分け ◆いずれも「完全な」という意味に用いられる．

頻度分析

	用例数		用例数
complete	5,585	❶ complete loss	311
		❷ complete remission	132
		❸ complete absence	119

形容詞

I

A

full	5,992	❶ full length	3,228
		❷ full thickness	113
perfect	236	❶ near perfect	28
		❷ perfect match	10
		❸ perfect correlation	7

⁑ complete（完全な）

❶ complete loss（完全な喪失）

Deletion of this motif or overexpression of a NUR77 dominant negative protein caused a complete loss of 20α-HSD promoter activation by $PGF_2\alpha$.（*J Biol Chem. 2000 275:37202*）

訳 NUR77のドミナントネガティブタンパク質の過剰発現は，$PGF_2\alpha$による20α-HSDプロモータの活性化の完全な喪失を引き起こした

⁑ full（完全な／十分な）

❶ full length ～（完全長の～）

Therefore, we cloned full-length cDNA for a human homolog of BGM, and we investigated the properties of its protein product, hsBG, to determine whether it had VLCS activity.（*J Biol Chem. 2000 275:35162*）

訳 われわれはヒトのBGMのホモログの完全長cDNAをクローニングした

perfect（完全な／完璧な）

❶ near perfect ～（ほとんど完全な～）

Near-perfect correlation existed between calculated scores and observed mortality, with higher scores associated with higher mortality.（*Circulation. 2001 104:263*）

訳 ほとんど完全な相関が，計算されたスコアと観察された死亡率の間に存在した

6. 十分な／適当な 【sufficient】

十分な	適当な／適正な	合理的な
sufficient adequate	appropriate adequate proper	reasonable

使い分け

◆ **sufficient** は「十分な」，**adequate** は「適当な／十分な」という意味に用いられる．

◆ **appropriate**, **proper** は「適当な／適正な」，**reasonable** は「合理的な」という意味で使われる．

頻度分析

	用例数		用例数
sufficient	5,582	❶ sufficient to	2,998
		❷ sufficient for	1,647
adequate	731	❶ adequate for	52
		❷ adequate to	46
appropriate	2,277	❶ appropriate for	153
		❷ appropriate to	59
proper	1,591	❶ for proper	448
		❷ proper development	69
		❸ proper folding	54
reasonable	358	❶ it is reasonable to	24

解説

◆ **sufficient**, **adequate**, **appropriate** は，後に to 不定詞や for を伴う用例が多い．

◆ **proper** は，for の後に使われることが多い．

◆ it is **reasonable** to（〜することは理にかなっている）の用例も多い．

★★ sufficient（十分な）

“sufficient to”“sufficient for”の用例が非常に多い．

❶ **sufficient to 〜（〜するのに十分な）**

In contrast, abrogation of retinoid signaling is sufficient to induce the expression of the chondroblastic phenotype in the presence of Noggin.（*J Cell Biol. 2000 148:679*）

訳 レチノイドシグナル伝達の抑止は，軟骨芽細胞様の表現型の発現を誘

導するのに十分である

❷ **sufficient for ～（～に十分な）**

The MEF2C DNA binding domain is sufficient for this interaction.（*J Biol Chem. 1998 273:26218*）

訳 MEF2C の DNA 結合ドメインはこの相互作用に十分である

★ adequate（適当な／十分な）

❶ **adequate for ～（～に適当な）**

Eighty-eight percent of biopsy specimens were adequate for histologic diagnosis.（*Transplantation. 2002 73:553*）

訳 生検試料の 88 %は組織学的診断に適当であった

❷ **adequate to ～（～するのに十分な）**

Interleukin 6（IL-6）levels were reduced but still adequate to support regeneration.（*Hepatology. 2001 33:915*）

訳 インターロイキン 6（IL-6）のレベルは低下したが，まだ再生を支持するのに十分である

★ appropriate（適当な／適した）

❶ **appropriate for ～（～に適当な）**

This method may be appropriate for use in the clinical laboratory.（*J Clin Microbiol. 2003 41:5683*）

訳 この方法は臨床検査室における利用に適当であるかもしれない

★ proper（適正な／適当な）

❶ **for proper ～（適正な～にとって）**

The Hox genes encode a group of transcription factors essential for proper development of the mouse.（*Dev Biol. 2002 249:96*）

訳 Hox 遺伝子はマウスの適正な発生に必須である一群の転写因子をコードする

reasonable（合理的な／適切な）

❶ **it is reasonable to ～（～することは理にかなっている）**

Emotions operate along the dimension of approach and aversion, and it is reasonable to assume that orienting behavior is intrinsically linked to emotionally involved processes such as preference decisions.（*Nat Neurosci. 2003 6:1317*）

訳 …ということ仮定することは理にかなっている

7. 明らかな 【clear】

明らかな	
clear	pronounced
apparent	obvious
evident	overt

使い分け
◆いずれの語も「明らかな」という意味に用いられる.
◆ pronounce は「発音する」という意味の動詞だが，**pronounced** は形容詞として用いられる場合が多い.

頻度分析

	用例数		用例数
clear	2,380	❶ not clear	558
		❷ clear that	247
		❸ clear whether	123
		❹ clear evidence	106
apparent	3,807	❶ no apparent	311
		❷ apparent in	212
		❸ apparent affinity	122
		❹ apparent molecular	122
evident	1,350	❶ evident in	437
pronounced	1,029	❶ more pronounced	289
		❷ pronounced in	177
obvious	512	❶ no obvious	194
overt	317	❶ no overt	47

解説
◆ **evident**, **pronounced** は，後に in を伴う用例が多い.
◆ **apparent**, **obvious, overt** は，前に no を伴う用例が多い.

★ clear（明らかな／明確な）

❶ not clear（明らかではない）

It is not clear whether p53 and RR can directly interact at the protein level to regulate DNA repair.（*Cancer Res. 2003 63:980*）

訳 p53 と RR がタンパク質レベルで直接相互作用できるかどうか明らかではない

★ apparent（明らかな／明白な）

❶ no apparent ～（明らかな～のない）

Although there was no apparent relationship between agouti mRNA levels and BMI, agouti mRNA levels were significantly elevated in subjects with type 2 diabetes.（*Diabetes. 2003 52:2914*）

訳 アグーチのメッセンジャーRNAレベルとBMIの間に明らかな関連性はなかった

★ evident（明らかな／明白な）

❶ evident in ～（～において明らかな）

In WAT, ZAG gene expression was evident in mature adipocytes and in stromal-vascular cells.（*Proc Natl Acad Sci USA. 2004 101:2500*）

訳 ZAG遺伝子発現は成熟した脂肪細胞において明らかであった

★ pronounced（明らかな／明白な／顕著な）

❶ more pronounced（より明らかな）

These skeletal abnormalities were more pronounced in mice deficient for both Dvl1 and Dvl2.（*Development. 2002 129:5827*）

訳 これらの骨格の異常が，Dvl1とDvl2の両方を欠損したマウスにおいてより明らかであった

★ obvious（明らかな／明白な／顕著な）

❶ no obvious ～（明らかな～のない）

There was no obvious correlation between functional status and donor age.（*Transplantation. 1999 68:1910*）

訳 機能的状態とドナーの年齢の間に明らかな相関はなかった

overt（明らかな／明白な）

❶ no overt ～（明らかな～のない）

In the absence of stress, MCIP1$^{-/-}$ animals exhibited no overt phenotype.（*Proc Natl Acad Sci USA. 2003 100:669*）

訳 MCIP1$^{-/-}$動物は明らかな表現型を示さなかった

8. 広い 【wide】

広い	
wide broad	extensive widespread

使い分け ◆いずれの語も「広い／広範の」という意味に用いられる．

頻度分析

	用例数		用例数
wide	3,856	❶ a wide range of	1,103
broad	2,179	❶ a broad range of	352
extensive	2,987	❶ more extensive	207
widespread	1,236	❶ widespread use	95

★ wide（広い）

"a wide range of" の用例が多い．

❶ a wide range of ～（広範囲の～）

Many BTB-containing proteins are transcriptional regulators involved in a wide range of developmental processes.（*Genetics. 2000 156:195*）

訳 多くのBTBを含むタンパク質は，広範囲の発生過程に関与する転写調節因子である

★ broad（広い）

❶ a broad range of ～（広範囲の～）

Using anti-Homer antibody we show that Homer is expressed in a broad range of tissues but is highly enriched in the CNS.（*J Neurosci. 2002 22:428*）

訳 ホーマーは広範囲の組織で発現する

★ extensive（広い／広範の）

❶ more extensive ～（より広範の～）

TIMP-3 immunoreactivity was more extensive in the SFD eye.（*Invest Ophthalmol Vis Sci. 2000 41:898*）

訳 TIMP-3の免疫反応性はSFDの目においてより広範であった

形容詞 I A

★ widespread（広い／広範の）

❶ widespread use（広範な使用）

These experiments may have implications for the widespread use of estrogenic substances in agriculture and the environment.（*Dev Biol. 2000 224:354*）

訳 これらの実験は，エストロゲン様物質の広範な使用のための意味を持つかもしれない

9. 全体の　【whole】

全体の	
whole	entire
overall	total

使い分け ◆いずれの語も「全体の」という意味に用いられる．

頻度分析

	用例数		用例数
whole	4,177	❶ whole cell	1,169
		❷ in whole	440
		❸ whole body	435
		❹ whole genome	418
overall	6,586	❶ overall survival	476
entire	2,214	❶ entire genome	64
		❷ entire length	52
total	9,409	❶ total number	335
		❷ total body	333

★ whole（全体の）

❷ in whole 〜（全体の〜において）

Changes in phosphorylation of multiple proteins were observed in whole cell lysates prepared from PMN-stimulated epithelial cells.（*J Immunol. 2002 169:476*）

訳 複数のタンパク質のリン酸化の変化は全細胞溶解液において観察された

★★ overall（全体の）

❶ overall survival ～（全体の生存～）

However, overall survival rates were similar in all three groups.（*N Engl J Med. 2003 349:2091*）

訳 全体の生存率は3つすべてのグループにおいて類似していた

★ entire（全体の）

❶ entire genome（ゲノム全体）

The entire genome of AAAV displays 56 to 65% identity at the nucleotide level with the other known AAVs.（*J Virol. 2003 77:6799*）

訳 AAAVのゲノム全体は，ヌクレオチドのレベルで他の既知のAAVと56から65％の同一性を示す

★★ total（全体の）

❶ total number（総数）

The number of founder cells was independent of the total number of cells injected into the host blastocysts.（*Dev Biol. 2004 275:192*）

訳 創始細胞の数は，ホストの胚盤胞に注入された総細胞数に依存しなかった

10. 多数の　【many】

多数の	
many numerous	large number of multitude of

使い分け ◆いずれの語も「多数の」という意味に用いられる．

頻度分析

	用例数		用例数
many	12,845	❶ many of	1,694
numerous	2,231	❶ numerous studies	120
large number of	1,205	❶ a large number of	697
multitude of	99	❶ a multitude of	85

⁑ many（多くの／多数の）

❶ many of ～（～の多く）

Additionally, VMAT2 and P-GP are inhibited by many of the same compounds.（*Brain Res. 2001 910:116*）

訳 VMAT2 と P-GP は同じ化合物の多くによって抑制される

★ numerous（多数の）

❶ numerous studies（多数の研究）

Numerous studies have demonstrated that estrogens induce rapid and transient activation of the Src/Erk phosphorylation cascade.（*Proc Natl Acad Sci USA. 2002 99:14783*）

訳 多数の研究が，…ということを実証してきた

★ large number of（多数の）

❶ a large number of ～（多数の～）

These data identified a large number of genes that are differentially expressed in *ob/ob* mice.（*Genes Dev. 2000 14:963*）

訳 これらのデータは，*ob/ob* マウスにおいて異なって発現する多数の遺伝子を同定した

multitude of（多数の）

❶ a multitude of ～（多数の～）

NF-κB is a transcription factor that regulates a multitude of genes, including those involved in immune, inflammatory, and anti-apoptotic responses.（*Biochemistry. 2002 41:7604*）

訳 NF-κB は多数の遺伝子を調節する転写因子である

11. いくつかの　【several】

いくつかの	
several	a number of

使い分け

◆どちらも「いくつかの」という意味に用いられる．

◆ **a number of** は「多数の」という意味に使われる場合もあるが，その意味では a large number of を用いるべきである．

頻度分析

	用例数		用例数
several	15,934	❶ several other	586
		❷ several of	539
		❸ several different	384
a number of	3,399	❶ a number of	3,399

★★ several（いくつかの）

❶ several other ～（いくつかの他の～）

Genetic studies have shown that KCBP is likely to interact with several other proteins.（*Plant Cell. 2004 16:185*）

訳 KCBPは，いくつかの他のタンパク質とおそらく相互作用しそうである

★ a number of（いくつかの／多数の）

❶ a number of ～（いくつかの～）

Although a number of genes involved in cancer development are regulated by c-myc, the actual mechanisms leading to Myc-induced neoplasia are not known.（*Mol Cell Biol. 2004 24:7538*）

訳 癌の発生に関与するいくつかの遺伝子はc-mycによって調節されているけれども

形容詞

I

A

12. かなりの 【substantial】

かなりの	
substantial	considerable

使い分け

◆ **substantial**は「かなりの／実質的な」，**considerable**は「かなりの」という意味に用いられる．

頻度分析

	用例数		用例数
substantial	2,840	❶ substantial increase in	95
		❷ substantial fraction	90
		❸ substantial proportion	78

considerable	1,260	❶ considerable interest	99
		❷ considerable evidence	54
		❸ considerable attention	45

★ substantial（かなりの／実質的な）

❶ **substantial increase in ～**（～のかなりの上昇）

In addition, exposure to IFN-γ resulted in a substantial increase in STAT 1 expression and a small increase in STAT 3 expression.（*J Biol Chem. 2001 276:7062*）

訳 IFN-γ への曝露は，STAT1 発現のかなりの上昇という結果になった

★ considerable（かなりの）

❶ **considerable interest**（かなりの関心）

There is considerable interest in understanding how cis-regulatory modifications drive morphological changes across species.（*Proc Natl Acad Sci USA. 2003 100:15666*）

訳 どのようにシス調節性の修飾が形態学的な変化を制御するかを理解することにかなりの関心がある

13. 中程度の 【moderate】

中程度の	大きくない
moderate	modest

使い分け ◆ **moderate** は「中程度の」，**modest** は「大きくない」という意味に用いられる．

頻度分析

	用例数		用例数
moderate	2,203	❶ moderate levels	81
		❷ moderate intensity	57
modest	1,204	❶ modest increase in	82
		❷ modest effect	51

★ moderate（中程度の／緩和する）

形容詞および動詞として用いられる．

❶ moderate levels（中程度のレベル）

Transiently transfected cells expressed moderate levels of ribozyme (approximately 50,000 molecules/cell) with predominant nuclear localization and a short half-life (23 min). (*J Biol Chem. 2002 277:25957*)

訳 一過性に遺伝子導入された細胞は，中程度のレベルのリボザイムを発現した

★ modest（あまり大きくない／わずかな）

❶ modest increase in ～（～の大きくない上昇）

Fibroblasts of Pms2 −/− mice exhibited only a modest increase in the frequency of clones with point mutations, such that mitotic recombination was still the primary cause of APRT deficiency. (*Oncogene. 2002 21:2840*)

訳 Pms2 −/− マウスの線維芽細胞は，点突然変異をもつクローンの頻度のわずかな上昇を示した

14. ほとんどない／わずかな／ない 【little】

ほとんどない	わずかな	少ない／小さい／弱い	ない
little few	a few subtle slight insignificant	less small weak	deficient free

使い分け

◆ **little** は，「ほとんどない」という意味に用いられる．
◆ **few** も，a がつかない場合は否定的な意味で使われる．
◆ **a few**, **subtle**, **slight**, **insignificant** は，「わずかな」という意味に用いられる．
◆ a little の用例は非常に少ない．
◆ **less** は「より少ない」，**small** は「小さい」，**weak** は「弱い」，**deficient** は「欠損した」という意味で使われる．
◆ **free** は，**free** of の形で「～のない」という意味に用いられる．
◆ **few** は数に，**little** は量に用いられる．

頻度分析

	用例数		用例数
little	6,142	❶ little is known	2,033
		❷ little effect	726
		❸ little or no	610
few	3,451	❶ a few	1,341
		❷ few studies	223
subtle	576	❶ subtle differences	74
		❷ subtle changes	72
slight	486	❶ a slight	299
		❷ slight increase	68
insignificant	108	❶ insignificant effects	8
less	11,007	❶ less than	2,677
small	13,322	❶ small number of	415
weak	1,871	❶ weak binding	72
deficient	9,599	❶ deficient mice	2,520
		❷ deficient in	1,669
free	9,209	❶ free of	325

解説
◆ **few** や **slight** は，前に a を伴う用例が多い．
◆ **less** は後に than を，**deficient** は後に in を伴う用例が多い．

★★ little（ほとんどない）

❶ little is known（ほとんど知られていない）

Little is known about the mechanisms by which these factors modulate transcription.（*J Biol Chem. 2000 275:3810*）

訳 それによってこれらの因子が転写を調節する機構についてはほとんど知られていない

★ few（わずかな／ほとんどない）

❶ a few ～（わずかな～）

TTAB is able to induce CYP26 mRNA expression in only a few of the lung carcinoma cell lines tested.（*Mol Pharmacol. 2000 58:483*）

訳 TTAB は，テストされた肺癌細胞株のうちのほんのわずかにおいてのみ CYP26 メッセンジャー RNA の発現を誘導できる

❷ few studies（…な研究はほとんどない）

Few studies have examined the separate effects of soy protein and isoflavones.（*Am J Clin Nutr. 2001 73:728*）

訳 大豆タンパク質とイソフラボンの別々の効果を調べた研究はほとんどない

★ subtle（わずかな）

❶ subtle differences（わずかな違い）

Finally, our findings also indicate that subtle differences in the pattern of core histone acetylation play a role in selective gene activation.（*Proc Natl Acad Sci USA. 2000 97:6340*）

訳 コアヒストンアセチル化のパターンのわずかな違いが，選択的な遺伝子活性化において役割を果たす

slight（わずかな）

❷ slight increase（わずかな上昇）

However, overexpression of APG14 led to only a slight increase in autophagy in nitrogen-rich medium.（*J Biol Chem. 2001 276:6463*）

訳 APG14の過剰発現は自己貪食のほんのわずかな上昇につながった

insignificant（わずかな／重要でない）

❶ insignificant effects（わずかな効果）

Mutations of Cys-822 and Cys-892 had insignificant effects on the $K_{i(\mathrm{app})}$, $K_{m(\mathrm{app})}$ or V_{max}, but mutations of Cys-813 to threonine and Cys-321 to alanine decreased the affinity for SCH28080.（*J Biol Chem. 2000 275:4041*）

訳 Cys-822とCys-892の変異は，$K_{i(\mathrm{app})}$，$K_{m(\mathrm{app})}$ あるいは V_{max} に対してわずかな効果しか持たなかった

★★ less（より少ない）

littleの比較級.「より少なく」という意味の副詞としても使われる.

❶ less than ～（～より少ない）

PGE_2 stimulated TSG-6 mRNA expression, but the magnitude of response was substantially less than that produced by TNF-α, and it was maximal only after 24 hours of incubation.（*Am J Pathol. 2002 160:1495*）

訳 しかし，反応の大きさはTNF-αによって産生されるそれよりも実質的に少なかった

★ small（小さい）

❶ **small number of ～**（少数の～／～の少数）

Expression analysis of this last class suggests that only a small number of these genes are expressed.（*Plant Mol Biol. 2002 48:319*）

訳 それらの遺伝子のほんの少数だけが発現している

★ weak（弱い）

❶ **weak binding**（弱い結合）

In addition, the weak binding of the heme in this protein resulted in conformational shifts at a location distant from the binding site.（*Biochemistry. 2001 40:4879*）

訳 このタンパク質におけるヘムの弱い結合は，結合部位から離れた位置における構造の転換という結果になった

★★ deficient（欠損した）

❷ **deficient in ～**（～を欠損した）

To test the latter, we generated mice deficient in both HIP1 and HIP1r.（*Mol Cell Biol. 2004 24:4329*）

訳 われわれはHIP1とHIP1rの両方を欠損したマウスを作製した

★★ free（ない／含まない）

❶ **free of ～**（～のない）

The 38,244 men who were free of diagnosed cardiovascular disease at that time were included in the analyses.（*Circulation. 2001 103:52*）

訳 そのときに診断された循環器疾患のない38,244名の男性が分析に含まれた

15. 普通の／一般的な【common】

共通の	一般的な	通常の	正常な
common	general	conventional	normal

使い分け

◆ **common**は「共通の／普通の」，**general**は「一般的な」，**conventional**は「通常の／従来の」，**normal**は「正常な」という意味に用いられる．

頻度分析

	用例数		用例数
common	9,325	❶ common in	645
		❷ common to	483
		❸ common cause	212
general	5,274	❶ general population	337
		❷ general transcription	180
		❸ general mechanism	162
conventional	2,899	❶ conventional methods	71
normal	20,497	❶ normal human	789
		❷ normal cells	450
		❸ normal levels	416
		❹ normal subjects	325

形容詞

★★ common（共通の／普通の）

❶ common in ～（～において共通な／～においてよくある）

Postoperative coagulation abnormalities were more common in patients with antibodies to human coagulation proteins.（*Ann Surg. 2001 233:88*）

訳 手術後の凝固異常は，…に対する抗体を持つ患者においてよりよくあった

I

★★ general（一般的な）

A

❸ general mechanism（一般的な機構）

These results also suggest that protein degradation may be a general mechanism for Cbl-mediated negative regulation of activated tyrosine kinases.（*Mol Cell Biol. 2000 20:851*）

訳 タンパク質の分解は，Cblに仲介される負の制御の一般的な機構かもしれない

★ conventional（通常の／従来の）

❶ conventional methods（通常の方法／従来の方法）

Th1 and Th2 populations were polarized *in vitro*, and their selective cytokine production was determined by conventional methods.（*Invest Ophthalmol Vis Sci. 2002 43:758*）

訳 それらの選択的なサイトカイン産生は通常の方法によって決定された

★★ normal（正常な／普通の）

❸ **normal levels**（正常なレベル）

However, Stat4α is required for normal levels of IL-12-induced interferon-γ production from Th1 cells.（*EMBO J. 2003 22:4237*）

訳 Stat4αは，正常なレベルのIL-12誘導性インターフェロン-γ産生に必要とされる

16. 特別の／特異的な【particular】

特別の／特異的な		
particular	special	specific

使い分け

◆ **particular**, **special** は「特別の／特に」という意味で，interest の前などに用いられる．

◆ **specific** は，「特異的な」という意味で使われる．

頻度分析

	用例数		用例数
particular	3,844	❶ of particular interest	163
		❷ particular importance	60
special	415	❶ of special interest	21
specific	40,963	❶ specific for	1,793
		❷ specific to	797
		❸ specific expression	656

★ particular（特別の／特に）

❶ **of particular interest**（特に興味深い）

This bifunctional enzyme is of particular interest because of its potential as a chemotherapeutic target.（*J Biol Chem. 2002 277:22168*）

訳 この二機能性の酵素は特に興味深い

special（特別の／特に）

❶ **of special interest**（特に興味深い）

Of special interest are the possible side effects of IL-12 therapy in patients with autoimmune diseases, especially those that are T cell mediated, such as rheumatoid arthritis (RA). (*Arthritis Rheum. 2000 43:461*)

訳 IL-12療法のありうる副作用は特に興味深い

⁑ specific（特異的な）

❶ specific for ～（～に対して特異的な）

Immunostaining was carried out with antibodies specific for 27 ribosomal proteins, two translation factors and one that specifically recognizes rRNA. (*Mol Cell. 2002 10:93*)

訳 免疫染色が，27個のリボソームタンパク質に対する特異的な抗体を使って行われた

形容詞

I

A

【I-B. 性質・状態】 性質・状態を表す形容詞

17. 可能な／ありそうな 【possible】

ありそうな／ほとんど確実な	可能な	潜在的な
probable likely	possible feasible	potential

使い分け

◆ **probable**, **likely** は「ありそうな／ほとんど確実な」, **possible**, **feasible** は「可能な／ありうる」, **potential** は「潜在的な／可能な」という意味に用いられる．

頻度分析

	用例数		用例数
probable	363	❶ a probable	58
likely	8,924	❶ likely to	4,026
		❷ more likely to	1,028
		❸ it is likely that	294
possible	6,809	❶ possible to	969
		❷ possible role	445
		❸ possible mechanism	236
feasible	500	❶ feasible to	52
potential	16,431	❶ potential role	618
		❷ potential target	239

解説

◆ **likely**, **possible**, **feasible** は，後に to 不定詞を伴う用例が多い．
◆ **likely** は，副詞としても用いられる．
◆ **possible**, **potential** は，role の前などで使われる．

probable（ありそうな／ほとんど確実な）

❶ **a probable ～**（ありそうな～／ほとんど確実な～）

Furthermore, hbl-1 is a probable target of microRNA regulation through its 3'UTR.（*Dev Cell. 2003 4:639*）

訳 hbl-1 はマイクロ RNA 調節のほとんど確実な標的である

★ likely（ありそうな）

形容詞および副詞として用いられる．"more likely" は，比較する対象がある場合に用いられる．

❶ **likely to ～**（おそらく～しそうな）

In contrast, the sequence amino-terminal to the Arm domain confers a negative regulatory function that is likely to be modulated by phosphorylation.（*Biochemistry. 2003 42:9195*）

訳 おそらくリン酸化によって調節されるであろう負の調節機能

❷ **more likely to ～**（おそらくより～しそうな）

Candidemic patients were more likely to have a history of underlying renal failure at baseline and to require dialysis at onset of septic shock.（*Crit Care Med. 2002 30:1808*）

訳 カンジダ血症の患者は，根底にある腎不全の病歴がおそらくよりありそうである

❸ **it is likely that ～**（～ということはありそうである）

Because the function of Rev is to bind HIV RNA and facilitate transport of singly spliced and unspliced RNA to the cytoplasm, it is likely that the nucleolus plays a critical role in HIV-1 RNA export.（*Proc Natl Acad Sci USA. 2000 97:8955*）

訳 核小体は，HIV-1 RNA搬出において決定的に重要な役割を果たしそうである

⁂ possible（ありうる／可能な）

❶ **possible to ～**（～することは可能な）

We show here that, although genes constitute only a small percentage of the maize genome, it is possible to identify them phenotypically as Ac receptor sites.（*Plant Cell. 2002 14:713*）

訳 それらをAc受容体部位として表現型的に同定することは可能である

❷ **possible role**（ありうる役割）

To investigate the possible role of recX in Gc, we cloned and insertionally inactivated the recX gene.（*Mol Microbiol. 2001 40:1301*）

訳 GcにおけるrecXのありうる役割を精査するために

★ feasible（可能な）

❶ **feasible to ～**（～することは可能な）

It is feasible to perform autotransplants solely with BM cells grown *ex vivo* in perfusion bioreactors from a small aliquot.（*Blood. 2000 95:2169*）

訳 BM細胞だけによる自家移植を実施することは可能である

★★ potential（潜在的な／可能な／可能性）

形容詞および名詞として用いられる．

❶ potential role（潜在的な役割）

To investigate the potential role of MK in tumor growth, we expressed MK in human SW-13 cells and studied receptor binding, signal transduction, and activity of MK.（*J Biol Chem. 2002 277:35990*）

訳 腫瘍の増殖における MK の潜在的な役割を精査するために

❷ potential target（潜在的な標的）

CAR is a potential target for the development of new drugs to treat neonatal, genetic, or acquired forms of jaundice.（*J Clin Invest. 2004 113:137*）

訳 CAR は新しい薬剤の開発のための潜在的な標的である

18. 重要な／決定的な【important】

重要な	決定的な	深刻な
important key vital	critical crucial definitive conclusive	serious

使い分け

◆ **important**, **key**, **vital** は，「重要な」という意味に用いられる．

◆ **critical**, **crucial**, **definitive**, **conclusive** は，「決定的な」という意味に使われる．

◆ **serious** は，「深刻な／重大な」という否定的な意味で用いられる．

頻度分析

	用例数		用例数
important	22,513	❶ important role	3,887
		❷ important for	3,871
		❸ important in	2,002
		❹ important implications	652
		❺ important mechanism	305
key	6,799	❶ key role	1,051
		❷ key regulator	230
		❸ key component	192

vital	641	❶ vital role	85
		❷ vital for	82
critical	11,413	❶ critical for	2,927
		❷ critical role	1,898
		❸ critical to	649
		❹ critical in	387
crucial	2,481	❶ crucial for	814
		❷ crucial role	546
		❸ crucial to	245
definitive	542	❶ definitive evidence	50
conclusive	85	❶ conclusive evidence	49
serious	783	❶ serious adverse events	57
		❷ serious complication	29

解説

- **important**, **vital**, **critical**, **crucial** は，後に for を伴う用例が多い．
- **important**, **key**, **vital**, **critical**, **crucial** は，role の前で用いられることも多い．
- **important** は，後に in を伴う用例も多い．
- **definitive**, **conclusive** は evidence の前，**serious** は events の前などで用いられる．

★★ important（重要な）

❶ important role（重要な役割）

Our observations suggest that Ang-2 may play an important role in regulating tumor angiogenesis.（*Am J Pathol. 2001 158:563*）

訳 Ang-2 は，腫瘍の血管新生を調節する際に重要な役割を果たすかもしれない

❷ important for ～（～にとって重要な）

Analysis of somatic mutations in V regions of Ig genes is important for understanding various biological processes.（*J Immunol. 2000 165:5122*）

訳 免疫グロブリン遺伝子の可変領域における体細胞変異の分析は，さまざまな生物学的過程を理解するために重要である

❸ important in ～（～の際に重要な）

The diagnosis is important in determining the need for antibiotic therapy, and the prognosis is important in determining the site of care.（*Ann Intern Med. 2003 138:109*）

訳 診断は，抗生物質治療の必要性を決定する際に重要である

形容詞

I

B

★★ key（重要な／鍵となる／鍵）

形容詞および名詞として用いられる．

❶ key role（重要な役割）

The signal transducer Stat5 plays a key role in the regulation of hematopoietic differentiation and hematopoietic stem cell function.（*Proc Natl Acad Sci USA. 2003 100:11904*）

訳 シグナルトランスデューサーStat5は，造血系の分化の調節において重要な役割を果たす

★ vital（非常に重要な／生命の）

❶ vital role（非常に重要な役割）

The EGFR plays a vital role in cell growth, and its overexpression can lead to transformation.（*Cancer Res. 2002 62:827*）

訳 EGFRは細胞増殖において非常に重要な役割を果たす

❷ vital for ～（～にとって非常に重要な）

The p53-MDM2 feedback loop is vital for cell growth control and is subjected to multiple regulations in response to various stress signals.（*Mol Cell Biol. 2004 24:7654*）

訳 p53-MDM2フィードバック・ループは，細胞増殖の調節にとって非常に重要である

★★ critical（決定的に重要な）

❶ critical for ～（～にとって決定的に重要な）

Taken together, our results show that a TFIIB conformational change is critical for the formation of activator-dependent transcription complexes.（*Nucleic Acids Res. 2004 32:1829*）

訳 TFIIBの構造的変化は，アクチベーター依存性の転写複合体の形成にとって決定的に重要である

❷ critical role（決定的に重要な役割）

Our genetic and biochemical studies demonstrate that Csk plays a critical role in mediating G protein signals to actin cytoskeletal reorganization.（*Dev Cell. 2002 2:733*）

訳 Cskは，Gタンパク質シグナルを仲介する際に決定的に重要な役割を果たす

★ crucial（決定的に重要な）

❶ crucial for ～（～にとって決定的に重要な）

Identification of Pax-6 target genes is crucial for understanding the gene regulatory network in these developmental processes.（*Gene. 2000 245:319*）

訳 Pax-6標的遺伝子の同定は，遺伝子調節ネットワークを理解するために決定的に重要である

❷ crucial role（決定的に重要な役割）

Ubiquitination plays a crucial role in regulating protein turnover.（*Mol Pharmacol. 2004 66:395*）

訳 ユビキチン化は，タンパク質の代謝回転を調節する際に決定的に重要な役割を果たす

★ definitive（決定的な）

❶ definitive evidence（決定的な証拠）

The results provide definitive evidence that iNOS is a key mediator of pulmonary pathology in sheep with ARDS resulting from combined burn and smoke inhalation injury.（*Am J Respir Crit Care Med. 2003 167:1021*）

訳 結果は，…という決定的な証拠を提供する

conclusive（決定的な／最終的な）

❶ conclusive evidence（決定的な証拠）

Our data provide conclusive evidence that FGFR1 signaling is necessary to maintain myoblast number and plays a role in myofiber organization.（*Dev Biol. 2000 218:21*）

訳 われわれのデータは，…という決定的な証拠を提供する

★ serious（深刻な／重大な）

❶ serious adverse events（深刻な有害事象）

No serious adverse events were observed.（*Radiology. 2003 228:457*）

訳 深刻な有害事象は観察されなかった

❷ serious complication（深刻な合併症）

Osteoporosis is a serious complication of kidney transplantation.（*Transplantation. 2001 72:83*）

訳 骨粗鬆症は腎臓移植の深刻な合併症である

19. 中心的な 【central】

中心的な	
central	pivotal

使い分け ◆ **central**, **pivotal** は「中心的な」という意味で，role の前などに用いられる．

頻度分析

	用例数		用例数
central	7,690	❶ central nervous	1,446
		❷ central role	856
		❸ central to	450
		❹ central region	205
pivotal	706	❶ pivotal role	414

★★ central（中心的な／主要な）

❷ central role（中心的な役割）

The binding of bacteria and platelets may play a central role in the pathogenesis of infective endocarditis.（*J Bacteriol. 2004 186:638*）

訳 細菌と血小板の結合は，感染性心内膜炎の病因において中心的な役割を果たすかもしれない

❸ central to ～（～にとって中心をなす）

Whether infection of endothelial cells is central to the pathogenesis of EBOV hemorrhagic fever（HF）remains unknown.（*Am J Pathol. 2003 163:2371*）

訳 内皮細胞の感染はEBOV出血熱の病因にとって中心をなす

★ pivotal（中心的な／きわめて重要な／中枢の）

❶ pivotal role（中心的な役割）

Since β-catenin plays a pivotal role in regulating cyclin D1 transcription, we studied whether PPARγ2-mediated inhibition of cyclin D1 transcription involved β-catenin.（*J Biol Chem. 2004 279:16927*）

訳 β-カテニンは，サイクリンD1の転写を調節する際に中心的な役割を果たす

20. 必要な 【necessary】

必要な	必須な
necessary required for	essential fundamental

使い分け

◆ **necessary**, **required** は「必要な」, **fundamental** は「不可欠な／基本的な」, **essential** は「必須な」という意味に用いられる.

頻度分析

	用例数		用例数
necessary	5,952	❶ necessary for	3,366
		❷ necessary to	948
required for	14,900	❶ required for	14,900
fundamental	1,558	❶ fundamental to	122
		❷ fundamental role	108
essential	11,770	❶ essential for	6,027
		❷ essential role	905
		❸ essential to	521

解説

◆ **necessary**, **required**, **essential** は, 後に for を伴う用例が非常に多い.

◆ **necessary**, **fundamental** は, 後に to を伴うことも多い.

⁑ necessary（必要な）

"necessary for" "necessary to" の用例が非常に多い.

❶ necessary for ～（～のために必要な）

We show that Sonic hedgehog is necessary for normal expression of both myf5 and myoD in adaxial slow muscle precursors, but not in lateral paraxial mesoderm.（*Dev Biol. 2001 236:136*）

訳 ソニック・ヘッジホックは, myf5 と myoD の両方の正常な発現のために必要である

❷ necessary to ～（～するために必要な）

These data show that clathrin function is necessary to maintain proper cellular distribution of early endosomes but does not play a prominent role in sorting and recycling events.（*Mol Biol Cell. 2001 12:2790*）

訳 クラスリン機能は適当な細胞分布を維持するために必要である

★★ required for（～のために必要とされる）

他動詞受動態で，"required for" の形で用いられることが非常に多い．

❶ is required for ～（…は，～のために必要とされる）

PAX6 is required for proper development of the eye, central nervous system, and nose.（*J Biol Chem. 2000 275:17306*）

訳 PAX6 は，目，中枢神経系および鼻の適正な発生のために必要とされる

★ fundamental（不可欠な／基本的な／必須な）

❶ fundamental to ～（～の基礎をなす／～に必要である）

The recognition of biologically distinct tumor subsets is fundamental to understanding tumorigenesis.（*Am J Pathol. 2003 163:1255*）

訳 生物学的にはっきり異なる腫瘍のサブセットの認識は腫瘍形成の理解に不可欠である

★★ essential（必須な）

"essential for" の用例が非常に多い．

❶ essential for ～（～にとって必須である）

Strikingly, the VMH is absent in newborn SF-1 knockout mice, suggesting that SF-1 is essential for the development of VMH neurons.（*J Comp Neurol. 2000 423:579*）

訳 SF-1 は VMH ニューロンの発生にとって必須である

21. 有用な 【useful】

有用な	
useful	available

使い分け ◆ **useful** は「有用な／役に立つ」，**available** は「利用できる／有用な」という意味に用いられる．

頻度分析

	用例数		用例数
useful	4,354	❶ useful for	1,088
		❷ useful in	967
		❸ available for	762
available	5,142	❶ available for	762
		❷ available at	299
		❸ available to	299

解説
◆ **useful** は，後に for または in を伴う用例が多い．
◆ **available** は，後に for を伴う用例が多い．

形容詞

★ useful（有用な／役に立つ）

"useful for" "useful in" の用例が多い．

❶ useful for ～（～のために有用な）

These results show that the yeast two-hybrid screen is useful for identifying compounds that can be exploited in mammalian cells. (*Oncogene. 2003 22:6151*)

訳 酵母ツーハイブリッドスクリーンは，…である化合物を同定する際に有用である

❷ useful in ～（～の際に有用な）

Thus, R-cadherin antagonists may be useful in the treatment of neovascular diseases in which circulating HSCs contribute to abnormal angiogenesis. (*Blood. 2004 103:3420*)

訳 R-カドヘリン拮抗剤は，新生血管疾患の治療の際に有用であるかもしれない

I

B

★★ available（利用できる／有用な）

"available for" の用例が多い．

❶ available for ～（～のために利用できる／～のために有用な）

After exclusions for insufficient data, 1,244 controls and 189 pancreatic cancer cases were available for analysis. (*Am J Epidemiol. 2004 159:693*)

訳 1,244 のコントロールと 189 の膵臓癌の症例が分析に利用できた

22. 効果的な／活発な 【effective】

効果的な	効率的な	活発な
effective efficacious	efficient	active

使い分け ◆ **effective**, **efficacious** は「効果的な」, **efficient** は「効率的な」, **active** は「活性のある／活発な」という意味に用いられる.

頻度分析

	用例数		用例数
effective	6,873	❶ effective in	1,036
		❷ effective than	319
		❸ effective at	209
		❹ effective for	209
efficacious	355	❶ efficacious in	71
efficient	4,390	❶ efficient in	175
		❷ efficient than	154
		❸ efficient at	104
active	17,070	❶ active site	4,444
		❷ active in	736
		❸ active form	417

解説 ◆いずれの語も，後に in を伴う用例が多い.

★★ effective（効果的な／効率的な）

“effective in” の用例が多い.

❶ effective in ～（～において効果的な）

In this population, hepatitis A vaccine was highly effective in preventing disease among recipients.（*JAMA. 2001 286:2968*）

訳 A 型肝炎ワクチンは疾患を予防する際に非常に効果的であった

efficacious（有効な）

❶ efficacious in ～（～において有効な）

Previous studies show that macrolide antibiotics may be efficacious in the treatment of panbronchiolitis and cystic fibrosis.（*Am J Respir Crit Care Med. 2003 168:121*）

訳 マクロライド抗生物質は，汎細気管支炎と嚢胞性線維症の治療において有効であるかもしれない

★ efficient（効率的な）

❶ efficient in ～（～において効率的な）

Cdc25A（T507A）was more efficient in binding to cyclin B1, activating cyclin B1–Cdk1, and promoting premature entry into mitosis.（*Mol Cell Biol. 2003 23:7488*）

訳 Cdc25A（T507A）はサイクリンB1への結合の際により効率的であった

★★ active（活性のある／活発な）

❶ active site（活性部位）

The 3-amidopyrrolidin-4-one inhibitors were bound in the active site of the enzyme in two alternate directions.（*J Med Chem. 2001 44:725*）

訳 3-amidopyrrolidin-4-one 阻害剤は酵素の活性部位に結合した

❷ active in ～（～において活性のある）

Insulin is a peptide growth factor that is active in most tissues, both during development and in adulthood.（*Invest Ophthalmol Vis Sci. 2001 42:2125*）

訳 インスリンはほとんどの組織において活性のあるペプチド成長因子である

23. 調節性の／誘導性の 【regulatory】

調節性の	誘導性の
regulatory modulatory	inducible

使い分け ◆ **regulatory**, **modulatory** は「調節性の」, **inducible** は「誘導性の」という意味に用いられる．

頻度分析

	用例数		用例数
regulatory	9,569	❶ regulatory elements	582
		❷ regulatory proteins	480
		❸ regulatory role	377

modulatory	333	❶ modulatory effects	52
inducible	3,546	❶ inducible expression	186
		❷ inducible genes	163
		❸ inducible by	99

⁑ regulatory（調節性の）

❶ regulatory elements（調節性のエレメント）

Phylogenetic footprinting is a method for the discovery of regulatory elements in a set of homologous regulatory regions, usually collected from multiple species.（*Nucleic Acids Res. 2003 31:3840*）

訳 系統発生的フットプリンティングは，一組の相同的な調節性領域にある調節性エレメントの発見のための方法である

modulatory（調節性の）

❶ modulatory effects（調節性の効果）

Taken together, these results show that the striatum of male mice is very sensitive to the modulatory effects of TMX upon MA-evoked DA output.（*Brain Res. 2004 1029:186*）

訳 オスのマウスの線条体は，…の調節性の効果にとても敏感である

＊ inducible（誘導性の）

❶ inducible expression（誘導性の発現）

Flow cytometric analyses indicated that the inducible expression of KLF4 caused a block in the G_1/S phase of the cell cycle.（*J Biol Chem. 2001 276:30423*）

訳 KLF4の誘導性の発現は，細胞周期のG_1/S期の遮断を引き起こした

24. 増加する【increasing】

増加する	
increasing	growing

使い分け ◆どちらも「増加する」という意味に用いられる．

頻度分析

	用例数		用例数
increasing	5,476	❶ increasing evidence	162
		❷ increasing concentrations	135
		❸ increasing number	122
growing	1,656	❶ growing cells	100
		❷ growing body	96
		❸ growing evidence	93

★★ increasing（増えている／増加する）

❶ increasing evidence（増えている証拠）

There is increasing evidence that chromatin components play an important part in plant epigenetics.（*Plant Mol Biol. 2000 43:221*）

訳 …という証拠が増えている

❷ increasing concentrations（増加する濃度）

GH3MOR cells were pretreated with increasing concentrations of β-funaltrexamine followed by functional testing after removal of unbound drug.（*J Biol Chem. 2001 276:37779*）

訳 GH3MOR細胞は，増加する濃度のβ-フナルトレキサミンによって前処理された

★ growing（増えている／増殖する）

❶ growing cells（増殖する細胞）

The population-doubling time of exponentially growing cells was 24 hours.（*Invest Ophthalmol Vis Sci. 2000 41:3898*）

訳 指数関数的に増殖する細胞の細胞集団倍加時間は24時間であった

25. 独特な／特有の 【unique】

独特な	特有の
unique	characteristic

使い分け

◆ **unique**は「独特な」，**characteristic**は「特有の／特徴的な」という意味に用いられる．

頻度分析

	用例数		用例数
unique	6,422	❶ unique to	556
		❷ unique in	165
		❸ unique among	134
characteristic	3,040	❶ characteristic of	1,442

解説 ◆ **unique** は後に to を，**characteristic** は後に of を伴う用例が多い．

★★ unique（独特な）

❶ **unique to ～**（～に独特な）

In addition to the shared residues, there are also residues that are unique to each target recognition.（*J Biol Chem. 2003 278:29901*）

訳 おのおのの標的の認識に独特である残基もある

★ characteristic（特有の／特徴的な／特徴）

形容詞として使われることが多いが，名詞の用例もある．

❶ **characteristic of ～**（～に特有な／～の特徴）

Thus, SurA recognizes a peptide motif that is characteristic of integral outer membrane proteins.（*J Biol Chem. 2003 278:49316*）

訳 SurA は，複合的な外膜タンパク質に特有であるペプチドモチーフを認識する

26. 異常な 【abnormal】

異常な		
abnormal	**aberrant**	**unusual**

使い分け ◆いずれの語も「異常な」という意味に用いられる．

頻度分析

	用例数		用例数
abnormal	2,793	❶ abnormal in	97

aberrant	1,515	❶ aberrant expression	94
		❷ aberrant methylation	94
unusual	1,696	❶ unusual in	90

★ abnormal（異常な）

❶ abnormal in ～（～において異常な）

Pulmonary function is frequently abnormal in patients with congestive heart failure (CHF), the mechanism of which has not been completely characterized.（*Circulation. 2002 106:1794*）

訳 肺機能はうっ血性心不全の患者においてしばしば異常である

★ aberrant（異常な）

❶ aberrant expression（異常な発現）

The accumulated evidence suggests that the aberrant expression of BCSG1 in breast carcinomas is caused by transcriptional activation of the BCSG1 gene.（*J Biol Chem. 2002 277:31364*）

訳 乳癌における BCSG1 の異常な発現は，BCSG1 遺伝子の転写の活性化によって引き起こされる

★ unusual（普通でない／異常な）

"unusual in that（～ということにおいて異常である）" の用例が多い．

❶ unusual in ～（～において異常な）

PADOX-1 is unusual in that it has a high affinity, inhibitory cyanide-binding site distinct from the distal heme face and the fatty acid site.（*J Biol Chem. 2004 279:29805*）

訳 PADOX-1 は，それが高親和性で抑制性のシアン化物結合部位を持つということにおいて異常である

27. 急速な 【rapid】

急速な		
rapid	fast	prompt

形容詞

I

B

使い分け ◆いずれの語も「急速な／迅速な」という意味に用いられる．

頻度分析

	用例数		用例数
rapid	7,706	❶ rapid increase	174
		❷ rapid degradation	128
		❸ rapid amplification	107
fast	2,445	❶ fast inactivation	62
prompt	143	❶ prompt treatment	10

★★ rapid（急速な）

❶ rapid increase（急速な上昇）

Oral dosing with DMP 777 caused a rapid increase in serum gastrin levels and severe hypochlorhydria.（*Gastroenterology. 2000 118:1080*）

訳 DMP 777の経口投与は，血清ガストリンレベルの急速な上昇を引き起こした

★ fast（急速な／速い）

❶ fast inactivation（急速な不活性化）

Apart from many similarities to HERG, however, the molecular mechanism of fast inactivation appears to be different.（*J Neurosci. 2000 20:511*）

訳 急速な不活性化の分子機構は異なるようである

prompt（迅速な）

❶ prompt treatment（迅速な処置）

Prompt treatment with streptomycin, gentamicin, doxycycline, or ciprofloxacin is recommended.（*JAMA. 2001 285:2763*）

訳 ストレプトマイシン，ゲンタマイシン，ドキシサイクリン，あるいはシプロフロキサシンによる迅速な処置が勧められる

28. 同時の 【simultaneous】

同時の	同期の／同時の
simultaneous concomitant concurrent coincident	synchronous

使い分け

◆ **simultaneous**, **concomitant**, **concurrent**, **coincident** は，「同時の」という意味に用いられる．

◆ **synchronous** は，「同期の／同時の」という意味で使われる．

頻度分析

	用例数		用例数
simultaneous	1,460	❶ simultaneous measurement	36
		❷ simultaneous activation	35
		❸ simultaneous detection	35
concomitant	1,420	❶ concomitant with	331
		❷ concomitant increase	121
concurrent	686	❶ concurrent with	108
coincident	520	❶ coincident with	401
synchronous	270	❶ synchronous with	20

解説

◆**concomitant**, **concurrent**, **coincident**, **synchronous** は，後に with を伴う用例が多い．

★ simultaneous（同時の／同時に起こる）

❷ simultaneous activation（同時の活性化）

The observed increment involves the simultaneous activation of annexin 7 by these two effectors.（*J Biol Chem. 2002 277:25217*）

訳 観察された増加はアネキシン7の同時の活性化を伴う

★ concomitant（同時の／随伴性の）

"concomitant with" の用例が多い．

❶ concomitant with ～（～と同時である）

The inhibition of apoptosis by COX-2 was concomitant with prevention of caspase 3 activation.（*Mol Cell Biol. 2000 20:8571*）

訳 COX-2によるアポトーシスの抑制はカスパーゼ3活性化の阻止と同時であった

★ concurrent（同時の／同時発生的な）

"concurrent with"の用例が多い．

❶ **concurrent with ～（～と同時である）**

Most interestingly, MKP-1 expression was potently induced by peptidoglycan, and this induction was concurrent with MAP kinase dephosphorylation.（*J Biol Chem. 2004 279:54023*）

訳 この誘導はMAPキナーゼ脱リン酸化と同時であった

★ coincident（同時の／一致する）

"coincident with"の用例が非常に多い．

❶ **coincident with ～（～と同時である）**

Phosphorylation of NSF is coincident with neurotransmitter release and requires an influx of external calcium.（*J Biol Chem. 2001 276:12174*）

訳 NSFのリン酸化は神経伝達物質の放出と同時である

synchronous（同期の／同時の）

❶ **synchronous with ～（～と同期した／～と同時である）**

These oscillations were synchronous with the firing of neurons, insensitive to transmitter receptor antagonists and disrupted by carbenoxolone, a gap junction blocker.（*J Neurosci. 2000 20:4091*）

訳 これらの周期的変動はニューロンの発火と同期した

29. 高齢の／～歳の 【elderly】

高齢の／～歳の	
elderly	old

使い分け

◆「高齢の」という意味には **elderly** が用いられることが多く，**old** はあまり使われない．

◆ **old** は，年齢を表すために用いられることが多い．

頻度分析

	用例数		用例数
elderly	805	❶ elderly patients	141
old	2,758	❶ -year-old	391

★ elderly（高齢の）

❶ elderly patients（高齢の患者）

NMDA receptor subunits are abnormally expressed in elderly patients with schizophrenia.（*Am J Psychiatry. 2001 158:1400*）

訳 NMDA受容体サブユニットは，統合失調症の高齢の患者において異常に発現する

★ old（～歳の／古い）

❶ -year-old（～歳の）

A 50-year-old woman underwent single lung transplantation for advanced chronic obstructive pulmonary disease.（*Transplantation. 2001 71:1859*）

訳 50歳の女性が片肺移植を受けた

【Ⅰ-C. 関係】 関係を表す形容詞

30. 責任ある／原因である 【responsible】

～に責任のある	原因である／原因となる
responsible for due to	causative causal

使い分け
◆ responsible, due は，それぞれ **responsible for**（～に責任がある／～の原因である），**due to**（～のせいである／～の原因である）の用例が圧倒的に多い．
◆ **causative** は「原因である」，**causal** は「原因となる」という意味で使われる．

頻度分析

	用例数		用例数
responsible for	6,082	❶ is responsible for	1,241
		❷ mechanisms responsible for	313
due to	8,564	❶ is due to	1,018
causative	484	❶ causative agent of	172
causal	458	❶ causal role	83

解説
◆ **due to** は，be 動詞の後や文頭などで使われる．

★★ responsible for（～に責任がある／～の原因である）

❶ **is responsible for ～**（…は，～に責任がある／…は，～の原因である）

Human serum albumin（HSA）is an abundant plasma protein that is responsible for the transport of fatty acids.（*J Mol Biol. 2000 303:721*）

訳 ヒト血清アルブミン（HSA）は，脂肪酸の輸送に責任がある豊富な血漿タンパク質である

❷ **mechanisms responsible for ～**（～の原因である機構／～に責任がある機構）

Wnts stimulate cell migration, although the mechanisms responsible for this effect are not fully understood.（*J Biol Chem. 2005 280:777*）

訳 この効果の原因である機構は完全には理解されていない

⁑ due to（～のせいである）

❶ **is due to ～**（…は，～のせいである）

We show here that this difference is due to the presence of a Mot3 binding site in OpA.（*Mol Cell Biol. 2000 20:7088*）

訳 この違いはOpAにおけるMot3結合部位の存在のせいである

causative（原因である）

❶ **causative agent of ～**（～の原因である媒介物）

Mycobacterium bovis is the causative agent of bovine tuberculosis (TB), and it has the potential to induce disease in humans.（*Infect Immun. 2003 71:4297*）

訳 *Mycobacterium bovis* はウシ結核の原因病原体である

causal（原因となる／原因である）

❶ **causal role**（原因となる役割）

Defective cardiac muscle relaxation plays a causal role in heart failure.（*J Clin Invest. 2001 107:191*）

訳 欠陥のある心筋弛緩は心不全において原因となる役割を果たす

31. 関連する 【related to】

関連する	
related to	relevant

使い分け ◆ **related to**, **relevant** to は，「～に関連する」という意味に用いられる．

頻度分析

	用例数		用例数
related to	4,901	❶ is related to	511
relevant	2,775	❶ relevant to	606
		❷ relevant for	119

形容詞 I C

★ related to（〜に関連する）

❶ is related to 〜（…は，〜に関連する）

We tested the hypothesis that response to prednisolone is related to the presence of eosinophilic airway inflammation.（*Lancet. 2000 356:1480*）

訳 プレドニゾロンへの反応は好酸性気道感染の存在と関連する

★ relevant（関連する）

"relevant to" の用例が非常に多い.

❶ relevant to 〜（〜に関連する）

Our results may be relevant to the mechanisms by which ducts proliferate in response to hepatic injury and to the hypercholeresis that occurs after experimentally induced bile duct proliferation.（*Am J Pathol. 2001 158:2079*）

訳 われわれの結果は，それによって…である機構に関連するかもしれない

32. 類似する／一致する 【similar】

似ている	同じ／相同の	一致する
similar analogous	same equal identical homologous	fit consistent

使い分け

◆ **similar** は「似ている」，**analogous** は「類似の」，**same** は「同じ」，**equal** は「等しい」，**identical** は「同一の」，**homologous** は「相同的な」，**fit** は「合う／一致する」，**consistent** は「一致する」という意味に用いられる.

頻度分析

	用例数		用例数
similar	21,777	❶ similar to	8,315
		❷ similar in	1,273
		❸ similar results	493
		❹ similar for	330
analogous	1,103	❶ analogous to	554

same	11,316	❶ at the same time	341
		❷ the same as	237
equal	1,272	❶ equal to	468
identical	4,107	❶ identical to	1,289
		❷ nearly identical	309
		❸ identical in	226
homologous	4,114	❶ homologous to	999
		❷ homologous recombination	852
fit	938	❶ fit to	130
consistent	11,964	❶ consistent with	10,971

解説

◆ **similar**, **analogous**, **equal**, **identical**, **homologous**, **fit** は，後に to を伴う用例が非常に多い．

◆ **same** は，time, as の前などで使われる．

◆ **consistent** は，後に with を伴う用例が多い．

★★ similar（似ている）

“similar to” の用例が非常に多い．

❶ **similar to ～**（～に似ている）

The mouse asporin gene structure is similar to that of biglycan and decorin with 8 exons.（*J Biol Chem. 2001 276:12212*）

訳 マウスのアスポリン遺伝子の構造はバイグリカンのそれに似ている

★ analogous（類似の）

“analogous to” の用例が非常に多い．

❶ **analogous to ～**（～に類似している）

The overall structure of the LH3 complex is analogous to that of the LH2 complex from *Rps. acidophila* strain 10050.（*Biochemistry. 2001 40:8783*）

訳 LH3 複合体の全体の構造は LH2 複合体のそれに類似している

★★ same（同じ）

❶ **at the same time**（同時に）

At the same time, myelin basic protein was apparent in the same restricted regions.（*J Comp Neurol. 2002 451:334*）

訳 同時に，ミエリン塩基性タンパク質が同じ限定された領域に明らかになった

❷ the same as ～（～と同じ）

The ELD1 gene sequence is the same as that of the KOBITO1 sequence.（*Plant Mol Biol. 2003 53:581*）

訳 ELD1 遺伝子配列は，KOBITO1 の配列のそれと同じである

★ equal（等しい／匹敵する）

“equal to” の用例が多い．

❶ equal to ～（～と等しい／～と同じ）

Fluconazole concentrations in the esophagus were greater than or equal to those in plasma.（*J Clin Microbiol. 2000 38:2369*）

訳 食道におけるフルコナゾール濃度は，血漿におけるそれらより大きいか同じであった

★ identical（同一の／同じ）

“identical to” の用例が多い．

❶ identical to ～（～と同一の）

This cDNA sequence is identical to EHD4, a recently described member of the EH domain family of proteins.（*J Biol Chem. 2001 276:43103*）

訳 この cDNA 配列は EHD4 と同一である

★ homologous（相同的な）

“homologous to” の用例が多い．

❶ homologous to ～（～と相同的な）

Schizosaccharomyces pombe Ddb1 is homologous to the mammalian DDB1 protein, which has been implicated in damaged-DNA recognition and global genomic repair.（*J Biol Chem. 2003 278:37006*）

訳 分裂酵母の Ddb1 は哺乳類の DDB1 タンパク質と相同的である

★ fit（合う／一致する／一致させる）

形容詞および他動詞として用いられる．

❶ fit to ～（～に一致する）

Therefore, the SPR kinetic data were fit to a conformational change model and kinetic rate constants determined.（*Biochemistry. 2004 43:9725*）

訳 SPR 動力学的データは高次構造の変化モデルに一致した

⁑ consistent（一致する）

"consistent with" の用例が圧倒的に多い．

❶ consistent with ～（～に一致する）

These results are consistent with the hypothesis that CaMKIIα accumulation at synapses is a memory trace of past synaptic activity.（*J Neurosci. 2004 24:9324*）

訳 これらの結果は，…という仮説に一致する

形容詞

33. 異なる／別々の／独立した 【different】

異なる	別個の／別々の	独立した
different distinct	separate discrete	independent

使い分け ◆ **different** は「異なる」，**distinct** は「はっきりと異なる」，**separate** は「別個の」，**discrete** は「別個の／別々の」，**independent** は「独立した」という意味に用いられる．

I

頻度分析

	用例数		用例数
different	23,456	❶ different from	1,348
		❷ different types	402
		❸ different cell	359
		❹ different between	309
distinct	11,169	❶ distinct from	1,640
		❷ distinct mechanisms	229
		❸ distinct roles	180
separate	2,340	❶ separate from	108
discrete	1,540	❶ discrete regions	54
independent	12,912	❶ independent of	3,936
		❷ is independent	1,028
		❸ independent manner	382

C

解説 ◆ **different**，**distinct**，**separate** は後に from を，**independent** は後に of を伴う用例が多い．

★★ different（異なる）

❶ different from ～（～と異なる）

The pH dependence of the mutant enzyme is not significantly different from that of the wild-type enzyme; this is most consistent with a requirement that the side chain of Tyr254 be uncharged for catalysis.（*Biochemistry. 2003 42:15208*）

訳 変異酵素のpH依存性は野生型酵素のそれと有意には異ならない

★★ distinct（はっきりと異なる）

❶ distinct from ～（～とはっきりと異なる）

Thus, the PGC-1α・ERRα interaction is distinct from that of other nuclear receptor PGC-1α partners, including PPARα, hepatocyte nuclear factor-4α, and estrogen receptor α.（*J Biol Chem. 2002 277:40265*）

訳 PGC-1α・ERRαの相互作用は，他の核内受容体PGC-1αのパートナーのそれとはっきりと異なる

★ separate（別個の／離れた／分ける）

形容詞および動詞として用いられる．

❶ separate from ～（～とは別個の／～から離れた）

These phenotypes indicate a disruption of striatal cell homeostasis by the mutant protein, via a mechanism that is separate from its normal activity.（*Hum Mol Genet. 2000 9:2799*）

訳 正常なそれの活性とは別個の機構によって

★ discrete（別々の）

❶ discrete regions（別々の領域）

Transcripts encoding Ucn II are expressed in discrete regions of the rodent central nervous system, including stress-related cell groups in the hypothalamus（paraventricular and arcuate nuclei）and brainstem（locus coeruleus）.（*Proc Natl Acad Sci USA. 2001 98:2843*）

訳 Ucn IIをコードする転写物は，げっ歯類の中枢神経系の別々の領域で発現する

★★ independent（独立した／非依存性の／無関係の）

"independent of"の用例が非常に多い．

❶ independent of ～（～とは無関係の／～とは独立した）

These data suggest that in myoblasts, SHP-2 represses myogenesis via a pathway that is independent of the Erks.（*Mol Cell Biol. 2002 22:3875*）

訳 SHP-2は，Erkとは無関係である経路によって筋形成を抑制する

34. 逆の 【opposite】

逆の	
opposite	inverse

使い分け ◆どちらも「逆の」という意味に用いられる．

頻度分析

	用例数		用例数
opposite	1,936	❶ opposite effects	139
		❷ opposite to	119
		❸ in opposite	92
inverse	826	❶ inverse correlation	126
		❷ inverse relationship	111
		❸ inverse association	92

★ opposite（逆の）

❶ opposite effects（逆の効果）

In this study, we show that the two sterol regulatory binding proteins（SREBPs）have opposite effects on the 12α-hydroxylase promoter.（*J Biol Chem. 2002 277:6750*）

訳 2つのsterol regulatory binding proteins（SREBPs）は，12α-水酸化酵素プロモータに対して逆の効果を持つ

★ inverse（逆の）

❷ inverse relationship（逆の関係）

There is an inverse relationship between serum bilirubin concentrations and risk of coronary artery disease.（*Am J Hum Genet. 2003 72:1029*）

訳 血清ビリルビン濃度と冠動脈疾患のリスクの間には逆の関係がある

【I-D. 頻度・時】 頻度・時に関係する形容詞

35. 連続の 【continuous】

連続の	
continuous	serial

使い分け
◆どちらも「連続の」という意味に用いられる．
◆**continuous** は切れ目のない連続性を意味し，**serial** は順番などの連続性を意味する．

頻度分析

	用例数		用例数
continuous	1,968	❶ continuous infusion	80
		❷ continuous flow	63
serial	910	❶ serial analysis	109
		❷ serial sections	59

★ continuous（連続の）

❶ continuous infusion（連続注入）

At 24 hrs after treatment, both groups of pigs were subjected to a 24-hr continuous infusion of lipopolysaccharide (LPS) at a rate of 80 ng/kg/min.（*Crit Care Med. 2000 28:2015*）

訳 両群のブタが，リポポリサッカライドの24時間連続注入に供された

★ serial（連続の）

❷ serial sections（連続切片）

Serial sections of 40 small (<3mm) sporadic colorectal adenomas were stained with H&E, MIB-1, and for β -catenin.（*Cancer Res. 2003 63:3819*）

訳 40個の小さな（＜ 3 mm）散発性結腸直腸腺腫の連続切片がH&Eで染色された

36. 次の／引き続いた 【following】

次の	引き続いた
following next	subsequent

使い分け ◆ **following**, **next** は「次の」, **subsequent** は「引き続いた」という意味に用いられる.

頻度分析

	用例数		用例数
following	9,375	❶ the following	1,262
next	1,135	❶ the next	575
		❷ next to	133
		❸ we next	130
subsequent	5,393	❶ and subsequent	1,436
		❷ subsequent to	212
		❸ subsequent activation	102

★★ following（次の）

動詞の進行形／動名詞だが，形容詞として用いられることも多い.

❶ the following ～（次の～）

The stability of these proteins was found to decrease in the following order: WT rat OM > rat OM I32L > rat OM A18S/L47R > rat OM A18S/L47R/I32L > bovine Mc cyt b5.（*Biochemistry. 2001 40:9469*）

訳 これらのタンパク質の安定性は次の順番に低下することが見つけられた

★ next（次の）

形容詞および副詞として用いられる.

❶ the next ～（次の～）

Consequently, by helping a relative reproduce, an individual passes its genes to the next generation, increasing their Darwinian fitness.（*Science. 2002 296:72*）

訳 個々はその遺伝子を次世代に伝える

形容詞

I

D

⁑ subsequent（引き続いた）

❶ and subsequent ～（…と引き続いた～）

Binding of Cdc42 to PLD1 and subsequent activation are GTP-dependent.（*J Biol Chem. 2000 275:15665*）

訳 Cdc42 の PLD1 への結合と引き続いた活性化は GTP 依存性である

37. まれな 【rare】

まれな		
rare	uncommon	infrequent

使い分け ◆いずれの語も「まれな」という意味に用いられ，後に in を伴うことが多い．

頻度分析

	用例数		用例数
rare	1,561	❶ rare in	115
uncommon	206	❶ uncommon in	32
infrequent	179	❶ infrequent in	20

⋆ rare（まれな）

❶ rare in ～（～においてまれな）

Our results showed that somatic BHD mutations are rare in sporadic renal tumors.（*Cancer Res. 2003 63:4583*）

訳 体細胞の BHD 変異は散発性の腎臓腫瘍においてまれである

uncommon（まれな）

❶ uncommon in ～（～においてまれな）

Atrial fibrillation（AF）is uncommon in children, and its mechanisms are unknown.（*Circulation. 2004 110:117*）

訳 心房細動（AF）は子供においてまれである

infrequent（まれな）

❶ **infrequent in ～**（～においてまれな）

Although p53 abnormalities are infrequent in SLVL, they underlie a more aggressive disease course and poor prognosis.（*Blood. 2001 97:3552*）

訳 p53の異常はSLVLにおいてまれである

38. 新しい【new】

新しい	最初の	最新の
new novel	initial	latest

使い分け ◆ **new**は「新しい」，**novel**は「新規の／新しい」，**initial**は「最初の」，**latest**は「最新の／最近の」という意味に用いられる．

頻度分析

	用例数		用例数
new	15,691	❶ new insights	432
		❷ new class	354
		❸ new method	309
		❹ new approach	291
novel	15,471	❶ novel mechanism	818
		❷ novel role	336
		❸ novel approach	304
initial	5,865	❶ initial step	141
		❷ initial rate	126
		❸ initial stages	101
latest	105	❶ the latest	82

★★ new（新しい）

❶ **new insights**（新しい洞察）

The results provide new insights into the roles of the FtsL and DivIB proteins in bacterial cell division.（*Mol Microbiol. 2000 36:278*）

訳 結果は，…の役割への新しい洞察を提供する

★★ novel（新規の）

❶ novel mechanism（新規の機構）

These results suggest a novel mechanism for the regulation of matrix protein degradation by GAGs.（*J Biol Chem. 2004 279:5470*）

訳 これらの結果は，…の調節の新規の機構を示唆する

★★ initial（最初の／初期の）

❶ initial step（最初のステップ）

The CYP7A1 gene encodes the enzyme cholesterol 7α-hydroxylase, which catalyzes the initial step in cholesterol catabolism and bile acid synthesis.（*J Clin Invest. 2002 110:109*）

訳 そして，それはコレステロール異化反応と胆汁酸合成における最初のステップを触媒する

latest（最新の／最近の）

lateの最上級．

❶ the latest ～（最新の～）

Remaining areas of uncertainty and research priorities are discussed in view of the latest findings.（*Science. 1990 247:166*）

訳 研究の優先度が最新の知見を考慮して議論される

39. 以前の／最近の 【previous】

以前の	最近の
previous	recent

使い分け

◆ **previous**は「以前の」，**recent**は「最近の」という意味で，studiesの前などに用いられる．

頻度分析

	用例数		用例数
previous	8,195	❶ previous studies	2,782
		❷ previous work	750
		❸ previous reports	392

recent	6,833	❶ recent studies	1,448
		❷ recent evidence	379
		❸ recent work	280

☆ previous（以前の）

❶ **previous studies**（以前の研究／先行研究）

Our previous studies have shown that PGP9.5 was highly expressed in primary lung cancers and lung cancer cell lines.（*Oncogene. 2002 21:3003*）

訳 われわれの以前の研究は，…ということを示した

☆ recent（最近の）

❶ **recent studies**（最近の研究）

Recent studies have demonstrated that CD47 plays an important role in regulating human neutrophil（PMN）chemotaxis.（*J Biol Chem. 2002 277:10028*）

訳 最近の研究は，…ということを実証した

40. 現在の 【present】

現在の	
present	**current**

使い分け

◆ **present**, **current** は「現在の」という意味で，study の前などに用いられる．

◆ **present** は，「存在する」という意味で用いられることも多い．

頻度分析

	用例数		用例数
present	20,359	❶ present in	6,157
		❷ present study	4,210
current	8,307	❶ current study	781
		❷ current in	246
		❸ current models	150

形容詞

I

D

⁑ present（現在の／存在する）

❷ present study（現在の研究）

In the present study, we examined the involvement of an intracortical mechanism in this functional modification.（*Proc Natl Acad Sci USA. 2004 101:5081*）

訳 現在の研究において，われわれはこの機能的修飾における皮質内機構の関与について調べた

⁑ current（現在の）

❶ current study（現在の研究）

In the current study, we used the yeast one-hybrid strategy to identify nuclear factors that bind to these three elements.（*J Biol Chem. 2004 279:8684*）

訳 現在の研究において，われわれは…する核因子を同定するために酵母ワンハイブリッド戦略を使った

第4章

副詞編

副詞は，主に動詞や形容詞を修飾する．また，副詞，前置詞句，節，文全体を修飾する場合もある．形容詞，副詞，前置詞句，節，文を修飾するときは，通常，その直前に置かれる．動詞を修飾するときもその直前に置かれる場合が多いが，直後に置かれることもある．目的語や補語の後で，文末付近に置かれる場合もある．用法は，それぞれの副詞ごとに決まったパターンがあるので，修飾する語の組み合わせも含めて個々の使い方をよく調べる必要がある．

【I-A. 程度】 程度を表す副詞

1. 著しく／非常に 【markedly】

有意に	著しく	劇的に	非常に	高度に
significantly	markedly remarkably strikingly	dramatically drastically	very extremely unusually	highly

使い分け

◆ **significantly** は，「（統計的に）有意に」という意味で用いられることが多いが，「著しく」という意味で使われることもある．

◆ **markedly**, **remarkably**, **strikingly** は「著しく」, **dramatically** は「劇的に」, **drastically** は「激烈に／顕著に」, **very** は「非常に」, **extremely** は「極度に／非常に」, **unusually** は「異常に／非常に」, **highly** は「高度に／非常に」という意味で用いられる．

頻度分析

	用例数		用例数
significantly	24,601	❶ significantly reduced	2,066
		❷ significantly higher	1,973
		❸ significantly increased	1,700
markedly	3,770	❶ markedly reduced	585
		❷ markedly increased	388
		❸ markedly decreased	194
remarkably	1,285	❶ remarkably similar	142
		❷ remarkably high	34
strikingly	859	❶ strikingly similar	105
		❷ strikingly different	67
dramatically	2,602	❶ dramatically reduced	281
		❷ dramatically increased	210
		❸ dramatically decreased	94
drastically	280	❶ drastically reduced	84
very	7,004	❶ very low	928
		❷ very similar	732
		❸ very high	514
extremely	1,183	❶ extremely low	144
		❷ extremely high	117
		❸ extremely sensitive	79

unusually	487	❶ unusually high	115
highly	12,763	❶ highly conserved	2,074
		❷ highly expressed	824
		❸ highly sensitive	368

解説

◆ **significantly**, **markedly**, **dramatically**, **drastically** は，reduced の前などで使われる．

◆ **remarkably**, **strikingly** は，similar の前などで用いられる．

◆ **very**, **extremely** は low の前，**unusually** は high の前，**highly** は conserved の前などで用いられる．

★★ significantly〔(統計的に) 有意に／著しく〕

用例数がきわめて多い．

❶ significantly reduced (有意に低下した)

Cytotoxic T-lymphocyte production was significantly reduced in C57BL/6 mice after a single intratracheal administration of modified vectors, and length of gene expression was extended from 4 to 42 days. (*J Virol. 2001 75:4792*)

訳 細胞傷害性Tリンパ球の産生はC57BL/6 マウスにおいて有意に低下した

❸ significantly increased (有意に上昇した)

Syndecan-4/PIP_2-dependent PKCα activity was significantly increased in PKCδ DN cells, while PKCδ overexpression was accompanied by decreased PKCα activity. (*J Biol Chem. 2002 277:20367*)

訳 シンデカン-4/PIP_2依存性のPKCα活性はPKCδ DN細胞において有意に上昇した

★★ markedly (顕著に／著しく)

❶ markedly reduced (顕著に低下した)

PPARγ expression was markedly reduced in lung and thoracic aorta after sepsis. (*J Immunol. 2003 171:6827*)

訳 PPARγ発現は肺において顕著に低下した

❷ markedly increased (顕著に上昇した)

Mortality was markedly increased in patients with versus without baseline RI both at 30 days (7.5% versus 0.8%, P<0.0001) and at 1 year (12.7% versus 2.4%, P<0.0001). (*Circulation. 2003 108:2769*)

訳 死亡率は，…の患者において顕著に上昇した

副詞

I

A

★ remarkably（著しく）

❶ **remarkably similar（著しく似た）**

We now show that DCP-1 has a substrate specificity that is remarkably similar to those of human caspase 3 and *Caenorhabditis elegans* CED-3, suggesting that DCP-1 is a death effector caspase.（*Mol Cell Biol. 2000 20:2907*）

訳 DCP-1は，…のそれらに著しく似た基質特異性を持つ

★ strikingly（著しく）

❶ **strikingly similar（著しく似た）**

These defects are strikingly similar to the phenotypes of IGF-1 receptor-deficient mice and suggest that Akt may serve as the most critical downstream effector of the IGF-1 receptor during development.（*Genes Dev. 2003 17:1352*）

訳 これらの欠損は，IGF-1受容体欠損マウスの表現型に著しく似ている

★ dramatically（劇的に）

❶ **dramatically reduced（劇的に減少した）**

Expression of GAL genes is dramatically reduced in gal4D strains and these strains are unable to grow on galactose as the sole carbon source.（*J Biol Chem. 2001 276:9825*）

訳 GAL遺伝子の発現はgal4D株において劇的に減少する

❷ **dramatically increased（劇的に上昇した）**

Epidermal growth factor receptor（EGFR）levels are dramatically increased in human keratinocytes（HKc）immortalized with full-length human papillomavirus type 16（HPV16）DNA（HKc/HPV16）, but increases in EGFR levels actually precede immortalization.（*Cancer Res. 2001 61:3837*）

訳 上皮細胞成長因子受容体（EGFR）レベルは，ヒトのケラチノサイトにおいて劇的に上昇する

drastically（激烈に／顕著に）

❶ **drastically reduced（激烈に低下した）**

Here we demonstrate that the expression level of six of the seven GA biosynthesis genes is drastically reduced in mutants lacking areA.（*Mol Microbiol. 2003 47:975*）

訳 7個のGA生合成遺伝子のうちの6個の発現レベルは，areAを欠損している変異体において激烈に低下する

★★ very（非常に）

❶ **very low（非常に低い）**

MDSP is expressed at very low levels in myotubes and early postnatal muscle.（*J Biol Chem. 2004 279:41404*）

訳 MDSP は筋管において非常に低いレベルで発現する

★ extremely（非常に／極度に）

❶ **extremely low（極度に低い）**

Western blot analysis revealed extremely low levels of CAR in the cytosols of female livers compared with male counterparts.（*Mol Pharmacol. 2001 59:278*）

訳 ウエスタンブロット分析は極度に低いレベルの CAR を明らかにした

❷ **extremely high（非常に高い）**

Adaptive inhibitors bind with extremely high affinity to a primary target within the family and maintain significant affinity against the remaining members.（*Biochemistry. 2003 42:8459*）

訳 適応抑制剤は，…に非常に高い親和性で結合する

★ unusually（非常に／異常に）

❶ **unusually high（非常に高い）**

Here we show that Pb^{2+} binds with unusually high affinity to the thrombin binding aptamer, d(GGTTGGTGTGGTTGG), inducing a unimolecular folded structure.（*J Mol Biol. 2000 296:1*）

訳 Pb^{2+} はトロンビン結合性アプタマーに非常に高い親和性で結合する

★★ highly（高度に／非常に）

❶ **highly conserved（高度に保存される）**

The type XXVII collagen gene codes for a novel vertebrate fibrillar collagen that is highly conserved in man, mouse, and fish（*Fugu rubripes*）.（*J Biol Chem. 2003 278:31067*）

訳 27 型コラーゲン遺伝子は，ヒト，マウスおよび魚において高度に保存されている新規の脊椎動物原線維コラーゲンをコードする

副詞

I

A

2. 強く／大きく 【strongly】

強く	大きく
strongly potently	greatly

使い分け ◆ **strongly** は「強く」，**potently** は「強力に」，**greatly** は「大きく」という意味に用いられる．

頻度分析

	用例数		用例数
strongly	5,633	❶ strongly suggest	512
		❷ strongly associated with	310
		❸ strongly inhibited	219
potently	759	❶ potently inhibited	134
greatly	2,468	❶ greatly reduced	485
		❷ greatly increased	198
		❸ greatly enhanced	194

★★ strongly（強く）

❶ strongly suggest ～（～を強く示唆する）

These results strongly suggest that the binding of the eIF4G HEAT domain to eIF1 and eIF5 is important for maintaining the integrity of the scanning ribosomal preinitiation complex.（*Mol Cell Biol. 2003 23:5431*）

訳 これらの結果は，…ということを強く示唆する

❷ strongly associated with ～（～と強く関連した）

The degree of viral replication was strongly associated with progression to CMV disease or viremia (risk ratio, 8.8 and 51.5 among patients with virus loads < or =2860 and >2860 copies/10^6 peripheral blood leukocytes, respectively).（*J Infect Dis. 2003 187:1801*）

訳 ウイルス複製の程度は，サイトメガロウイルス疾患への進行と強く関連した

★ potently（強力に）

❶ potently inhibited（強力に抑制される／強力に抑制した）

We report here that SMRT function is potently inhibited by a mitogen-activated protein kinase (MAPK) kinase kinase (MAPKKK) cascade that operates downstream of this growth factor receptor.（*Mol Cell Biol. 2000 20:6612*）

訳 SMRT機能は，…によって強力に抑制される

★ greatly（大きく）

❶ greatly reduced（大きく減少した）

Also, JNK phosphorylation in response to 2-5A was greatly reduced in RNase L −/− mouse cells.（*J Biol Chem. 2004 279:1123*）

訳 2-5Aに応答したJNKリン酸化は，RNase L −/− マウス細胞において大きく減少した

❸ greatly enhanced（大きく増強される）

This alkylating agent induces liver carcinogenesis and its effect is greatly enhanced by TGF-α.（*Am J Pathol. 2002 160:1555*）

訳 その効果はTGF-αによって大きく増強される

3. 主に 【predominantly】

主に	主に／第一に	もっぱら	大部分は
predominantly mainly mostly	primarily principally	exclusively	largely

使い分け

◆ **predominantly**, **mainly**, **mostly** は「主に」, **primarily**, **principally** は「主に／第一に」, **exclusively** は「もっぱら」という意味で，inの前などに用いられる．

◆ **largely** は「大部分は」という意味で，unknownの前で使われることが多い．

頻度分析

	用例数		用例数
predominantly	2,807	❶ predominantly in	429
mainly	1,662	❶ mainly in	244

mostly	845	❶ mostly in	69
primarily	4,007	❶ primarily in	531
principally	282	❶ principally by	33
exclusively	1,645	❶ exclusively in	423
largely	3,236	❶ largely unknown	605

★ predominantly（主に）

❶ predominantly in ～（主に～において）

Uncoupling protein-3（UCP-3）is a recently identified member of the mitochondrial transporter superfamily that is expressed predominantly in skeletal muscle.（*Nature. 2000 406:415*）

訳 脱共役タンパク質-3（UCP-3）は，主に骨格筋において発現するミトコンドリア・トランスポーター・スーパーファミリーの最近同定されたメンバーである

★ mainly（主に）

❶ mainly in ～（主に～において）

GFAT1 is ubiquitous, whereas GFAT2 is expressed mainly in the central nervous system.（*Diabetes. 2001 50:2419*）

訳 一方，GFAT2は主に中枢神経系において発現する

★ mostly（主に／ほとんど）

❶ mostly in ～（主に～において）

The mutant rhodopsin was found mostly in perinuclear locales（endoplasmic reticulum; ER）as evidenced by colocalization using the antibodies Rho1D4 and calnexin-NT.（*Invest Ophthalmol Vis Sci. 2001 42:826*）

訳 変異ロドプシンは主に核周囲の部位において見つけられた

★★ primarily（主に／一次的に）

❶ primarily in ～（主に～において）

The cpk gene is expressed primarily in the kidney and liver and encodes a hydrophilic, 145-amino acid protein, which we term cystin.（*J Clin Invest. 2002 109:533*）

訳 cpk遺伝子は主に腎臓と肝臓において発現する

principally（主に／第一に）

❶ principally by ～（主に～によって）

The oscillations persisted even in perfusions of zero calcium-EGTA Krebs solution suggesting that the calcium oscillation is mediated principally by intracellular calcium release-reuptake mechanisms.（*J Physiol. 2005 562:553*）

訳 カルシウムの周期的変動は，主に細胞内カルシウムの放出と再取り込みの機構によって仲介されている

★ exclusively（もっぱら／排他的に）

❶ exclusively in ～（もっぱら～において）

Mouse PTPRO mRNA is expressed exclusively in two cell types: neurons and kidney podocytes.（*J Comp Neurol. 2003 456:384*）

訳 マウスのPTPROメッセンジャーRNAは，もっぱら2つのタイプの細胞において発現する

★★ largely（大部分は／ほとんど）

❶ largely unknown（ほとんど知られていない）

Neuronal activity influences myelination of the brain, but the molecular mechanisms involved are largely unknown.（*Neuron. 2002 36:855*）

訳 しかし，関与する分子機構はほとんど知られていない

4. 完全に／十分に 【completely】

完全に	十分に	全く
completely entirely fully	sufficiently enough fully	quite

使い分け
- ◆ **completely**, **entirely** は，「完全に」という意味に用いられる．
- ◆ **sufficiently**, **enough** は，「十分に」という意味で使われる．
- ◆ **fully** は「完全に／十分に」，**quite** は「全く／完全に」という意味に用いられる．

頻度分析

	用例数		用例数
completely	4,321	❶ completely blocked	387
		❷ completely inhibited	274
		❸ completely abolished	270
entirely	721	❶ almost entirely	95
		❷ entirely dependent	36
fully	3,634	❶ not fully understood	231
		❷ fully functional	153
		❸ fully active	109
sufficiently	491	❶ sufficiently high	49
enough	583	❶ enough to	372
quite	638	❶ quite different	122

解説

◆ **completely** は blocked の前，**entirely** は dependent の前，**fully** は functional の前，**sufficient** は high の前，**enough** は to の前，**quite** は different の前などで用いられる．

◆ **fully** は，not **fully** understood などの否定形の用例も多い．

★★ completely（完全に）

❶ **completely blocked**（完全にブロックされた／完全にブロックした）

However, ADP caused Akt phosphorylation in Gαq- and P2Y1-deficient platelets, which was completely blocked by AR-C69931MX.（*J Biol Chem. 2004 279:4186*）

訳 そして，それは AR-C69931MX によって完全にブロックされた

❸ **completely abolished**（完全に消滅させられた／完全に消滅させた）

This effect was completely abolished by EGFR-specific tyrosine kinase inhibitors providing evidence for the role of the EGFR in this response.（*J Biol Chem. 2003 278:35451*）

訳 この効果は，EGFR 特異的チロシンキナーゼ阻害剤によって完全に消滅させられた

★ entirely（完全に／全体的に）

❶ **almost entirely**（ほとんど全体的に）

The predicted HetL protein is composed almost entirely of pentapeptide repeats with a consensus of A(D/N)L*X, where * is a polar amino acid.（*J Bacteriol. 2002 184:6873*）

訳 予想される HetL タンパク質は，ほとんど全体的に A(D/N)L*X のコンセンサスのペンタペプチドの繰り返しで構成されている

❷ **entirely dependent**（完全に依存している）

Such binding was entirely dependent on the formation of gp120-CD4-CXCR4 tricomplexes since staining was absent with SDF-treated or coreceptor-negative target cells.（*J Virol. 2001 75:11096*）

訳 そのような結合は，…の形成に完全に依存した

⋆⋆ fully（十分に／完全に）

❶ **not fully understood**（十分には理解されていない）

Cytokinesis in eukaryotic organisms is under the control of small GTP-binding proteins, although the underlying molecular mechanisms are not fully understood.（*EMBO J. 2001 20:3705*）

訳 根底にある分子機構は十分には理解されていないけれども

❷ **fully functional**（完全に機能的である）

However, it is unclear whether AR is fully functional in recurrent prostate cancer after androgen withdrawal.（*Cancer Res. 2003 63:4552*）

訳 ARが再発性の前立腺癌において完全に機能的であるかどうかは明らかでない

⋆ sufficiently（十分に）

❶ **sufficiently high**（十分に高い）

Levels of induced IP-10 protein were sufficiently high to induce chemotaxis of peripheral blood lymphocytes.（*J Virol. 2000 74:9214*）

訳 誘導されたIP-10タンパク質のレベルは，末梢血白血球の走化性を誘導するのに十分高かった

⋆ enough（十分に）

通常の副詞と違って，修飾する形容詞の後に置かれる．

❶ **enough to ～**（～するのに十分な）

Some of these pores should be large enough to allow cytochrome c to permeate.（*J Biol Chem. 2000 275:38640*）

訳 これらの細孔のいくつかは，チトクロムcを透過させるのに十分なほど大きいに違いない

⋆ quite（全く／完全に）

❶ **quite different**（全く異なる）

This sequence is quite different from the consensus Dun1

副詞

I

A

phosphorylation sequence reported previously from peptide library studies.（*J Biol Chem. 2004 279:11293*）

訳 この配列はコンセンサスのDun1リン酸化配列とは全く異なる

5. 明らかに／明確に 【clearly】

明確に／明らかに	決定的に
clearly **apparently** **distinctly** **unambiguously** **evidently** **obviously** **definitively**	**conclusively**

使い分け

◆ **clearly**, **distinctly**, **unambiguously**, **evidently**, **obviously**, **definitively**は，「明確に／明らかに」という意味に用いられる．

◆ **apparently**は「明らかに」という意味だけでなく，「見かけ上」という意味で用いられることも多い．

◆ **conclusively**は「決定的に／明確に」という意味で使われる．

頻度分析

	用例数		用例数
clearly	1,637	❶ clearly defined	117
		❷ clearly demonstrate	112
apparently	1,463	❶ apparently due to	42
distinctly	240	❶ distinctly different	130
unambiguously	178	❶ unambiguously identified	19
evidently	95	❶ Evidently,	34
obviously	57	❶ not obviously	27
definitively	132	❶ to definitively	25
conclusively	120	❶ conclusively demonstrate	14

解説

◆ **clearly**はdefinedやdemonstrateの前，**distinctly**はdifferentの前などで用いられる．

◆ **evidently**は，文頭で用いられることが多い．

◆ **obviously**は，否定形で使われることが多い．

★ clearly（明確に／明らかに）

❶ clearly defined（明確に規定される）

Platelet/Endothelial Cell Adhesion Molecule-1 (PECAM-1 or CD31) is thought to be a vascular-specific protein, but its function has not been clearly defined.（*Dev Biol. 2001 234:317*）

訳 しかし，その機能は明確に規定されてはいない

❷ clearly demonstrate（明確に実証する）

Our results clearly demonstrate that Tsg101 is not a primary tumor suppressor in mouse embryonic fibroblasts.（*J Biol Chem. 2002 277:43216*）

訳 われわれの結果は，…ということを明確に実証する

★ apparently（明らかに／見かけ上）

❶ apparently due to ～（明らかに～のせいで）

The cross-reactivity of gp130 is apparently due to a chemical plasticity evident in the amphipathic gp130 cytokine-binding sites.（*Science. 2001 291:2150*）

訳 gp130の交差反応性は明らかに化学的可塑性のせいである

distinctly（明らかに／はっきりと）

❶ distinctly different（明らかに異なる）

This role for Cod1p is distinctly different from that of the well-characterized Ca^{2+} P-type ATPase Pmr1p which is neither required for Hmg2p degradation nor its control.（*J Cell Biol. 2000 148:915*）

訳 このCod1pの役割は，よく特徴づけられているCa^{2+} P型ATP分解酵素Pmr1pのそれとは明らかに異なる

unambiguously（明らかに）

❶ unambiguously identified（明らかに同定される）

The Sindbis virus was unambiguously identified by either approach.（*Anal Chem. 2002 74:2529*）

訳 シンドビスウイルスは，どちらのアプローチによっても明らかに同定された

evidently（明らかに）

❶ Evidently, ～（明らかに，～）

Evidently, manganese uptake in Synechocystis 6803 cells occurs in two

副詞

I

A

steps.（*Biochemistry. 2002 41:15085*）

訳 明らかに，Synechocystis 6803 細胞におけるマンガンの取り込みは 2 段階で起こる

obviously（明らかに）

❶ not obviously ～（明かに～でない）

Since the YciA, YciB, and YciC proteins are not obviously related to any known transporter family, they may define a new class of metal ion uptake system.（*J Bacteriol. 2002 184:6508*）

訳 YciA，YciB および YciC タンパク質は，どの既知のトランスポーター・ファミリーにも明らかに関連していないので

definitively（明確に／はっきりと）

❶ to definitively ～（明確に～すること）

Therefore, the purpose of this study was to definitively characterize the human immunodeficiency virus type 1（HIV-1）quasispecies transmitted in utero and intrapartum.（*J Virol. 2001 75:2194*）

訳 この研究の目的は，1 型ヒト免疫不全症ウイルス（HIV-1）疑似種を明確に特徴づけることであった

conclusively（決定的に／明確に）

❶ conclusively demonstrate（決定的に実証する）

Our data conclusively demonstrate that loss of p53 function is a requirement for progression of Friend erythroleukemia *in vivo.*（*Oncogene. 2001 20:2946*）

訳 われわれのデータは，…ということを決定的に実証する

6. 広く／普遍的に 【widely】

広く／広範に	普遍的に／遍在性に	構成的に／恒常的に
widely broadly extensively	universally ubiquitously	constitutively

使い分け ◆ **widely, broadly** は「広く／広範に」，**extensively** は「広範に」という意味に用いられる．

◆ **universally** は「普遍的に」，**ubiquitously** は「遍在性に」，**constitutively** は「構成的に／恒常的に」という意味で使われる．

頻度分析

	用例数		用例数
widely	2,508	❶ widely used	622
		❷ widely expressed	336
		❸ widely distributed	237
broadly	585	❶ broadly expressed	71
extensively	904	❶ extensively studied	161
universally	181	❶ universally conserved	56
ubiquitously	451	❶ ubiquitously expressed	332
constitutively	2,876	❶ constitutively active	1,245
		❷ constitutively expressed	326
		❸ constitutively activated	290

解説

◆ **widely**, **broadly**, **ubiquitously** は，expressed の前などで用いられる．

◆ **extensively** は studied の前，**universally** は conserved の前，**constitutively** は active の前などで使われる．

副詞

I

A

★ widely（広く／広範に）

❶ widely used（広く使われる）

Microtubule-damaging agents（MDA）are potent antineoplastic drugs that are widely used in clinical treatment for a variety of cancers.（*J Biol Chem. 2004 279:39431*）

訳 微小管損傷剤（MDA）は，さまざまな癌の臨床治療に広く使われる強力な抗癌剤である

❷ widely expressed（広範に発現する）

BCL-XL is a key anti-apoptotic BCL-2 family protein that is widely expressed in human cancer cells and is induced in response to diverse survival signals.（*Oncogene. 2000 19:5534*）

訳 BCL-XL は，ヒトの癌細胞において広範に発現する重要な抗アポトーシス性 BCL-2 ファミリータンパク質である

★ broadly（広く／広範に）

❶ broadly expressed（広範に発現する）

Northern blot analyses indicate that OSBP1 is broadly expressed in human and monkey tissues.（*J Biol Chem. 2001 276:18570*）

訳 OSBP1 は，ヒトおよびサルの組織において広範に発現する

★ extensively（広範に）

❶ extensively studied（広範に研究される）

The role of oncogenes in the induction of angiogenesis has been extensively studied in benign and malignant tumors.（*Am J Pathol. 2003 163:1321*）

訳 血管形成の誘導における癌遺伝子の役割は良性および悪性の腫瘍において広範に研究されてきた

universally（普遍的に／広く）

❶ universally conserved（普遍的に保存される）

To test for base-pairing interactions between 23S rRNA and aminoacyl tRNA, site-directed mutations were made at the universally conserved nucleotides U2552 and G2553 of 23S rRNA in both *E. coli* and *B. stearothermophilus* ribosomal RNA and incorporated into ribosomes.（*Mol Cell. 1999 4:859*）

訳 部位特的な変異が，大腸菌と *B. stearothermophilus* の両方のリボソーム RNA において普遍的に保存されている 23S リボソーム RNA 塩基 U2552 と G2553 につくられた

★ ubiquitously（遍在性に）

❶ ubiquitously expressed（遍在性に発現する）

The gene is ubiquitously expressed in human and murine tissues, although the expression pattern is more restricted during mouse development.（*Proc Natl Acad Sci USA. 2003 100:10358*）

訳 その遺伝子は，ヒトおよびマウスの組織において遍在性に発現する

★ constitutively（構成的に／恒常的に）

❶ constitutively active（構成的に活性のある）

Furthermore, overexpression of a constitutively active form of Notch1 inhibited the proliferation of various prostate cancer cells, including DU145, LNCaP, and PC3 cells.（*Cancer Res. 2001 61:7291*）

訳 構成的活性型の Notch1 の過剰発現は，さまざまな前立腺癌細胞の増殖を抑制した

❷ constitutively expressed（構成的に発現する）

The LST1 gene is constitutively expressed in leukocytes and dendritic cells, and it is characterized by extensive alternative splicing. (*J Immunol. 2000 164:3169*)

訳 LST1 遺伝子は，白血球および樹状細胞において構成的に発現する

7. はるかに／もっと 【far】

はるかに／もっと	
far	much

使い分け ◆ **far** は「はるかに」，**much** は「もっと／はるかに」という意味で，more の前などで用いられる．

頻度分析

	用例数		用例数
far	1,901	❶ far red	147
		❷ far more	141
		❸ far from	132
much	5,135	❶ much more	761
		❷ much less	508

★ far（はるかに）

❷ far more（はるかにもっと）

The histogram method is far more sensitive than the 10-HU threshold method for diagnosis of adrenal adenomas at enhanced CT, with specificity maintained at 100%. (*Radiology. 2003 228:735*)

訳 ヒストグラム法は，…の診断にとって 10-HU 閾値法よりもはるかに敏感である

★★ much（もっと／はるかに）

❶ much more（はるかにもっと）

Peak 1 IIGlc migrated much more slowly than peak 2 IIGlc. (*J Bacteriol. 2004 186:8453*)

訳 ピーク 1 の IIGlc は，ピーク 2 の IIGlc よりもはるかにゆっくりと移動した

副詞

I

A

8. ほとんど 【nearly】

ほとんど	実質的に
nearly almost	virtually

使い分け
- ◆ **nearly**, **almost** は，「ほとんど」という意味に用いられる．
- ◆ **virtually** は，「実質的に／ほとんど」という意味で使われる．

頻度分析

	用例数		用例数
nearly	2,749	❶ nearly all	363
		❷ nearly identical	309
		❸ nearly complete	156
almost	2,335	❶ almost all	359
		❷ almost completely	240
		❸ almost exclusively	225
virtually	921	❶ virtually all	257
		❷ virtually identical	140
		❸ virtually no	94

解説
- ◆ **nearly**, **almost**, **virtually** は，all の前などで用いられる．

★ nearly（ほとんど）

❶ nearly all ～（ほとんどすべての～）

Several types of epidermal growth factor receptor (EGFR) gene mutations have been reported in glioblastomas, and in nearly all cases the alterations have been reported in tumors with EGFR amplification. (*Cancer Res. 2000 60:1383*)

訳 ほとんどすべての症例において，変化がEGFRの増幅を伴う腫瘍において報告されている

❷ nearly identical（ほとんど同じ）

The structure of the SHY protein is nearly identical to other extracellular matrix glycoproteins that are composed of LRRs, such as the polygalacturonase inhibitor proteins (PGIP) of plants. (*Plant J. 2004 39:643*)

訳 SHYタンパク質の構造は，他の細胞外基質糖タンパク質とほとんど同じである

★ almost（ほとんど）

❶ almost all ～（ほとんどすべての～）

The t(9;22) chromosomal translocation is found in almost all patients with chronic myelogenous leukemia.（*Proc Natl Acad Sci USA. 2000 97:2093*）

訳 t(9;22)染色体転座は，慢性骨髄性白血病のほとんどすべての患者において見つけられる

❷ almost completely（ほとんど完全に）

This decrease was almost completely blocked by a PARP inhibitor.（*Nucleic Acids Res. 2003 31:e104*）

訳 この減少はPARP抑制剤によってほとんど完全にブロックされた

★ virtually（実質的に／ほとんど）

❶ virtually all ～（実質的にすべての～）

Immunodepletion experiments indicate that virtually all of the endogenous GRSP1 protein exists as a complex with GRP1 in lung.（*J Biol Chem. 2001 276:40065*）

訳 実質的にすべての内在性GRSP1タンパク質が，GRP1との複合体として存在する

❷ virtually identical（実質的に同じ）

The induction was virtually identical to that observed in −2100-FAS-CAT transgenic mice and to the endogenous FAS mRNA.（*J Biol Chem. 2000 275:10121*）

訳 その誘導は，…において観察されたそれと実質的に同じであった

副詞

I

A

9. かなり　【substantially】

かなり	ますます
substantially considerably fairly reasonably appreciably	increasingly

使い分け

◆ **substantially**, **considerably**, **fairly**, **reasonably**, **appreciably** は，「かなり」という意味に用いられる．

◆ **increasingly** は，「ますます」という意味で使われる．

頻度分析

	用例数		用例数
substantially	2,369	❶ substantially reduced	219
		❷ substantially increased	119
		❸ substantially higher	113
considerably	733	❶ considerably more	108
		❷ considerably reduced	30
fairly	181	❶ fairly well	23
reasonably	158	❶ reasonably well	41
appreciably	125	❶ not appreciably	51
increasingly	816	❶ increasingly important	82

解説

◆ **substantially**, **considerably** は，reduced の前などで用いられる．

◆ **considerably** は，more など比較級の形容詞を修飾することも多い．

◆ **fairly**, **reasonably** は，well の前などで使われる．

◆ **appreciably** は，否定形の用例が多い．

◆ **increasingly** は，important の前などで用いられる．

★ substantially（実質的に／かなり）

❶ substantially reduced（かなり低下した）

While nuclear BER protein levels and activities were generally not altered in p^0 cells, AP endonuclease activity was substantially reduced in nuclear and in whole cell extracts.（*Cancer Res. 2001 61:2015*）

訳 AP エンドヌクレアーゼ活性は，核および全細胞抽出液においてかなり

低下した

★ considerably（相当に／かなり）

比較級の形容詞を修飾して強調する表現が多い.

❶ considerably more ～（相当により～）

In particular, bleomycin-induced DSB repair is considerably more sensitive to inhibition by increased ionic strength than repair of bluntended DNA.（*Nucleic Acids Res. 2001 29:E78*）

訳 ブレオマイシンに誘導されるDSBの修復は，…に対して相当により敏感である

❷ considerably reduced（かなり低下した）

Our observations show that the mean B cell life span is considerably reduced in HEL-Ig Tgn compared with Ig Tgn mice, but also demonstrate that some HEL–Ig Tgn B cells survive to maturity.（*J Immunol. 2000 164:3035*）

訳 平均B細胞寿命は，Ig Tgnマウスと比べてHEL-Ig Tgnマウスにおいてかなり低下する

fairly（かなり）

❶ fairly well（かなりよく）

Although the gross morphology of amyloid fibrils is fairly well understood, very little is known about how the constituent polypeptides fold within the amyloid folding motif.（*Biochemistry. 2001 40:11757*）

訳 アミロイド原線維の肉眼的形態はかなりよく理解されているけれども

reasonably（かなり／適度に）

❶ reasonably well（かなりよく）

Although biochemical aspects of these metabolic pathways are reasonably well understood, the identification and characterization of genes encoding nucleotide sugar interconversion enzymes is still in its infancy.（*Plant Mol Biol. 2001 47:95*）

訳 これらの代謝経路の生化学的側面はかなりよく理解されているけれども

appreciably（かなり）

否定形の用例が多い.

副詞

I

A

❶ **not appreciably**（あまり〜でない）

Asn-382 and Thr-383 mutations did not appreciably alter the K_m value for arachidonate, the cyclooxygenase product profile, or the Tyr-385 radical spectroscopic characteristics, confirming the structural integrity of the cyclooxygenase site.（*J Biol Chem. 2004 279:4084*）

訳 アスパラギン382とスレオニン383の変異は，アラキドン酸に対する K_m 値をあまり変えなかった

⋆ increasingly（ますます）

❶ **increasingly important**（ますます重要な）

Haplotype analysis has become increasingly important for the study of human disease as well as for reconstruction of human population histories.（*Am J Hum Genet. 2000 67:518*）

訳 ハプロタイプ分析は，…の研究にとってますます重要になってきた

10. 中程度に／部分的に 【moderately】

中程度に	部分的に／いくらか
moderately	partially somewhat

使い分け

◆ **moderately** は「中程度に」，**partially** は「部分的に」，**somewhat** は「いくらか」という意味に用いられる．

頻度分析

	用例数		用例数
moderately	639	❶ only moderately	59
partially	4,318	❶ partially inhibited	205
		❷ partially purified	153
		❸ partially blocked	142
somewhat	494	❶ somewhat more	47

解説

◆ **moderately** は，前に only を伴うことも多い．

◆ **partially** は inhibited の前，**somewhat** は more や less の前などで用いられる．

★ moderately（中程度に）

❶ only moderately（中程度にしか）

The inhibitors were typically potent against collagenase-3（MMP-13）and gelatinase A（MMP-2）, while they spared collagenase-1（MMP-1）and only moderately inhibited stromelysin（MMP-3）.（*J Med Chem. 2001 44:1060*）

訳 一方，それらはコラゲナーゼ-1（MMP-1）を容赦し，そしてストロメライシン（MMP-3）を中程度にしか抑制しなかった

★★ partially（部分的に）

❶ partially inhibited（部分的に抑制される）

In contrast, TRAF2-mediated NFκB induction was partially inhibited by DTRAF1.（*J Biol Chem. 2000 275:12102*）

訳 TRAF2に仲介されるNFκBの誘導は，DTRAF1によって部分的に抑制された

★ somewhat（いくらか）

❶ somewhat more（いくらかもっと）

The mixed HOPO ligand is somewhat more effective than 3,4,3-LI(1,2-HOPO) when given orally, and the enhanced reduction of liver Pu by the mixed ligand is statistically significant.（*J Med Chem. 2002 45:3963*）

訳 混合HOPOリガンドは，3,4,3-LI(1,2-HOPO）よりもいくらかもっと効果的である

11. 比較的／選択的に 【relatively】

比較的	比較できるほど	選択的に
relatively comparatively	comparably	selectively

使い分け ◆ **relatively**, **comparatively** は「比較的」, **comparably** は「比較できるほど」, **selectively** は「選択的に」という意味に用いられる.

頻度分析

	用例数		用例数
relatively	4,100	❶ relatively high	351
		❷ relatively low	341
		❸ relatively small	254
comparatively	132	❶ comparatively little	22
comparably	114	❶ comparably to	15
selectively	3,252	❶ selectively expressed	130
		❷ selectively inhibited	89

解説 ◆ **relatively** は high の前，**comparatively** は little の前， **comparably** は to の前，**selectively** は expressed の前などで用いられる．

★★ relatively（比較的）

❶ relatively high（比較的高い）

Ape1/ref-1 was expressed at relatively high levels in the tumor cells of nearly all sections.（*Cancer Res. 2001 61:2220*）

訳 Ape1/ref-1 は腫瘍細胞において比較的高いレベルで発現した

comparatively（比較的）

❶ comparatively little（比較的わずかしか）

However, comparatively little is known about how IL-17 expression is controlled.（*J Biol Chem. 2004 279:52762*）

訳 どのように IL-17 発現が調節されるかについては比較的わずかしか知られていない

comparably（比較できるほど／匹敵するほど）

❶ comparably to ～（～に匹敵するほど）

The method is applied to estimate amino acid probabilities based on observed counts in an alignment and is shown to perform comparably to previous methods.（*Bioinformatics. 2001 17:S65*）

訳 そして，従来の方法に匹敵するほど機能することが示される

★ selectively（選択的に）

❶ selectively expressed（選択的に発現される）

Here, we demonstrate that LPLA2 is selectively expressed in alveolar macrophages but not in peritoneal macrophages, peripheral blood

monocytes, or other tissues. (*J Biol Chem. 2004 279:42605*)
訳 LPLA2は肺胞のマクロファージにおいて選択的に発現される

12. わずかに／単に 【slightly】

わずかに／不十分に	弱く	単に／唯一の
slightly **modestly** **marginally** **poorly**	**weakly**	**simply** **only**

使い分け

◆ **slightly**, **modestly** は「わずかに」, **marginally** は「わずかに／かろうじて」, **poorly** は「不十分に」, **weakly** は「弱く」, **simply** は「単に」, **only** は「唯一の」という意味に用いられる.

頻度分析

	用例数		用例数
slightly	1,690	❶ slightly higher	136
		❷ slightly more	113
		❸ slightly lower	98
modestly	388	❶ only modestly	96
marginally	216	❶ marginally significant	30
poorly	3,096	❶ poorly understood	1,582
		❷ poorly defined	239
		❸ poorly characterized	130
weakly	996	❶ only weakly	173
		❷ weakly to	64
		❸ weakly with	59
simply	560	❶ simply by	62
only	26,891	❶ only in	2,401
		❷ only a	1,898
		❸ only the	1,674

解説

◆ **slightly** は higher の前, **marginally** は significant の前, **poorly** は understood の前などで使われる.

◆ **weakly** は only の後, **simply** は by の前, **only** は in の前などで用いられる.

副詞

I

A

★ slightly（わずかに）

❶ slightly higher（わずかにより高い）

Elastic energy storage values calculated using the molecular model were slightly higher than those obtained from collagen fibers, but display the same increases in slope as the fiber data.（*J Theor Biol. 2004 229:371*）

訳 分子モデルを使って計算された弾性エネルギー格納値は，コラーゲン線維から得られたそれらよりわずかに高い

modestly（中程度に）

❶ only modestly（中程度にしか）

The trafficking of transferrin was only modestly affected by these treatments.（*Mol Biol Cell. 2004 15:3758*）

訳 トランスフェリンの輸送は，それらの処置によって中程度にしか影響を受けなかった

marginally（わずかに／かろうじて）

❶ marginally significant（かろうじて有意な）

We observed a marginally significant reduction in the risk of breast cancer among ^{194}Trp carriers.（*Cancer Res. 2003 63:8536*）

訳 われわれは，乳癌のリスクのかろうじて有意な減少を観察した

★★ poorly（不十分に／ほとんど～でない）

❶ poorly understood（ほとんど理解されていない）

Obesity contributes to the development of type 2 diabetes, but the underlying mechanisms are poorly understood.（*Science. 2004 306:457*）

訳 しかし，根底にある機構はほとんど理解されていない

★ weakly（弱く）

❶ only weakly（かろうじて弱く）

Our data show that RCP interacts only weakly with Rab4 *in vitro* and does not play the role of coupling Rab11 and Rab4 *in vivo*.（*Mol Biol Cell. 2004 15:3530*）

訳 RCP は Rab4 とかろうじて弱く相互作用する

★ simply（単純に／単に）

❶ simply by ～（単に～によって）

These effects could not be explained simply by changes in locomotor activity or general arousal.（*Behav Neurosci. 2000 114:320*）

訳 これらの効果は単に運動活性の変化によってでは説明できない

★★ only（唯一の／～のみ）

❶ only in ～（～においてのみ）

Expression of the IL-6 gene was found only in the testes and spleen of infected animals.（*J Immunol. 2001 167:4527*）

訳 IL-6遺伝子の発現は，感染した動物の精巣と脾臓においてのみ見つけられた

副詞

13. 通常／一般に 【commonly】

I

通常／一般に	通常／普通
commonly **generally** **in general**	**usually**

A

使い分け

◆ **commonly**, **generally**, **in general** は，「一般に／通常」という意味に用いられる．

◆ **usually** は「通常／普通」という意味で使われる．

頻度分析

	用例数		用例数
commonly	2,035	❶ commonly used	562
		❷ commonly found	101
		❸ commonly associated with	92
generally	2,628	❶ generally accepted	92
		❷ generally applicable	78
		❸ generally thought	71
in general	826	❶ In general,	326
usually	1,404	❶ usually associated with	53

解説

◆ **commonly** は，used の前で用いられることが非常に多い．

◆ **in general** は，文頭で使われることが多い．

★ commonly（一般に／通常）

❶ commonly used（通常使われる）

Antibodies to the amyloid precursor protein（APP）are commonly used to detect traumatic axonal injury（TAI).（*Brain Res. 2000 871:288*）

訳 アミロイド前駆タンパク質（APP）に対する抗体は，外傷性の軸索損傷（TAI）を検出するために通常使われる

★ generally（一般に／通常）

❶ generally accepted（一般に受け入れられる）

It is generally accepted that the ability of cocaine to inhibit the dopamine transporter（DAT）is directly related to its reinforcing actions.（*Proc Natl Acad Sci USA. 2004 101:372*）

訳 …ということは一般に受け入れられている

★ in general（一般に／通常は／概して）

❶ In general, ～（概して，～）

In general, the magnitude of zinc induction was dependent on the concentration of zinc in the culture medium, but independent of the amount of MTF-1 expression.（*Nucleic Acids Res. 2002 30:3130*）

訳 概して，亜鉛による誘導の大きさは培地中の亜鉛の濃度に依存した

★ usually（通常／普通）

❶ usually associated with ～（通常～と関連する）

Finally, while cell spreading is usually associated with cell migration, xPTP-PESTr promotes ectodermal cell spreading on fibronectin but also reduces cell migration in response to activin-A, suggesting an adverse effect on cell translocation.（*Dev Biol. 2004 265:416*）

訳 細胞伸展は，通常，細胞遊走と関連する

14. 特に／特異的に 【particularly】

特に	特異的に
particularly especially in particular notably	specifically

使い分け

◆ **particularly**, **especially**, **in particular**, **notably** は，「特に」という意味に用いられる．

◆ **specifically** は，「特異的に」という意味で使われる．

頻度分析

	用例数		用例数
particularly	3,360	❶ particularly in	544
		❷ particularly important	161
especially	2,284	❶ especially in	482
		❷ especially when	104
in particular	1,796	❶ In particular,	954
notably	935	❶ Notably,	483
specifically	6,922	❶ specifically to	596
		❷ specifically in	417

解説

◆ **particularly**, **especially** は，in の前で用いられることが多い．

◆ **in particular**, **notably** は，文頭で使われることが多い．

◆ **specifically** は，to や in の前で用いられることが多い．

★★ particularly（特に）

❶ particularly in ～（特に～において）

However, resistance has been reported, particularly in patients with advanced-stage disease.（*Cancer Res. 2002 62:7149*）

訳 特に進行したステージの疾患の患者において

❷ particularly important（特に重要な）

Residue β D305 would not tolerate substitution with Val or Ser and had extremely low activity as β D305E, suggesting that this residue is particularly important for synthesis and hydrolysis activity.（*J Biol Chem. 2003 278:51594*）

訳 この残基は合成と加水分解活性にとって特に重要である

★ especially（特に）

❶ especially in ～（特に～において）

Left ventricular（LV）mass is an important predictor of morbidity and mortality, especially in patients with systemic hypertension.（*Circulation. 2004 110:1814*）

訳 特に高血圧症の患者において

❷ especially when ～（特に～のとき）

Systematic dissection of the CO revealed that the region of contiguous DNA bases was the active component in the repair process, especially when the single-stranded ends were protected against nuclease attack.（*Nucleic Acids Res. 2001 29:4238*）

訳 特に一本鎖の末端がヌクレアーゼの攻撃から保護されたとき

★ in particular（特に）

文頭で用いられることが多い．

❶ In particular, ～（特に，～）

In particular, we found that BRCA1 expression attenuated p53-mediated cell death in response to γ-irradiation.（*Oncogene. 2003 22:3749*）

訳 特に，われわれは…ということを見つけた

★ notably（特に）

文頭で用いられることが多い．

❶ Notably, ～（特に，～）

Notably, the antibodies did not react with Env cells when treated with a covalent cross-linker either alone or during fusion with target cells.（*J Virol. 2001 75:11096*）

訳 特に，…で処理されたとき抗体が Env 細胞と反応しなかった

★★ specifically（特異的に）

❶ specifically to ～（～に特異的に）

Mecp2 is an X-linked gene encoding a nuclear protein that binds specifically to methylated DNA and functions as a general transcriptional repressor by associating with chromatin-remodeling complexes.（*Nat Genet. 2001 27:327*）

訳 Mecp2 は，メチル化した DNA に特異的に結合する核タンパク質をコードする X 染色体連鎖遺伝子である

❷ **specifically in ～**（～において特異的に）

Whereas myocardin is expressed specifically in cardiac and smooth muscle cells, MRTF-A and -B are expressed in numerous embryonic and adult tissues.（*Proc Natl Acad Sci USA. 2002 99:14855*）

訳 myocardinは心筋および平滑筋細胞において特異的に発現する

15. 実際に 【actually】

実際に	
actually	practically

使い分け ◆ **actually**, **practically** は，be動詞の後などで「実際に」という意味で用いられる．

頻度分析

	用例数		用例数
actually	444	❶ is actually	59
practically	57	❶ is practically	6

★ actually（実際に）

❶ **is actually ～**（…は，実際に～である）

We show here that PCP is actually converted to tetrachlorobenzoquinone, which is subsequently reduced to tetrachlorohydroquinone by PcpD, a protein that had previously been suggested to be a PCP hydroxylase reductase.（*J Bacteriol. 2003 185:302*）

訳 PCPは実際にテトラクロロベンゾキノンに変換される

practically（実際に）

❶ **is practically ～**（…は，実際に～である）

Therefore, this chemokine is practically inactive without pGlu1.（*Biochemistry. 2000 39:14075*）

訳 このケモカインは実際にpGlu1なしには不活性である

16. およそ 【approximately】

約／およそ	おおよそ
approximately about ca.	roughly

使い分け
◆いずれの語も「約／およそ」という意味に用いられる．
◆**roughly** は，「おおまかに」という意味合いが強い．
◆**about** は副詞だけでなく，「〜について」という意味の前置詞としても用いられる．

頻度分析

	用例数		用例数
approximately	16,785	❶ of approximately	2,189
about	9,087	❶ of about	558
ca.	328	❶ of ca.	46
roughly	381	❶ of roughly	25

⁂ approximately（約／およそ）

❶ **of approximately 〜**（約〜の…）

In Western blot analyses, the purified polyclonal antibody recognized a specific signal with a molecular mass of approximately 40 kDa in RL and HT lymphomas.（*Proc Natl Acad Sci USA. 2004 101:15160*）

訳 ウエスタンブロット分析において，精製されたポリクローナル抗体は約 40 kDa の分子量を持つ特異的なシグナルを認識した

⁂ about（約／およそ）

❶ **of about 〜**（約〜の…）

MPCBP has a molecular mass of about 72 kDa and contains three tetratricopeptide repeats（TPR）suggesting that it is a member of the TPR family of proteins.（*J Biol Chem. 2000 275:35457*）

訳 MPCBP は約 72 kDa の分子量を持つ

ca.（約）

❶ of ca.～（約～の…）

A peptide corresponding to the last 22 residues of UL54 was sufficient to bind specifically to UL44 in a 1:1 complex with a dissociation constant of ca. 0.7 microM.（*J Virol. 2004 78:158*）

訳 UL54の最後の22残基に相当するペプチドは，約0.7 μMの解離定数を持つ1：1の複合体でUL44に特異的に結合するのに十分であった

roughly（おおよそ／およそ）

❶ of roughly ～（おおよそ～の…）

Western blot analysis of the supernatant using α-H1 and α-ubiquitin antibodies detected the same band of roughly 46 kDa; this band was absent from the control supernatant.（*Biochemistry. 2004 43:16203*）

訳 α-H1およびα-ユビキチン抗体を使った上清のウエスタンブロット分析は，おおよそ46 kDaの同一のバンドを検出した

【Ⅰ-B. 様態】 様子・状態・関係などを表す副詞

17. おそらく／もしかしたら 【probably】

おそらく	たぶん	もしかしたら／潜在的に
probably presumably likely	perhaps	possibly potentially

使い分け

- **probably**, **presumably**, **likely** は「おそらく」，**perhaps** は「たぶん」，**possibly**, **potentially** は「もしかしたら」という意味に用いられる．
- **probably**, **presumably**, **likely** は確率がかなり高く，**perhaps** はやや低く，**possibly**, **potentially** はもっと低い．
- most **likely**, most **probably** は，それぞれ **likely**, **probably** より確率が高い場合に用いられる．

頻度分析

	用例数		用例数
probably	2,310	❶ probably due to	110
		❷ probably by	91
		❸ probably because	76
presumably	1,213	❶ presumably by	134
		❷ presumably because	84
		❸ presumably due to	81
likely	8,924	❶ most likely	1,077
		❷ will likely	95
perhaps	1,144	❶ and perhaps	361
		❷ perhaps by	101
possibly	2,193	❶ and possibly	724
		❷ possibly by	195
		❸ possibly through	127
potentially	3,081	❶ potentially important	211
		❷ potentially useful	123
		❸ potentially involved	68

★ probably（おそらく）

❶ probably due to ～（おそらく～のせいで）

This increase is probably due to the activation of the gene encoding 204 (Ifi204) by Smad transcription factor, including Smad1, -4, and -5.（*J Biol Chem. 2005 280:2788*）

訳 この上昇は，おそらく…をコードする遺伝子の活性化のせいである

★ presumably（おそらく）

❶ presumably by ～（おそらく～によって）

In summary, *in vitro* selection for a strain of EIAV that rapidly killed cells resulted in the generation of a virus that was able to superinfect these cells, presumably by the use of a novel mechanism of cell entry.（*J Virol. 2003 77:2385*）

訳 おそらく細胞侵入の新規の機構の利用によって

❷ presumably because ～（おそらく～のせいで）

However, ectopic expression of both transcripts together severely reduces viability, presumably because of the formation of inappropriate gap junctions.（*Mol Biol Cell. 2000 11:2459*）

訳 おそらく不適当なギャップ結合の形成のせいで

★★ likely（ありそうな／おそらく）

❶ most likely（おそらく）

This lack of stability is most likely due to the fact that the wild-type yeast TyrRS misaminoacylates the *E. coli* proline tRNA.（*Proc Natl Acad Sci USA. 2001 98:2268*）

訳 この安定性の欠如は，おそらく…という事実のせいである

❷ will likely ～（おそらく～であろう）

The function of jeltraxin will likely be related to its calcium-dependent lectin properties.（*Biochemistry. 2003 42:12761*）

訳 jeltraxinの機能は，カルシウム依存性レクチンの性質におそらく関連するであろう

★ perhaps（たぶん／もしかしたら）

❶ ～ and perhaps（～そしてたぶん）

Thus, these data suggest that DHT and perhaps other androgenic hormones may protect normal females against the risk of dofetilide-induced arrhythmia.（*Circulation. 2002 106:2132*）

訳 DHT そしてたぶん他の男性ホルモンは，ドフェチリド誘導性の不整脈のリスクから正常の女性を保護するかもしれない

❷ **perhaps by ～（たぶん～によって）**

The crystal structure of the Vβ8.2-SEC3 complex suggests that the CDR2 mutations act by disrupting Vβ main chain interactions with SEC3, perhaps by affecting the conformation of CDR2.（*J Exp Med. 2000 191:835*）

訳 たぶん CDR2 の構造に影響を与えることによって

★ possibly（もしかしたら／たぶん）

❷ **possibly by ～（もしかしたら～によって）**

It has been hypothesized that β-catenin may also regulate cell migration and cell shape changes, possibly by regulating the microtubule cytoskeleton via interactions with APC.（*Dev Biol. 2001 235:33*）

訳 もしかしたら APC との相互作用によって微小管細胞骨格を調節することによって

❸ **possibly through ～（もしかしたら～によって）**

At higher concentrations, openers reduced ATPase activity, possibly through stabilization of MgADP at the channel site.（*FASEB J. 2000 14:1943*）

訳 もしかしたらチャネル部位における MgADP の安定化によって

★★ potentially（潜在的に／もしかすると）

❶ **potentially important（もしかすると重要な）**

This finding is potentially important in elucidating the physiological function of Trp.（*J Biol Chem. 2000 275:11934*）

訳 この知見は，もしかすると Trp の生理学的機能を解明するのに重要かもしれない

18. 容易に／うまく 【readily】

容易に	うまく
readily easily	successfully

使い分け

◆ **readily**, **easily** は「容易に」という意味で，detected の前などに用いられる．

◆ **successfully** は「うまく」という意味で，used の前などで使われる．

副詞

頻度分析

	用例数		用例数
readily	1,970	❶ readily detected	112
		❷ readily available	97
		❸ readily detectable	95
easily	846	❶ easily detected	37
successfully	1,194	❶ successfully used	68
		❷ successfully to	59
		❸ successfully applied	52

I

B

★ readily（容易に／すぐに）

❶ readily detected（容易に検出される）

In contrast, NOS2 mRNA and enzyme activity was readily detected in the spleens of infected mice.（*J Immunol. 2001 166:1912*）

訳 NOS2 メッセンジャー RNA と酵素活性は，感染したマウスの脾臓において容易に検出された

★ easily（容易に）

❶ easily detected（容易に検出される）

This virus was easily detected in the same cells by immunostaining and PCR.（*Cancer Res. 1999 59:6103*）

訳 このウイルスは，免疫染色と PCR によって同じ細胞において容易に検出された

★ successfully（うまく）

❶ **successfully used**（うまく使われる）

MALDI-TOF mass spectrometry was successfully used to measure these probes simultaneously.（*Anal Chem. 2001 73:2126*）

訳 MALDI-TOF 質量分析法は，これらのプローブを同時に測定するためにうまく使われた

19. 効率的に／活発に 【efficiently】

効率的に	活発に
efficiently effectively	actively

使い分け

◆ **efficiently** は「効率的に」，**effectively** は「効果的に／効率的に」，**actively** は「活発に」という意味に用いられる．

頻度分析

	用例数		用例数
efficiently	2,861	❶ more efficiently	339
effectively	1,951	❶ more effectively	169
actively	729	❶ actively transcribed	49

★ efficiently（効率的に）

❶ **more efficiently**（もっと効率的に）

E6/E7 induced immortalization of EGF-R wild-type cells 5-fold more efficiently than null cells.（*Cancer Res. 2000 60:4397*）

訳 E6/E7 は，EGF-R 野生型細胞の不死化をヌル細胞より 5 倍効率的に誘導する

★ effectively（効果的に／効率的に）

❶ **more effectively**（もっと効果的に／もっと効率的に）

The three altered proteins defective in single-stranded DNA binding cannot mediate the annealing of homologous DNA, whereas gp2.5-Δ26C mediates the reaction more effectively than does wild-type.（*J Biol Chem. 2003 278:7247*）

訳 gp2.5-Δ26C は野生型よりもっと効率的に反応を仲介する

★ actively（活発に）

❶ **actively transcribed**（活発に転写される）

While much is known about the roles of histone methyltransferases (HMTs) in the establishment of heterochromatin, little is known of their roles in the regulation of actively transcribed genes.（*Mol Cell Biol. 2003 23:5972*）

訳 活発に転写される遺伝子の調節におけるそれらの役割についてはほとんど知られていない

副詞

20. 適切に 【properly】

適切に	正確に
properly adequately appropriately	correctly

使い分け

◆ **properly**, **adequately**, **appropriately** は，「適切に」という意味に用いられる．

◆ **correctly** は，「正確に」という意味で使われる．

I

頻度分析

	用例数		用例数
properly	450	❶ properly folded	37
adequately	258	❶ adequately described	11
appropriately	346	❶ appropriately to	26
correctly	719	❶ correctly identified	84

B

properly（適切に／正しく）

❶ **properly folded**（適切に折りたたまれた）

We show that both proteins were secreted at high levels and that the purified proteins were properly folded.（*J Mol Biol. 2002 324:165*）

訳 精製されたタンパク質は適切に折りたたまれた

adequately（適切に／適当に）

❶ **adequately described**（適切に記述される）

The rate of decline was adequately described by a simple linear function.（*Invest Ophthalmol Vis Sci. 2000 41:325*）

訳 減少の速度は単純な一次関数によって適切に記述された

appropriately（適切に）

❶ **appropriately to ～**（～に適切に）

During animal development, cells have to respond appropriately to localized secreted signals.（*Nature. 2000 403:789*）

訳 細胞は局在化された分泌シグナルに適切に応答しなければならない

★ correctly（正確に／正しく）

❶ **correctly identified**（正確に同定される／正確に同定した）

Using the LRP NAP test, 47（94%）out of 50 isolates were correctly identified as tuberculosis complex.（*J Clin Microbiol. 2001 39:3883*）

訳 50の分離株のうち47（94%）は結核複合体として正確に同定された

21. 一様に【uniformly】

一様に	
uniformly	evenly

使い分け ◆ **uniformly**, **evenly** は distributed の前などで,「一様に」という意味に用いられる.

頻度分析

	用例数		用例数
uniformly	505	❶ uniformly distributed	66
evenly	123	❶ evenly distributed	56

★ uniformly（一様に）

❶ **uniformly distributed**（一様に分布する）

PMR1 is uniformly distributed throughout the cytoplasm on polysomes

and in lighter complexes and does not colocalize in cytoplasmic foci with Dcp1. (*Mol Cell. 2004 14:435*)
訳 PMR1は細胞質でくまなく一様に分布する

evenly（一様に）

❶ **evenly distributed**（一様に分布する）

In cells expressing E-cadherin lacking the cytoplasmic region, β-catenin was evenly distributed in the cytoplasm. (*Cancer Res. 2000 60:7057*)
訳 β-カテニンは細胞質で一様に分布した

22. 緊密に／直接 【closely】

緊密に	直接
closely tightly	directly

使い分け ◆ **closely**, **tightly** は「緊密に」, **directly** は「直接」という意味で使われる.

頻度分析

	用例数		用例数
closely	3,286	❶ closely related	1,574
		❷ closely linked	189
		❸ closely associated	185
tightly	1,453	❶ tightly regulated	274
		❷ tightly to	137
		❸ tightly bound	125
directly	8,471	❶ directly to	719
		❷ directly with	581
		❸ directly from	297

解説 ◆ **closely** は related の前, **tightly** は regulated の前, **directly** は to の前などで用いられる.

⁑ closely（緊密に）

❶ **closely related**（緊密に関連した）

Analysis of this cDNA reveals that Hunk is most closely related to the SNF1 family of serine/threonine kinases and contains a newly described SNF1 homology domain.（*Genomics. 2000 63:46*）

訳 Hunkは，SNF1ファミリーのセリン／スレオニンキナーゼにもっとも緊密に関連している

* tightly（しっかりと／緊密に）

❶ **tightly regulated**（しっかりと調節される）

Taken together, our results support the concept that ganglioside biosynthesis is tightly regulated by the formation of glycosyltransferase complexes in the ER and/or Golgi.（*Biochemistry. 2002 41:11479*）

訳 ガングリオシド生合成は糖転移酵素複合体の形成によってしっかりと調節される

⁑ directly（直接）

❶ **directly to ～**（～に直接）

We demonstrated with recombinant proteins that the PPARγ/RXRα heterodimer binds directly to this PPAR response element.（*J Biol Chem. 2004 279:23908*）

訳 PPARγ/RXRαヘテロ二量体はこのPPAR応答エレメントに直接結合する

23. 速く／手短に　【rapidly】

速く	手短に
rapidly quickly	briefly

使い分け

◆ **rapidly**, **quickly** はmoreの後などで，「速く／急速に」という意味に用いられる．

◆ **briefly** は文頭やdiscussedの前などで，「手短に」という意味で使われる．

頻度分析	用例数		用例数
rapidly	4,901	❶ more rapidly	417
quickly	425	❶ more quickly	63
briefly	172	❶ briefly discussed	20

★★ rapidly（速く／急速に）

❶ more rapidly（より速く）

VC＋ tumors grew more rapidly than mock-transfected tumors and exhibited parallel increases in tumor angiogenesis.（*Cancer Res. 2001 61:2404*）

訳 VC＋腫瘍はニセの遺伝子導入された腫瘍よりも速く増殖した

★ quickly（速く／急速に）

❶ more quickly（より速く）

In vitro studies show that ARF6 T157A can spontaneously bind and release GTP more quickly than the wild-type protein suggesting that it is a fast cycling mutant.（*J Biol Chem. 2002 277:40185*）

訳 ARF6 T157Aは，野生型タンパク質よりも速く自然にGTPを結合および放出する

briefly（手短に／短時間に）

❶ briefly discussed（手短に議論される）

A mechanism for copper translocation is briefly discussed.（*Biochemistry. 2000 39:7337*）

訳 銅の移行の機構が手短に議論される

24. 同時に 【simultaneously】

同時に		
simultaneously	concomitantly	concurrently

使い分け ◆ **simultaneously**，**concomitantly**，**concurrently** は with の前などで，「同時に」という意味に用いられる．

副詞 I B

頻度分析

	用例数		用例数
simultaneously	1,800	❶ simultaneously with	143
		❷ to simultaneously	123
		❸ simultaneously in	103
concomitantly	331	❶ concomitantly with	78
concurrently	291	❶ concurrently with	67

★ simultaneously（同時に）

❶ simultaneously with ～（～と同時に）

These data also show that the NLS of IFNγ（95-132）can interact simultaneously with IFNGR-1 and the nuclear import machinery.（*Biochemistry. 2004 43:5445*）

訳 IFNγ（95-132）の NLS は，IFNGR-1 と核移行機構に同時に相互作用できる

concomitantly（同時に）

❶ concomitantly with ～（～と同時に）

An increase in cdk4 kinase activity occurs concomitantly with the increase in cyclin D mRNA.（*Mol Cell Biol. 2002 22:4863*）

訳 cdk4 リン酸化酵素活性の上昇は，サイクリン D メッセンジャー RNA の上昇と同時に起こる

concurrently（同時に）

❶ concurrently with ～（～と同時に）

We found that cleavage at Asp720 occurred concurrently with caspase 3 activation and the increased production of total secreted Aβ and Aβ1-42 in association with staurosporine- and etoposide-induced apoptosis.（*J Biol Chem. 2003 278:46074*）

訳 Asp720 における切断は，カスパーゼ 3 の活性化と同時に起こった

25. 同様に／同等に　【similarly】

同様に	～に加えて	同等に
similarly identically	as well as not only ～ but also	equally equivalently

使い分け
- ◆**similarly**, **identically**, **as well as** は「同様に」という意味に用いられる．
- ◆**as well as** は「～に加えて／同様に」，**not only ～ but also** は「～だけでなく…も」という意味で使われる．
- ◆**equally** は「同じぐらい／等しく」，**equivalently** は「同等に」という意味に用いられる．

頻度分析

	用例数		用例数
similarly	1,489	❶ similarly to	221
identically	95	❶ identically to	21
as well as	12,698	❶ as well as	12,698
not only ～ but also	1,001	❶ not only ～ but also in	131
equally	919	❶ equally well	105
		❷ equally effective	92
equivalently	70	❶ equivalently in	14

★ similarly（同様に）

❶ similarly to ～（～と同様に）

We have recently demonstrated that S6K2 is regulated similarly to S6K1 by the mammalian target of rapamycin pathway and by multiple PI3-K pathway effectors *in vivo*.（*J Biol Chem. 2001 276:7892*）

訳 S6K2 は S6K1 と同様に…によって調節される

identically（同様に）

❶ identically to ～（～と同様に）

Functionally, xHP1α behaves identically to human HP1α.（*EMBO J. 2003 22:3164*）

訳 xHP1α はヒトの HP1α と同様に働く

副詞　I　B

⁑ as well as（～に加えて／同様に）

❶ as well as ～（～に加えて／～と同様に）

We discovered that, in these cells, YY1 activated endogenous Msx2 gene expression as well as Msx2 promoter-luciferase fusion gene activity.（*Nucleic Acids Res. 2002 30:1213*）

訳 YY1は，Msx2プロモーター・ルシフェラーゼ融合遺伝子活性に加えて内在性のMsx2遺伝子発現も活性化した

★ not only ～ but also（～だけでなく…も）

❶ not only ～ but also in（～だけでなく…においても）

This transgene is expressed not only in peripheral T cells, but also in immature thymocytes and thymocytes undergoing positive selection, in agreement with endogenous IL-2 expression.（*J Immunol. 2001 166:1730*）

訳 この導入遺伝子は末梢のT細胞においてだけでなく，未成熟な胸腺細胞においても発現する

★ equally（同じぐらい／等しく）

❶ equally well（同じぐらいよく）

MARCKS-(151-175) binds equally well to $PI_{4,5}P_2$ and $PI_{3,4}P_2$.（*J Biol Chem. 2001 276:5012*）

訳 MARCKS-(151-175) は，$PI_{4,5}P_2$ と $PI_{3,4}P_2$ に同じぐらいよく結合する

equivalently（同等に）

❶ equivalently in ～（～において同等に）

By contrast, both alleles were transcribed equivalently in Sp3-rich hepatocytic HepG2 cells.（*J Immunol. 2001 167:5838*）

訳 両方のアレルは，Sp3に富む肝細胞のHepG2細胞において同等に転写された

26. 異なって／別々に／独立に 【differently】

異なって／別々に	個々に／独立に
differently separately	individually independently

使い分け
◆ **differently** は「異なって」，**separately** は「別々に」，**individually** は「個々に／それぞれ」，**independently** は「独立的に」という意味に用いられる．

頻度分析

	用例数		用例数
differently	454	❶ differently to	64
separately	604	❶ separately from	48
individually	771	❶ individually and	69
independently	2,966	❶ independently of	1,019

★ differently（異なって／別々に）

❶ **differently to ～**（～に異なって）

We find that the two Q cells respond differently to EGL-20 because they have different response thresholds.（*Mol Cell. 1999 4:851*）

訳 2つのQ細胞はEGL-20に異なって応答する

★ separately（別個に／別々に）

❶ **separately from ～**（～とは別個に）

A cell-free system was used to evaluate the activation of phosphatase separately from MEK inactivation.（*Dev Biol. 2001 236:244*）

訳 無細胞系は，MEKの不活性化とは別個にホスファターゼの活性化を評価するために使われた

★ individually（個々に／それぞれ）

❶ **individually and ～**（それぞれ，そして～）

In undifferentiated odontoblasts, Nrf1 and C/EBPβ repress DSPP promoter activity individually and synergistically by cooperatively interacting with each other.（*J Biol Chem. 2004 279:45423*）

訳 Nrf1 と C/EBPβ は，DSPP プロモータ活性を個々にそして相乗的に抑制する

★ independently（独立的に）

“independently of” の用例が非常に多い．

❶ independently of ～（～とは独立して）

However, appropriate tissue-specific repression of the cyclin A1 promoter occurs independently of CpG methylation.（*Mol Cell Biol. 2000 20:3316*）

訳 サイクリン A1 プロモータの適切な組織特異的抑制は CpG メチル化とは独立して起こる

27. 逆に 【inversely】

逆に	不利に
inversely oppositely	adversely

使い分け

◆ **inversely** は correlated の前などに用い，関係が逆であることを意味することが多い．

◆ **adversely** は「不利に」という意味で，affect の前で用いられことが多い．

頻度分析

	用例数		用例数
inversely	804	❶ inversely correlated	188
		❷ inversely correlated with	158
		❸ inversely associated	147
oppositely	75	❶ oppositely charged	30
adversely	234	❶ adversely affect	98

解説

◆ **oppositely** は，**oppositely** charged の用例が多い．

★ inversely（逆に）

❷ inversely correlated with ～（～と逆に相関する）

The level of phosphorylated Akt was inversely correlated with the endogenous level of PTEN protein and overexpression of PTEN-blocked Akt phosphorylation in all cells analysed.（*Hum Mol Genet. 2001 10:251*）

訳 リン酸化されたAktのレベルは，PTENタンパク質の内在性のレベルと逆に相関した

oppositely（逆に）

❶ oppositely charged（逆に荷電した）

We identify a pair of oppositely charged residues, E153 and R169, that comprise an intermolecular salt bridge within a functional Int multimer.（*Proc Natl Acad Sci USA. 2004 101:2770*）

訳 われわれは，一組の逆に荷電した残基E153とR169を同定する

adversely（不利に／逆に）

❶ adversely affect（不利な影響を与える）

Incubation with SCCA did not adversely affect cell viability（typically >98%).（*J Periodontol. 2001 72:1059*）

訳 SCCAとのインキュベーションは，細胞の生存度には不利な影響を与えなかった

【I-C. 頻度・時】 頻度・時を表す副詞

28. 連続的に 【continuously】

連続的に	進行性に／連続的に	持続的に	引き続いて
continuously **serially**	**progressively**	**persistently**	**subsequently**

使い分け

- ◆ **continuously**, **serially** は「連続的に」という意味に用いられる.
- ◆ **continuously** は時間の連続性を示す場合に, **serially** は連続的に何かが行われるときに用いられる.
- ◆ **progressively** は「進行性に／連続的に」, **persistently** は「持続的に」, **subsequently** は「引き続いて」という意味で使われる.

頻度分析

	用例数		用例数
continuously	582	❶ continuously for	24
serially	178	❶ serially sectioned	15
progressively	720	❶ progressively more	46
		❷ progressively increased	45
persistently	257	❶ persistently infected	64
subsequently	2,175	❶ and subsequently	563

★ continuously（連続的に）

❶ continuously for ～（～の間連続的に）

The CCD image of the waveguide was monitored continuously for 25 min.（*Anal Chem. 2001 73:5518*）

訳 導波管の CCD イメージが 25 分間連続的にモニターされた

serially（連続的に）

❶ serially sectioned（連続的に切片にされる）

After infusion, rats were euthanized, and their brains were removed and serially sectioned.（*Cancer Res. 2002 62:6552*）

訳 それらの脳は取り出され，そして連続的に切片にされた

★ progressively（進行性に／連続的に）

❶ **progressively more ～**（進行性により～）

Although transthoracic echocardiography is often sufficient for this purpose initially, visualization of the coronary arteries becomes progressively more difficult as children grow.（*Circulation. 2002 105:908*）

訳 子供が成長するにつれて，冠動脈の可視化が進行性により困難になる

❷ **progressively increased**（連続的に上昇した）

In bats and technological sonars, the gain of the receiver is progressively increased with time after the transmission of a signal to compensate for acoustic propagation loss.（*Nature. 2003 423:861*）

訳 レシーバーのゲインは時間とともに連続的に上昇した

persistently（持続的に）

❶ **persistently infected**（持続的に感染した）

Examination of lymph nodes from the animals at 18 weeks postchallenge had shown that all six animals were persistently infected with challenge virus.（*J Virol. 2000 74:10489*）

訳 6匹の動物すべてが曝露されたウイルスに持続的に感染していた

★ subsequently（引き続いて）

❶ **～ and subsequently**（～そして引き続いて）

Rabbits were immunized against α-toxin and subsequently challenged with *S. aureus* strain Newman.（*Invest Ophthalmol Vis Sci. 2002 43:1109*）

訳 ウサギはα-毒素に対して免疫され，そして引き続いてブドウ球菌株 Newman に曝露された

29. しばしば 【frequently】

しばしば	
frequently	often

使い分け ◆ **frequently**, **often** は in や associated の前などで，「しばしば」という意味に用いられる．

頻度分析

	用例数		用例数
frequently	2,583	❶ frequently in	246
		❷ frequently associated	105
		❸ frequently observed	99
often	4,263	❶ often associated	160
		❷ often associated with	157
		❸ often in	107
		❹ often used	90

★ frequently（頻繁に／しばしば）

❶ frequently in ～（～において頻繁に）

We found that Myc activations occurred more frequently in p27 −/− lymphomas than in p27 +/+ tumors.（*Proc Natl Acad Sci USA. 2002 99:11293*）

訳 Mycの活性化はp27 −/−リンパ腫において，p27 +/+腫瘍においてより頻繁に起こった

★★ often（しばしば）

❷ often associated with ～（～としばしば付随する）

Up-regulation of CD44 is often associated with morphogenesis and tumor invasion.（*J Biol Chem. 2003 278:8661*）

訳 CD44の上方制御は形態形成と腫瘍浸潤にしばしば付随する

30. まれに 【rarely】

まれに	
rarely	**infrequently**

使い分け ◆ **rarely**はobservedの前，**infrequently**はoccursの後などで，「まれに」という意味に用いられる．

頻度分析

	用例数		用例数
rarely	552	❶ rarely observed	31
infrequently	101	❶ occurs infrequently	9

★ rarely（まれに／めったにない）

❶ rarely observed（まれに観察される）

Although this triad of symptoms is rarely observed in a single patient, a three-generation kindred with autosomal-dominant transmission of these three disorders has been reported as "PAPA syndrome"（MIM 604416).（*Am J Hum Genet. 2000 66:1443*）

訳 この症状の三徴候はまれに１人の患者において観察される

infrequently（まれに）

❶ occurs infrequently（まれに起こる）

Reinfarction occurs infrequently after fibrinolysis but confers increased risk of 30-day and 1-year mortality.（*Circulation. 2001 104:1229*）

訳 再梗塞は線維素溶解の後まれに起こる

31. 早期に／最初に／すぐに【early】

早期に／最初に	新たに	すぐに
early initially originally	newly	immediately soon readily

使い分け

- **early** は in の前などで，「早期に」という意味に用いられる．
- **initially**, **originally** は identified の前などで，「最初に」という意味で使われる．
- **newly** は synthesized の前などで，「新たに」という意味に用いられる．
- **immediately**, **soon**, **readily** は，「すぐに」という意味で使われる．

頻度分析

	用例数		用例数
early	14,953	❶ early in	990
		❷ early stages	560
		❸ early as	397
		❹ as early as	387
initially	1,874	❶ initially identified	89
originally	668	❶ originally identified	144

newly	2,329	❶ newly synthesized	409
		❷ newly identified	307
		❸ newly diagnosed	218
immediately	1,596	❶ immediately after	454
		❷ immediately adjacent	120
		❸ immediately upstream	118
soon	294	❶ soon after	185
readily	1,972	❶ readily available	97

★★ early（早期に／初期に／初期の）

副詞および形容詞として用いられる．

❶ early in ～〔～の初期（早期）に〕

These defects occurred early in development prior to the onset of measurable neurological or mitochondrial abnormalities.（*Mol Cell Biol. 2004 24:8195*）

訳 これらの欠損は発生の初期に起こった

❹ as early as ～（早ければ～／～と同じぐらい早期に）

High grade PIN-like lesions resembling early human prostate cancer were detected as early as 10 weeks of age.（*Oncogene. 2002 21:4099*）

訳 初期のヒト前立腺癌に似た高悪性度の PIN 様病変は早ければ 10 週齢で検出された

★ initially（最初に／初期に）

❶ initially identified（最初に同定された）

Caveolin-1 was initially identified as a phosphoprotein in Rous sarcoma virus-transformed cells.（*J Biol Chem. 2001 276:35150*）

訳 Caveolin-1 は，ラウス肉腫ウイルスに感染した細胞におけるリン酸化タンパク質として最初に同定された

★ originally（最初に）

❶ originally identified（最初に同定された）

p120-catenin（p120）was originally identified as a tyrosine kinase substrate, and subsequently shown to regulate cadherin-mediated cell-cell adhesion.（*Biochemistry. 2003 42:9195*）

訳 p120-catenin（p120）はチロシンリン酸化酵素の基質として最初に同定された

★ newly（新たに）

❶ newly synthesized ～（新たに合成された～）

Transport of newly synthesized proteins is not impaired.（*J Biol Chem. 2001 276:11461*）

訳 新たに合成されたタンパク質の輸送は障害されない

★ immediately（すぐに）

❶ immediately after ～（～の後すぐに）

Loading doses were administered immediately after the procedure, and the drugs were prescribed for 2 weeks.（*Circulation. 2001 104:539*）

訳 初回量が処置の後すぐに投薬された

soon（すぐに）

❶ soon after ～（～の後すぐに）

The jmj mutants died soon after birth, apparently as a result of respiratory insufficiency caused by rib and sternum defects in addition to the heart defects.（*Circ Res. 2000 86:932*）

訳 jmj変異体は出生の後すぐに死亡した

readily（すぐに／容易に）

❶ readily available（すぐに利用できる）

The apparatus required for the delivery of inhaled prostacyclin is simple, inexpensive, and readily available in most hospitals.（*Crit Care Med. 2002 30:2762*）

訳 吸引プロスタサイクリンの投与のために必要とされる装置は，単純，安価で，そしてほとんどの病院ですぐに利用できる

32. 以前に 【previously】

以前に	すでに	最近	今まで
previously before formerly	already	recently	to date

使い分け

- **previously** は reported の前などで，「以前に～した」という場合に用いられる．
- **before** は「前に」という意味で，days や years の後などで用いられる．**before** and after の用例も多い．
- **formerly** は known の前などで，以前の名前や説を示すときに使われる．
- **already** は known の前などで，「すでに～」という意味に用いられる．
- **recently** は identified の前などで，「最近～した」という場合に使われる．
- **to date** は reported や identified の後などで，「今まで」という意味で用いられる．

頻度分析

	用例数		用例数
previously	16,945	❶ previously reported	1,714
		❷ previously shown	1,227
		❸ previously described	992
before	7,681	❶ before and after	862
		❷ days before	172
formerly	209	❶ formerly known as	30
already	765	❶ already known	36
recently	7,052	❶ recently identified	552
		❷ recently reported	387
		❸ recently described	343
to date	888	❶ to date	888

☆ previously（以前に）

❶ previously reported（以前に報告した／以前に報告された）

We previously reported that expression of tumor necrosis factor-α

(TNFα) was attenuated in macrophages exposed to febrile range temperatures. (*J Biol Chem. 2000 275:9841*)

訳 われわれは，…ということを以前に報告した

❷ **previously shown**（以前に示した／以前に示された）

We have previously shown that the SCF^{Skp2}-mediated ubiquitination pathway plays an important role in Cdt1 degradation. (*J Biol Chem. 2004 279:17283*)

訳 われわれは，…ということを以前に示した

⁑ before（前に／以前に）

❶ **before and after ～**（～の前と後で）

Net protein balance was determined before and after treatment with oxandrolone. (*Ann Surg. 2003 237:422*)

訳 ネットのタンパク質バランスが，オキサンドロロンによる処置の前と後で決定された

❷ **days before ～**（～の…日前）

Animals were studied by NMR 2 days before therapy and 1 and 5 days after therapy. (*Cancer Res. 2001 61:2002*)

訳 動物はNMRによって治療の2日前，1日後および5日後に調査された

formerly（以前に）

❶ **formerly known as ～**（以前には～として知られていた）

We investigated the influence and mechanism of action of HMGB-1/-2 (formerly known as HMG-1/-2) on estrogen receptor α (ERα) and ERβ. (*J Biol Chem. 2004 279:14763*)

訳 以前にはHMG-1/-2として知られていた

* already（すでに／以前に）

❶ **already known**（すでに知られている）

Although BAFF is already known to bind two receptors, BCMA and TACI, we have identified a third receptor for BAFF that we have termed BAFF-R. (*Science. 2001 293:2108*)

訳 BAFFは2つのレセプターに結合するとすでに知られている

⁑ recently（最近）

❶ **recently identified**（最近同定された／最近同定した）

Sef was recently identified as a negative regulator of fibroblast growth

副詞

I

C

factor（FGF）signaling in a genetic screen of zebrafish and subsequently in mouse and humans.（*J Biol Chem. 2004 279:38099*）

訳 Sef は，線維芽細胞成長因子（FGF）シグナル伝達の負の調節因子として最近同定された

❷ **recently reported**（最近報告した／最近報告された）

We recently reported that the Krüppel-like zinc finger transcription factor KLF15 can induce adipocyte maturation and GLUT4 expression.（*J Biol Chem. 2003 278:2581*）

訳 われわれは，…ということを最近報告した

★ to date（今まで）

❶ **to date**（今まで）

Although in most plant species no more than two annexin genes have been reported to date, seven annexin homologs have been identified in *Arabidopsis*, Annexin Arabidopsis 1-7（AnnAt1-AnnAt7）.（*Plant Physiol. 2001 126:1072*）

訳 ほとんどの植物種において，2つ以上のアネキシン遺伝子は今まで報告されていない

33. 現在 【currently】

現在	今	今日	まだ
currently presently at present	now	today	still

使い分け

◆ **currently**, **presently** は，「現在」という意味で使われる．

◆ **at present** は，文頭などで「現在では」という意味に用いられる．

◆ **now** は「今」という意味で，we **now** report ／ show ／ demonstrate that の用例が多い．

◆ **today** は，「今日（こんにち）」という意味で，文頭などで用いられる．

◆ **still** は，「まだ」という意味に用いられる．

頻度分析

	用例数		用例数
currently	1,959	❶ currently available	239
		❷ currently used	130
		❸ currently unknown	95
presently	267	❶ presently unknown	32
at present	208	❶ At present,	89
now	4,528	❶ now report	593
		❷ now show	501
		❸ now demonstrate	273
today	189	❶ Today,	20
still	2,934	❶ still unclear	124
		❷ still not	102
		❸ still poorly	95

★ currently（現在）

❶ currently available（現在利用できる）

No specific vaccine for West Nile virus（WNV）is currently available for human use.（*J Infect Dis. 2004 190:2104*）

訳 ウエストナイルウイルス（WNV）に対する特異的なワクチンで，現在，ヒトに使うことができるものはない

presently（現在）

❶ presently unknown（現在知られていない）

Although the precise mechanism of action of ethanol（EtOH）is presently unknown, studies suggest that it acts, in part, by interfering with normal cerebellar functioning.（*J Neurosci. 2004 24:3746*）

訳 エタノール（EtOH）の作用の正確な機構は現在知られていない

at present（現在では）

文頭で用いられることが多い．

❶ At present, ～（現在では，～）

At present, however, such information is difficult to obtain directly through laboratory experiments.（*Nature. 2001 411:934*）

訳 しかし，現在では，そのような情報は研究室での実験によって直接得ることが困難である

副詞

I

C

★★ now（今）

❶ now report ～（今，～を報告する）

We now report that Rac and Cdc42 play distinct roles in regulating this asymmetry.（*J Cell Biol. 2003 160:375*）

訳 われわれは，今，…ということを報告する

❷ now show ～（今，～を示す）

We now show that the Rho kinase effector LIM kinase is responsible for this effect.（*Dev Cell. 2003 5:273*）

訳 われわれは，今，…ということを示す

today（今日）

文頭で用いられることも多い．

❶ Today, ～（今日，～）

Today, non-mammalian nervous systems continue to provide ideal platforms for the study of fundamental problems in neuroscience.（*Nature. 2002 417:318*）

訳 今日，非哺乳類の神経系は，…に対する理想的なプラットフォームを提供し続けている

★ still（まだ）

❶ still unclear（まだはっきりしない）

It is still unclear whether HIV-1 kills infected cells directly or indirectly.（*FASEB J. 2001 15:5*）

訳 HIV-1 が感染した細胞を直接的あるいは間接的に殺すかどうかは，まだはっきりしない

34. 将来　【prospectively】

予見的に	今後は	その後
prospectively	hereafter	thereafter

使い分け

◆ **prospectively** は studied や evaluated の前などで，「予見的に」という意味に用いられる．

◆ **hereafter** は，referred の前などで「今後は（～と呼ぶ）」という形で使われる．

◆ **thereafter** は，「その後」という意味に用いられる．

頻度分析	用例数		用例数
prospectively	687	❶ prospectively evaluated	48
hereafter	44	❶ hereafter referred to as	12
thereafter	369	❶ Thereafter,	59

★ prospectively（予見的に）

❶ prospectively evaluated ～（～を予見的に評価した／予見的に評価された）

We prospectively evaluated nondietary and dietary determinants of iron stores.（*Am J Clin Nutr. 2003 78:1160*）

訳 われわれは，鉄貯蔵の非食餌性および食餌性の決定基を予見的に評価した

hereafter（今後は／これから先）

❶ hereafter referred to as ～（今後は～と呼ぶ）

We examined the relationship between mitogen-activated MEK (mitogen and extracellular signal-regulated protein kinase kinase) and phosphorylation of the gene product encoded by retinoblastoma (hereafter referred to as Rb) in vascular smooth muscle cells.（*J Biol Chem. 2004 279:24899*）

訳 今後は Rb と呼ぶ

thereafter（その後）

❶ Thereafter, ～（その後，～）

Thereafter, 25 tissue samples of each dog were collected for further analysis.（*J Clin Microbiol. 2000 38:2191*）

訳 その後，それぞれのイヌの 25 の組織サンプルがさらなる解析のために集められた

第5章

接続詞・接続語編

英語論文の書き方に関する本を読むと，よく「長い文は避けよ」と書いてあるのを目にする．確かにそのとおりなのだろうが，実際の英語論文の抄録を調べてみると，結構，長い文が多いのがわかる．われわれの調査では，英文抄録中の一文あたりの平均単語数は23.7語であった．つまり，「長い文は避けよ」ではなく，「長い文を書け」と言いたくなるぐらいだ．なぜ短い文が好まれないのかを考えてみると，1つはスタイルの問題であり，もう1つは文を短く刻むとそれぞれの文の意味はわかりやすくても，文章全体の意味がつかみにくくなるという問題があるからだろう．短い文をつないで意味の通じる文章をわかりやすく書くためには接続詞・接続語をいかにうまく使うかが大きなポイントになるのではないだろうか．本章では，文を接続するために論文でよく使われる単語・連語をまとめてあるので，ぜひ活用していただきたい．

【Ⅰ-A．逆説・比較】 逆説や比較の意味を持つ接続詞，副詞，前置詞，句

1．しかし／だが一方／にもかかわらず／それどころか 【however】

しかし	だが一方	にもかかわらず	それどころか
however but	whereas while	nevertheless nonetheless	on the contrary

使い分け

◆ **however** は前の文の内容に対して，「しかし」という意味で用いられることが非常に多い．

◆ **but** も「しかし」という意味で用いられるが，文頭には通常使われず，後半の文の先頭で用いられることが多い．

◆ **whereas, while** は，通常，文頭には用いられず，対立・対照・譲歩などの副詞節を導いて「だが一方」という意味で使われることが多い．

◆ **while** は，文頭で「～だけれども／～の間に」という意味に用いられることもある．

◆ **nevertheless, nonetheless** は前の文の内容を受けて，「にもかかわらず」という意味で用いられることが非常に多い．

◆ **on the contrary** は前に述べたことを否定して，「それどころか／それに比べて」という意味で用いられる．

頻度分析

	用例数		用例数
however	31,602	❶ However,	22,507
		❷ there is, however,	24
but	67,204	❶ but the	4,718
whereas	17,426	❶ , whereas	15,027
while	12,692	❶ , while	6,749
nevertheless	884	❶ Nevertheless,	663
nonetheless	367	❶ Nonetheless,	267
on the contrary	65	❶ On the contrary,	55

★ however（しかし）

文頭に使われることが非常に多い．文頭以外にも，副詞句の後の文の先頭，主語と述語動詞の間，be 動詞の後などで用いられる．文頭以外の場合は，前後にコンマを伴う．

❶ **However, ～（しかし，～）**

However, the mechanism by which the integrin β6 promotes oral tumor progression is not well understood.（*J Biol Chem. 2003 278:41646*）

訳 しかし，それによってインテグリンβ6が口腔癌の進行を促進する機構は，よく理解されてはいない

❷ **there is, however, ～（しかし，～がある）**

There is, however, little information related to the efficacy of this agent beyond the normal 6-mo assessment period.（*J Nucl Med. 2003 44:1*）

訳 しかし，…の効率に関連する情報はほとんどない

⁑ but（しかし）

文だけでなく，語句をつなぐ場合にも用いられる．文頭に使われることは通常ない．

❶ **but the ～（しかし～）**

Reduced serotonin transporter（5-HTT）expression is associated with abnormal affective and anxiety-like symptoms in humans and rodents, but the mechanism of this effect is unknown.（*Science. 2004 306:879*）

訳 しかし，この効果の機構は知られていない

⁑ whereas（だが一方～）

文頭に用いられることは非常に少ない．

❶ **, whereas ～（だが一方～）**

One degron leads to ubiquitylation on internal lysine(s), whereas the other is independent of ubiquitylation.（*J Biol Chem. 2004 279:23851*）

訳 だが一方，他方はユビキチン化とは無関係である

⁑ while（だが一方～／～だけれども／～の間に）

❶ **, while ～（だが一方～）**

Initiation of DNA unwinding *in vitro* appears to require a dimeric UvrD complex in which one subunit is bound to the ssDNA/dsDNA junction, while the second subunit is bound to the 3' ssDNA tail.（*J Mol Biol. 2003 325:913*）

訳 だが一方，2番目のサブユニットは3'の一本鎖DNA尾部に結合している

★ nevertheless（にもかかわらず）

文頭に用いられることが多い．

❶ **Nevertheless, ～**（にもかかわらず，～）

Nevertheless, we found that Bcl-x_L expression can stimulate cell respiration in cells with mitochondrial DNA.（*J Biol Chem. 2000 275:7087*）

訳 にもかかわらず，われわれはBcl-x_L 発現が細胞呼吸を刺激しうることを見つけた

nonetheless（にもかかわらず）

文頭に用いられることが多い．

❶ **Nonetheless, ～**（にもかかわらず，～）

Nonetheless, the role of these leukocytes remains poorly understood.（*Science. 2004 305:1773*）

訳 にもかかわらず，これらの白血球の役割はあまりよく理解されないままである

on the contrary（それどころか／それに比べて）

文頭に用いられることが多い．

❶ **On the contrary, ～**（それどころか，～）

On the contrary, we found that IFN-γ suppresses EAM.（*Circulation. 2001 104:3145*）

訳 それどころか，われわれはIFN-γがEAMを抑制することを見つけた

2. 対照的に／一方では／代わりに 【in contrast】

対照的に／逆に	～とは対照的に	一方では	代わりに
in contrast by contrast conversely	in contrast to as opposed to	on the other hand meanwhile	instead alternatively instead of

使い分け

◆ **in contrast**, **by contrast** は「対照的に」，**conversely** は「逆に」という意味に用いられる．

◆ **in contrast to**, **as opposed to** は，「～とは対照的に」という意味で使われる．

◆ **on the other hand**, **meanwhile** は，「一方では」という意味に用いられる．

◆ **instead**, **alternatively** は「その代わりに」，**instead of** は「～の代わ

りに」という意味で使われる．

頻度分析

	用例数		用例数
in contrast	13,631	❶ In contrast,	10,008
by contrast	1,247	❶ By contrast,	1,112
conversely	1,434	❶ Conversely,	1,286
in contrast to	3,394	❶ In contrast to	1,981
as opposed to	304	❶ , as opposed to	117
on the other hand	842	❶ On the other hand,	658
meanwhile	30	❶ Meanwhile,	25
instead	1,410	❶ Instead,	822
alternatively	859	❶ Alternatively,	276
instead of	537	❶ instead of	537

解説 ◆ **in contrast**, **by contrast**, **conversely**, **in contrast to**, **on the other hand**, **meanwhile**, **instead**, **alternatively** は，文頭で用いられることが多い．

接続詞

I

A

★★ in contrast（対照的に）

文頭に用いられることが多い．

❶ **In contrast, 〜**（対照的に，〜）

In contrast, expression of RAMPs 1, 2, and 3 was unaffected by low oxygen tension.（*FASEB J. 2003 17:1499*）

訳 対照的に，RAMP1，2，3の発現は低酸素圧によって影響を受けなかった

★ by contrast（対照的に）

文頭に用いられることが多い．

❶ **By contrast, 〜**（対照的に，〜）

By contrast, the most severely disruptive mutation（G392E）resulted, at age 13 years, in progressive myoclonus epilepsy, with many inclusions present in almost all neurons.（*Lancet. 2002 359:2242*）

訳 対照的に，もっとも重篤な破壊的な変異（G392E）は，13歳で，進行性ミオクローヌスてんかんを引き起こした

★ conversely（逆に）

文頭に用いられることが多い．

❶ **Conversely, ～（逆に，～）**

Conversely, overexpression of GSK-3α or GSK-3β enhances Thr-58 phosphorylation and ubiquitination of c-Myc.（*J Biol Chem. 2003 278:51606*）

訳 逆に，GSK-3α あるいは GSK-3β の過剰発現は，c-Myc の Thr-58 リン酸化およびユビキチン化を亢進する

☆☆ in contrast to（～とは対照的に）

文頭に用いられることが多い．

❶ **In contrast to ～（～とは対照的に）**

In contrast to previous reports, we find that SL4 binds weakly to NC（K_d =（± 14 μM）), suggesting an alternative function.（*J Mol Biol. 2001 314:961*）

訳 以前の報告とは対照的に

as opposed to（～とは対照的に）

文頭に用いられることは少ない．

❶ **, as opposed to ～（～とは対照的に）**

Only two isoforms of the adenine nucleotide translocase（Ant）protein have been identified in mouse, as opposed to the three in humans.（*Gene. 2000 254:57*）

訳 ヒトにおける 3 つとは対照的に，ただ 2 つのアイソフォームのアデニンヌクレオチド転位酵素（Ant）タンパク質だけがマウスにおいて同定されている

☆ on the other hand（一方では／他方では）

文頭に用いられることが多い．対応する表現として“on（the）one hand”が用いられることもあるが，頻度は少ない．

❶ **On the other hand, ～（一方では，～）**

On the other hand, the level of RGS9 mRNA was not affected by the knockout.（*Nature. 2001 411:843*）

訳 一方では，RGS9 メッセンジャー RNA のレベルはノックアウトによって影響されなかった

meanwhile（一方では）

文頭に用いられることが多い．

❶ **Meanwhile, ～（一方では，～）**

Meanwhile, genetic and reverse genetic approaches are providing evidence for the importance of natural products in host defence.（*Nature. 2001 411:843*）

訳 一方では，遺伝学的および逆遺伝学的アプローチは，…の重要性の証拠を提供している

★ instead（代わりに）

文頭に用いられることが多い．

❶ **Instead, ～（その代わりに，～）**

Instead, we found that recombinant proIL-18 was cleaved into smaller fragments by the complex of surface-associated and released amebic proteinases.（*Infect Immun. 2003 71:1274*）

訳 その代わりに，われわれは組換え型 proIL-18 がより小さな断片に切断されることを見つけた

★ alternatively（代わりに）

文頭に用いられることも多い．

❶ **Alternatively, ～（その代わりに，～）**

Alternatively, the additional channel isoforms may be present only during early development, when they may serve to strengthen collectively presynaptic release during critical periods of synaptogenesis.（*J Neurosci. 2001 21:412*）

訳 その代わりに，付加的なチャネルアイソフォームが初期の発生の間だけに存在するかもしれない

★ instead of（～の代わりに）

❶ **instead of ～（～の代わりに）**

Carboxymethylated transferrin is used instead of casein as a substrate for assaying rat MMP-7.（*Anal Biochem. 2001 293:38*）

訳 C 末端のメチル化されたトランスフェリンがカゼインの代わりに基質として使われる

3. ～だけれども／～にもかかわらず／～と違って 【although】

～だけれども	～にもかかわらず	～と違って／～に反して
although though even though while	despite in spite of albeit	unlike contrary to

使い分け

- **although**, **though** は副詞節を導き，「～だけれども」という意味に用いられる．
- **though** は，**even though** の形で用いられることが多い．
- **while** は，文頭で「～だけれども／～の間に」という意味に用いられることもあるが，後半の節を導いて「だが一方」という意味で使われることの方が多い．
- **despite**, **in spite of** は副詞句を導いて，「～にもかかわらず」という意味に用いられる．
- **albeit** は省略節を導く接続詞で，「～ではあるが」という意味で使われる．
- **unlike** は文頭で副詞句を導いて，「～と違って」という意味に用いられることが多い．
- **contrary to** は副詞句を導いて，「～に反して／～と違って」という意味に用いられることが多い．

頻度分析

	用例数		用例数
although	18,485	❶ Although	11,787
though	1,550	❶ Though	135
even though	900	❶ even though	900
while	12,692	❶ While	2,502
despite	6,073	❶ Despite	2,836
		❷ Despite this,	94
in spite of	163	❶ in spite of	163
albeit	403	❶ albeit at	88
		❷ albeit with	70
unlike	2,659	❶ Unlike	1,357
		❷ unlike in	54
contrary to	480	❶ Contrary to	285

⁑ although（〜だけれども／〜にもかかわらず）

文頭で副詞節を導くことが多い．

❶ Although 〜（〜だけれども／〜にもかかわらず）

Although the mechanism of this transition remains elusive, glycosylation has been proposed to impede the PrPC to PrPSc conversion.（*Proc Natl Acad Sci USA. 2003 100:7593*）

訳 この遷移の機構はわかりにくいままだけれども

* though（〜だけれども／〜にもかかわらず）

単独で用いるよりも，"even though" の用例の方が多い．

❶ Though 〜（〜だけれども／〜にもかかわらず）

Though the method has broad applicability, we focus our analysis on proteins and demonstrate its usefulness by examining protein-metal, protein-protein, protein-DNA, and protein-RNA interactions.（*Anal Chem. 2003 75:3281*）

訳 その方法は広い適用性を持っているけれども

* even though（〜だけれども）

"even if（たとえ〜だとしても）" と混同しないように注意が必要である．

❶ even though 〜（〜だけれども）

Conversely, staurosporine, which alters kinase domain structure, disrupted receptor binding, even though the catalytic activity of Jak3 is dispensable for receptor binding.（*Mol Cell. 2001 8:959*）

訳 Jak3 の触媒活性は受容体結合に不必要だけれども

⁑ while（〜だけれども／〜の間に／だが一方〜）

❶ While 〜（〜だけれども）

While expression of GSH-FDH often increases in the presence of metabolic or exogenous sources of formaldehyde, little is known about the factors that regulate this response.（*J Bacteriol. 2004 186:7914*）

訳 GSH-FDH の発現は代謝性あるいは外来性由来のホルムアルデヒド存在下でしばしば上昇するけれども

⁑ despite（〜にもかかわらず）

文頭で副詞句を導くことが多い．

❶ Despite 〜（〜にもかかわらず）

Despite the fact that the entire nucleotide sequence of its genome has

接続詞

I

A

recently become available, its mechanisms of pathogenicity are poorly understood.（*J Bacteriol. 2001 183:2384*）

訳 そのゲノムの全塩基配列が最近利用できるようになったという事実にもかかわらず

❷ Despite this, ～（これにもかかわらず，～）

Despite this, little is known about their pathogenesis.（*Hum Mol Genet. 2003 12:1875*）

訳 これにもかかわらず，それらの病因についてはほとんど知られていない

in spite of（～にもかかわらず）

文頭に用いられることは比較的少ない．

❶ in spite of ～（～にもかかわらず）

GIF did not bind human GM-CSF or IL-2 in spite of the fact that orf virus is a human pathogen.（*J Virol. 2000 74:1313*）

訳 オルフウイルスはヒトの病原体であるという事実にもかかわらず

★ albeit（～ではあるが／～にもかかわらず）

接続詞．省略節に用いられる．

❶ albeit at ～（～においてではあるが）

Similar incubation with 8-Cl-cAMP also resulted in accumulation of 8-Cl-ATP in the cells, albeit at a lower level.（*Cancer Res. 2001 61:5474*）

訳 より低いレベルにおいてではあるが

❷ albeit with ～（～によってではあるが）

A wide range of cell types are capable of initiating translation of c-myc by internal ribosome entry, albeit with different efficiencies.（*Nucleic Acids Res. 2000 28:687*）

訳 異なる効率によってではあるが

★ unlike（～と違って）

文頭に用いられることが多い．前置詞，接続詞，形容詞として用いられる．

❶ Unlike ～（～と違って）

Unlike wild type mice, hypoE mice were susceptible to diet-induced hypercholesterolemia, which was fully reversed within 3 weeks after resumption of a chow diet.（*J Biol Chem. 2002 277:11064*）

訳 野生型マウスと違って

❷ **unlike in ～**（～の場合とは違って）

However, unlike in human patients, there is no evidence of neuron loss or astrogliosis in the striatum.（*Hum Mol Genet. 2002 11:347*）

訳 ヒトの患者の場合とは違って，…の証拠はない

★ contrary to（～に反して／～と違って）

文頭に用いられることが多い．

❶ **Contrary to ～**（～に反して）

Contrary to previous reports, we find that MEK inhibitors dose-dependently inhibit OC differentiation.（*J Biol Chem. 2002 277:6622*）

訳 以前の報告に反して

4. ～と比較して 【compared with】

～と比較して	
compared with compared to	in comparison with in comparison to relative to

使い分け

◆ **compared with**, **compared to**, **in comparison with**, **in comparison to**, **relative to** は，「～と比較して」という意味に用いられる．

頻度分析

	用例数		用例数
compared with	15,619	❶ compared with	15,619
		❷ as compared with	1,364
compared to	5,508	❶ compared to	5,508
		❷ as compared to	559
in comparison with	447	❶ in comparison with	447
in comparison to	409	❶ in comparison to	409
relative to	4,093	❶ relative to	4,093

★★ compared with（～と比較して／～と比べて）

“as compared with”の用例もあるが，asは省略されることが多い．

❶ compared with ～（～と比較して）

Retinal lesions（acellular capillaries and pericyte ghosts）were not significantly（$P > 0.05$）present at 3 months in any experimental groups compared with the control group.（*Invest Ophthalmol Vis Sci. 2001 42:2964*）

訳 網膜の病変は，コントロール群と比較してどの実験群にも3カ月で有意には（$P > 0.05$）存在しなかった

❷ as compared with ～（～と比較して）

CCR4 was significantly elevated in bleomycin-treated mice as compared with control mice.（*J Immunol. 2004 173:4692*）

訳 CCR4はコントロールマウスと比較して，ブレオマイシン処置されたマウスにおいて有意に上昇した

★★ compared to（～と比較して／～と比べて）……………………

“as compared to”の用例もあるが，asは省略されることが多い．

❶ compared to ～（～と比較して）

Biochemical studies in mouse epidermis showed that cdk6 activity increased twofold in cdk4-deficient mice compared to wild-type siblings.（*Am J Pathol. 2002 161:405*）

訳 cdk6活性は野生型の同胞と比較して，cdk4欠損マウスにおいて2倍上昇した

❷ as compared to ～（～と比較して）

Using the array, 37 genes were up-regulated and 28 were down-regulated in FL cells as compared to normal GC B cells.（*Blood. 2002 99:282*）

訳 正常なGCのB細胞と比較して，FL細胞において28遺伝子が下方制御された

★ in comparison with（～と比較して／～と比べて）……………

❶ in comparison with ～（～と比較して）

Cisplatin mediates its cytotoxicity by forming covalent adducts on DNA, and we find that Δnhp6a/b mutants are hypersensitive to cisplatin in comparison with the wild-type strain.（*Biochemistry. 2002 41:5404*）

訳 Δnhp6a/b変異体は野性株と比較してシスプラチンに高感受性である

★ in comparison to（〜と比較して／〜と比べて）………………

❶ in comparison to 〜（〜と比較して）

In addition, after lead exposure renal function was significantly diminished in MT-null mice in comparison to WT mice.（*Am J Pathol. 2002 160:1047*）

訳 腎機能は，野生型マウスと比較してMT欠損マウスにおいて有意に低下した

★★ relative to（〜と比較して／〜と比べて）…………………………

❶ relative to 〜（〜と比較して）

It was previously shown that the V-ATPase must possess at least 5-10% activity relative to wild type to undergo *in vivo* dissociation in response to glucose withdrawal.（*J Biol Chem. 2003 278:12985*）

訳 V-ATPaseは，野生型と比較して少なくとも5〜10％の活性を持つに違いない

【I-B．肯定】 肯定の意味を持つ接続詞，副詞，前置詞，句

5. ～なので／～のせいで 【because】

～なので	～のせいで
because since as	due to because of owing to

使い分け

◆ **because**，**since** は従属節を導いて，「～なので」という意味で用いられることが多い．

◆ **since** は，「～以来」という意味で使われることも多い．

◆ **as** は従属節を導いて「～なので」という意味に用いられることも多いが，原因・理由以外にも時／比例／対照の意味でも用いられる．誤解をさけるために，**because** や **since** を使うことが望ましい．

◆ **due to**，**because of**，**owing to** は，「～のせいで／～の原因で」という意味に用いられる．

頻度分析

	用例数		用例数
because	14,543	❶ Because	5,218
since	4,762	❶ Since	2,192
as	129,146	❶ As the	436
due to	8,564	❶ due to	8,564
because of	3,860	❶ because of	3,860
owing to	379	❶ owing to	379

★★ because（～なので）

文頭に用いられることが比較的多い．

❶ **Because ～**（～なので）

Because the majority of human tumors possess mutant p53, it is important to know the molecular mechanism by which mutant p53 regulates RR and to what extent.（*Cancer Res. 2003 63:6583*）

訳 大多数のヒトの腫瘍は変異した p53 を持つので

★ since（～なので／～以来）

文頭に用いられることもかなり多い．

❶ **Since ～**（～なので）

Since it is known that DNA MTase levels are regulated by the ras-mitogen-activated protein kinase (MAPK) pathway, this study sought to determine whether decreased ras-MAPK signaling could account for the DNA hypomethylation in lupus T cells.（*Arthritis Rheum. 2001 44:397*）

訳 …ということが知られているので

★ as（～なので／～につれて／～であるとき／～のように）

asは，接続詞や前置詞などとして非常に多様な意味で用いられる．接続詞としても多義語なので，この意味ではbecauseやsinceを使う方が望ましい．

❶ **As the ～**（～なので）

As the p53-dependent apoptosis pathway is not well understood, we sought to identify apoptosis-specific p53 target genes using a subtractive cloning strategy.（*Genes Dev. 2000 14:704*）

訳 p53依存性のアポトーシス経路はよく理解されていないので

★★ due to（～のせいで／～の原因で）

副詞句を導いて「～のせいで」という意味で用いられるが，be動詞や名詞のあとに用いられることも多い．

❶ **due to ～**（～のせいで）

Due to the presence of two double bonds, formation of four different isomers is possible.（*J Med Chem. 2001 44:2270*）

訳 2つの二重結合の存在のせいで

★★ because of（～のせいで）

副詞句を導いて「～のせいで」という意味で用いられる．

❶ **because of ～**（～のせいで／～の理由で）

The exact mechanism by which EHEC induces disease remains unclear because of the lack of a natural animal model for the disease.（*J Infect Dis. 2002 186:1682*）

訳 その疾患に対する自然の動物モデルの欠如のせいで

owing to（～のせいで／～の原因で）

副詞句を導いて「～のせいで」という意味で用いられる．

❶ **owing to ～**（～のせいで）

This down-regulation of STAT4 is specific for IL-12 signaling, presumably owing to the prolonged activation of STAT4 induced by IL-12.（*Blood. 2001 97:3860*）

訳 IL-12によって誘導されるSTAT4の延長した活性化のせいで

6. 従って／それゆえ【therefore】

従って／それゆえ	従って／このように
therefore hence accordingly consequently	thus

使い分け ◆ **therefore**, **hence**, **accordingly**は「従って／それゆえ」, **consequently**は「従って／結果的に」, **thus**は「従って／このように」という意味で，文頭に用いられることが多い．

頻度分析

	用例数		用例数
therefore	8,602	❶ Therefore,	3,644
hence	1,556	❶ Hence,	700
accordingly	638	❶ Accordingly,	520
consequently	1,118	❶ Consequently,	614
thus	19,406	❶ Thus,	12,166

★★ therefore（従って／それゆえ）

文頭に用いられることが多い．

❶ **Therefore, ～**（従って，～）

Therefore, we examined whether p53 regulates the expression of genes required for global genomic repair.（*Proc Natl Acad Sci USA. 2002 99:12985*）

訳 従って，われわれは…かどうかを調べた

★ hence（従って／それゆえ）

文頭に用いられることが多い．

❶ **Hence, ～**（従って，～）

Hence, we hypothesized that HNO can exert selective toxicity to cells subjected to acidosis.（*J Biol Chem. 2003 278:42761*）

訳 従って，われわれはHNOは…に対する選択的な毒性を発揮できると仮定した

★ accordingly（従って／それゆえ）

文頭に用いられることが非常に多い．

❶ **Accordingly, ～**（従って，～）

Accordingly, we tested the hypothesis that the hemostatic system contributes to liver injury in LPS/RAN-treated rats.（*Hepatology. 2004 40:1342*）

訳 従って，われわれは…という仮説をテストした

★ consequently（従って／結果的に）

文頭に用いられることが多い．

❶ **Consequently, ～**（従って，～）

Consequently, we propose that E7.16 can directly target cdc25A transcription and maintains cdc25A gene expression by disrupting Rb/E2F/HDAC-1 repressor complexes.（*J Virol. 2002 76:619*）

訳 従って，われわれは…ということを提案する

★★ thus（従って／このように）

文頭に用いられることが非常に多い．

❶ **Thus, ～**（従って，～）

Thus, we conclude that the binding of Axin to LRP-5 is an important part of the Wnt signal transduction pathway.（*Mol Cell. 2001 7:801*）

訳 従って，われわれは…ということを結論する

7. それから／それによって 【then】

それから／今度は	そして	それによって／その結果
then in turn	and	thereby so that

使い分け
◆ **then** は「それから」，**in turn** は「今度は」，**and** は「そして」，**thereby** は「それによって」という意味に用いられる．
◆ **so that** は前にコンマを伴って，「その結果／それで」という意味で使われる．

頻度分析

	用例数		用例数
then	6,555	❶ We then	526
in turn	1,398	❶ , which in turn	628
and	902,291	❶ and the	50,395
thereby	3,124	❶ , thereby	2,105
		❷ and thereby	804
so that	893	❶ , so that	255

⁑ then（それから）

❶ We then ～（それから，われわれは～）

We then used this information to design a multiplex PCR assay based on the simultaneous amplification of fragments of these genes.（*J Clin Microbiol. 2003 41:3526*）

訳 それから，われわれは…するためにこの情報を使った

★ in turn（今度は／順に）

“, which in turn” の用例が非常に多い．

❶ , which in turn ～〔そして（それは）今度は～〕

The apoptosome recruits and activates caspase-9, which in turn activates caspase-3 and-7, which then kill the cell by proteolysis.（*J Biol Chem. 2001 276:34244*）

訳 アポプトソームはカスパーゼ9を動員して活性化し，そして（それは）今度はカスパーゼ3と7を活性化する

⁑ and（そして）

文だけでなく，語句をつなぐ場合にも用いられる．文頭に使われることは通常ない．

❶ and the ～（そして～）

Human PEG1 is transcribed from two promoters; the transcript from promoter P1 is derived from both parental alleles, and the transcript from P2 is exclusively from the paternal allele.（*J Biol Chem. 2002 277:13518*）

訳 プロモータP1からの転写は両親のアレルに由来し，そしてP2からの転写はもっぱら父系性のアレルに由来する

★★ thereby（それによって）

❶ , thereby ～（それによって～）

Production of AT would inactivate TRAP, thereby increasing trp operon expression.（*Mol Cell. 2004 13:703*）

訳 AT の産生は TRAP を不活性化し，それによって trp オペロンの発現を上昇させる

❷ and thereby ～（そしてそれによって～）

NSAIDs inhibit cyclooxygenase activity and thereby reduce prostaglandin synthesis; prostaglandins stimulate aromatase gene expression and thereby stimulate estrogen biosynthesis.（*JAMA. 2004 291:2433*）

訳 NSAID はシクロオキシゲナーゼ活性を抑制し，そしてそれによってプロスタグランジン合成を低下させる

★ so that（その結果／それで／～するように）

“so that”は，前にコンマがある場合には「その結果／それで」という意味に用いられる．コンマがない場合は，「～するために／～にするように」という意味で使われることが多い．

❶ , so that ～（その結果～／それで～）

We have proposed a new model of rat intestinal sugar absorption in which high glucose concentrations promote rapid insertion of GLUT2 into the apical membrane, so that absorptive capacity is precisely regulated to match dietary intake.（*J Physiol. 2004 560:281*）

訳 その結果，吸収能が食物摂取に釣り合うように正確に調節される

8. さらに／そのうえ／加えて　【furthermore】

さらに／そのうえ	～に加えて
furthermore	in addition to
further	besides
moreover	
in addition	
additionally	

使い分け ◆ **furthermore, moreover** は，文頭で「さらに」という意味に用いられることが非常に多い．

◆ **further** は，文頭などで「さらに」という意味の副詞として使われるだけでなく，「さらに進んだ」という意味の形容詞として使われる．
◆ **additionally**, **in addition** は，文頭で「そのうえ」という意味に用いられることが多い．
◆ **in addition to**, **besides** は，「～に加えて」という意味に用いられる．

頻度分析	用例数		用例数
furthermore	10,848	❶ Furthermore,	10,570
further	14,222	❶ Further,	1,178
moreover	5,999	❶ Moreover,	5,817
in addition	13,369	❶ In addition,	9,460
additionally	2,213	❶ Additionally,	1,855
in addition to	3,600	❶ in addition to	3,600
besides	258	❶ Besides	134

⁑ furthermore（さらに／そのうえ）

文頭に使われることが多い．

❶ Furthermore, ～（さらに，～）

Furthermore, we show that the interaction between BRLF1 and CBP is important for BRLF1-induced activation of the early lytic EBV gene SM in Raji cells.（*J Virol. 2001 75:6228*）

訳 さらに，われわれはBRLF1とCBPの間の相互作用が…にとって重要であるということを示す

⁑ further（さらに／そのうえ／さらに進んだ）

副詞および形容詞として用いられる．

❶ Further, ～（さらに，～）

Further, we demonstrate that NF1 is required for both the association of BRG1 chromatin remodeling complex and the GR on the promoter *in vivo*.（*Mol Cell Biol. 2003 23:887*）

訳 さらに，われわれはNF1が…に必要とされるということを実証する

⁑ moreover（さらに）

文頭に使われることが圧倒的に多い．

❶ Moreover, ～（さらに，～）

Moreover, we found that activation of p38 is required for caspase 3 and

9 cleavage, suggesting that potassium currents enhancement is required for caspase activation. (*J Neurosci. 2001 21:3303*)

訳 さらに，われわれはp38の活性化がカスパーゼ3と9の切断に必要とされるということを見つけた

★★ in addition（そのうえ／加えて）

文頭に用いられることが多い．

❶ In addition, ～（そのうえ，～）

In addition, we have identified an interaction between hREV7 and hMAD2 but not hMAD1. (*J Biol Chem. 2000 275:4391*)

訳 そのうえ，われわれは…の間の相互作用を同定した

★ additionally（そのうえ／さらに）

文頭に用いられることが非常に多い．

❶ Additionally, ～（さらに，～）

Additionally, we show that the movo1 promoter is activated by the lymphoid enhancer factor 1 (LEF1)/β-catenin complex, a transducer of wnt signaling. (*Proc Natl Acad Sci USA. 2002 99:6064*)

訳 さらに，われわれは…ということを示す

★★ in addition to（～に加えて）

❶ in addition to ～（～に加えて）

Cytohesin-1, in addition to its role in cell adhesion, is a guanine nucleotide-exchange protein activator of ARF GTPases. (*Proc Natl Acad Sci USA. 2002 99:2625*)

訳 細胞接着における役割に加えて

besides（～に加えて）

文頭に用いられることが多い．

❶ Besides ～（～に加えて）

Besides its key role in regulating complement-mediated cell lysis, Factor H also appears to play a role when "hijacked" by invading organisms in enabling cellular evasion of complement. (*J Biol Chem. 2000 275:16666*)

訳 補体に仲介される細胞溶解を調節する際の重要な役割に加えて

9. ～に一致して／～に従って 【in agreement with】

～に一致して	～に従って
in agreement with coincident with in accordance with in accord with	according to

使い分け

◆ **in agreement with**, **coincident with**, **in accordance with**, **in accord with** は，副詞句を導いて「～に一致して」という意味に用いられる．

◆ **according to** は，副詞句を導いて「～に従って」という意味で使われる．

頻度分析

	用例数		用例数
in agreement with	572	❶ In agreement with	188
coincident with	401	❶ , coincident with	98
in accordance with	164	❶ In accordance with	49
in accord with	150	❶ In accord with	34
according to	1,653	❶ According to	211

⋆ in agreement with（～に一致して）

❶ In agreement with ～（～に一致して）

In agreement with previous studies, there was no evidence for a similar preferential infection of CD4+ naive lymphocytes.（*J Virol. 2001 75:4091*）

訳 以前の研究に一致して，…の証拠はなかった

⋆ coincident with（～に一致して）

❶ , coincident with ～（～に一致して）

Viral antigens disappeared from the blood as early as 7 days after transfection, coincident with the appearance of antiviral antibodies.（*Proc Natl Acad Sci USA. 2002 99:13825*）

訳 抗ウイルス抗体の出現に一致して，早ければトランスフェクションの後７日でウイルス抗原が血液から消失した

in accordance with（～に一致して）

❶ **In accordance with ～**（～に一致して）

In accordance with this model, GCN2 bound several deacylated tRNAs with similar affinities, and aminoacylation of tRNAphe weakened its interaction with GCN2.（*Mol Cell. 2000 6:269*）

訳 このモデルに一致して，GCN2 はいくつかの脱アセチル化された tRNA に結合した

in accord with（～に一致して）

❶ **In accord with ～**（～に一致して）

In accord with these findings, survivin and COX-2 were frequently upregulated and co-expressed in human lung cancers *in situ*.（*FASEB J. 2004 18:206*）

訳 これらの知見に一致して，サバイビンと COX-2 はしばしば上方制御された

★ according to（～に従って）

❶ **According to ～**（～に従って）

According to this model, ArcA-P plays a central role in cydAB regulation by antagonizing H-NS repression of cydAB transcription when oxygen becomes limiting.（*Mol Microbiol. 2000 38:1061*）

訳 このモデルに従って，ArcA-P は cydAB の調節において中心的な役割を果たす

10. 実際に 【indeed】

実際に	
indeed	in fact

使い分け ◆ **indeed**, **in fact** は，文頭で「実際に」という意味に用いられることが多い．

頻度分析

	用例数		用例数
indeed	1,278	❶ Indeed,	676
in fact	548	❶ In fact,	257

★ indeed（実際に）

文頭に使われることが多い．

❶ **Indeed, ～**（実際に，～）

Indeed, we found that BMP2 treatment of MCF-7 cells decreased the association of PTEN with two proteins in the degradative pathway, UbCH7 and UbC9.（*Hum Mol Genet. 2003 12:679*）

訳 実際に，われわれは…ということを見つけた

★ in fact（実際に）

文頭に使われることが多い．

❶ **In fact, ～**（実際に，～）

In fact, we observed signal enhancement proportional to the amount of RCA product formed.（*Anal Biochem. 2004 333:246*）

訳 実際に，われわれは形成されたRCA産物の量に比例したシグナルの増強を観察した

11. たとえば／すなわち【for example】

たとえば	すなわち
for example for instance e.g.	namely i.e.

使い分け
◆ **for example**, **for instance**, **e.g.**は「たとえば」，**namely**, **i.e.**は「すなわち」という意味に用いられる．

頻度分析

	用例数		用例数
for example	1,269	❶ For example,	727
for instance	112	❶ For instance,	67
e.g.	2,175	❶ e.g.,	1,528
namely	716	❶ , namely	663
i.e.	2,841	❶ i.e.,	2,014

解説
◆ **for example**, **for instance**は，文頭で使われることが多い．
◆ **e.g.**, **namely**, **i.e.**は，文中で挿入的に語句の言い換えなどに用いられる．

★ for example（たとえば）

文頭に使われることが多い.

❶ For example, ～（たとえば，～）

For example, the Patched receptor limits the range of its ligand Hedgehog.（*Dev Biol. 2001 235:467*）

訳 たとえば，Patched受容体はそのリガンドであるHedgehogの範囲を限定する

for instance（たとえば）

文頭に使われることが多い.

❶ For instance, ～（たとえば，～）

For instance, PS1 binds to β-catenin and modulates β-catenin signaling.（*J Biol Chem. 2001 276:38563*）

訳 たとえば，PS1はβ-カテニンに結合し，そしてβ-カテニンシグナル伝達を調節する

★ e.g.（たとえば）

❶ e.g., ～（たとえば，～）

This suggests that T cell activation in hypoxic conditions *in vivo* may lead to different patterns of lymphokine secretion and accumulation of cytokines（e.g., vascular endothelial growth factor）affecting endothelial cells and vascular permeabilization.（*J Immunol. 2001 167:6140*）

訳 低酸素状態におけるT細胞の活性化は，異なるパターンのリンホカイン分泌とサイトカイン（たとえば，血管内皮増殖因子）の蓄積につながるかもしれない

★ namely（すなわち）

❶ , namely ～（すなわち～）

Three late assembly domain consensus motifs, namely PTAP, PPPY, and LYPXL, have been identified in different retroviruses.（*J Virol. 2004 78:6636*）

訳 3つのlate assemblyドメインのコンセンサスモチーフ，すなわちPTAP，PPPYおよびLYPXLは，異なるレトロウイルスにおいて同定されている

★ i.e.（すなわち）

❶ i.e., ～（すなわち，～）

Male rats underwent laparotomy（i.e., soft tissue trauma）and were bled to and maintained at a blood pressure of 40 mm Hg until 40% of shed blood volume was returned in the form of lactated Ringer's solution.（*Crit Care Med. 2000 28:2837*）

訳 雄のラットは開腹術（すなわち，軟組織外傷）を受けた

12. 同様に 【similarly】

同様に		
similarly	likewise	correspondingly

使い分け

◆ **similarly**, **likewise**, **correspondingly** は，「同様に」という意味に用いられる．

◆ **correspondingly** は，「対応して」という意味も持つ．

頻度分析

	用例数		用例数
similarly	2,886	❶ Similarly,	1,325
likewise	560	❶ Likewise,	384
correspondingly	173	❶ Correspondingly,	78

★ similarly（同様に）

文頭に用いられることが多い．

❶ Similarly, ～（同様に，～）

Similarly, in the developing rat aorta, we found increased PTEN activity associated with increased perlecan deposition and decreased SMC replication rates.（*Circ Res. 2004 94:175*）

訳 同様に，発生中のラット大動脈においてわれわれはPTEN活性の上昇を見つけた

★ likewise（同様に）

文頭に用いられることが多い．

❶ **Likewise, ～**（同様に，～）

Likewise, expression of a tetracycline-regulated wild-type p53 cDNA in p53-null fibroblasts caused a reduction in 53BP2 protein levels.（*Mol Cell Biol. 2000 20:8018*）

訳 同様に，p53 ヌル線維芽細胞におけるテトラサイクリン調節性の野生型 p53 cDNA の発現は 53BP2 タンパク質レベルの減少を引き起こした

correspondingly（同様に／対応して）

文頭に用いられることが多い．

❶ **Correspondingly, ～**（同様に，～）

Correspondingly, enzyme activity of AMACR increases approximately 4-fold in PCa in comparison with adjacent normal prostate.（*Cancer Res. 2003 63:7365*）

訳 同様に，AMACR の酵素活性は，…と比較して PCa において約 4 倍上昇する

13. まとめると【taken together】

まとめると	
taken together	**in summary**
collectively	**in conclusion**

使い分け

◆ **taken together**, **collectively** は，文頭で「まとめると」という意味に用いられることが多い．

◆ **in summary** は，文頭で「要約すると」という意味で使われる．

◆ **in conclusion** は，文頭で「まとめると／結論として」という意味に用いられる．

頻度分析

	用例数		用例数
taken together	3,060	❶ Taken together,	2,786
collectively	1,305	❶ Collectively,	985
in summary	894	❶ In summary,	884
in conclusion	1,436	❶ In conclusion,	1,426

★★ taken together（まとめると）

文頭に用いられることが多い.

❶ Taken together, ～（まとめると，～）

Taken together, these results suggest that the TGF-β-1-mediated inhibition of the flk-1/KDR gene is mediated by a 5'-untranslated region palindromic GATA site.（*J Biol Chem. 2001 276:5395*）

訳 まとめると，これらの結果は…ということを示唆する

★ collectively（まとめると）

文頭に用いられることが多い.

❶ Collectively, ～（まとめると，～）

Collectively, these data indicate that ZBP-89 regulates cell proliferation in part through its ability to directly bind the p53 protein and retard its nuclear export.（*Mol Cell Biol. 2001 21:4670*）

訳 まとめると，これらのデータは…ということを示している

★ in summary（要約すると）

文頭に用いられることが非常に多い.

❶ In summary, ～（要約すると，～）

In summary, we have identified a novel mechanism for NGF-induced activation of atypical PKC involving tyrosine phosphorylation by c-Src.（*Mol Cell Biol. 2001 21:8414*）

訳 要約すると，われわれは非定型プロテインキナーゼCのNGF誘導性の活性化のための新規の機構を同定した

★ in conclusion（まとめると／結論として）

文頭に用いられることが非常に多い.

❶ In conclusion, ～（まとめると，～）

In conclusion, we have identified KLF4 as a novel regulator of u-PAR expression that drives the synthesis of u-PAR in the luminal surface epithelial cells of the colon.（*J Biol Chem. 2004 279:22674*）

訳 まとめると，われわれは…を同定した

14. もし～なら／たとえ～だとしても 【if】

もし～なら	たとえ～でも	～でない限りは	いったん～すると
if Given	if any even if	unless	once

使い分け
- ◆科学論文では，条件節が使われることは比較的少ない．
- ◆文頭などの **if** は，「もし～なら」という意味に用いられる．
- ◆文頭などの **Given** は，「～を考慮に入れれば／～を考慮に入れて」という意味で使われる．
- ◆ **if any** は「たとえあったとしても」，**even if** は「たとえ～だとしても」，**unless** は「～でない限りは」という意味に用いられる．
- ◆文頭などの **once** は，「いったん～すると」という意味で使われる．

頻度分析

	使用回数		使用回数
if	6,835	❶ If	1,125
Given	870	❶ Given the	493
		❷ Given that	195
if any	273	❶ little, if any,	44
even if	168	❶ even if	168
unless	367	❶ unless	367
once	1,354	❶ Once	328

★★ if（もし～なら／～かどうか）

❶ If ～（もし～なら）

If this is the case, DNA repair and genome stability might be compromised in quiescent lymphocytes with potentially negative consequences.（*Oncogene. 2004 23:1911*）

訳 もしその場合は

★ Given（～を考えれば／～を考慮に入れて）

❶ Given the ～（～を考えれば／～を考慮に入れて）

Given the importance of BRG-1 in RB function, germ line BRG-1 mutations in tumorigenesis may be tantamount to RB inactivation.（*J Biol*

Chem. 2002 277:4782）

訳 RB機能におけるBRG-1の重要性を考えれば

❷ **Given that ～**（～ということを考えれば／～ということを考慮に入れて）

Given that the primary structure of all known proteases will soon be available, an important challenge is to define the structure-activity relationships that govern substrate hydrolysis.（*Anal Biochem. 2001 294:176*）

訳 すべての既知のプロテアーゼの一次構造がすぐに利用可能になることを考えれば

if any（たとえあったとしても）

❶ **little, if any,**（たとえあったとしても，ほとんどない）

Glucocorticoids have little, if any, effect on mRNA stability.（*Biochemistry. 2004 43:10851*）

訳 グルココルチコイドは，たとえあったとしても，メッセンジャーRNAの安定性に対する効果をほとんど持たない

even if（たとえ～だとしても）

"even though（～だけれども）" と混同しないように注意が必要である．

❶ **even if ～**（たとえ～だとしても）

Seven of 120（6%）physicians and 4 of 108 family members would not intubate or perform CPR even if there was a chance of recovery.（*Am J Respir Crit Care Med. 2002 166:1430*）

訳 たとえ回復の見込みがあったとしても

unless（～でない限りは）

❶ **unless ～**（～でない限りは）

Anonymous ESTs are of limited value unless they are connected to function.（*Bioinformatics. 2003 19:249*）

それらが機能に結びつかない限りは

★ once（いったん～すると／一度）

文頭に用いられることが多い．

❶ **Once ～**（いったん～すると）

Once activated, they dimerize and translocate to the nucleus and modulate the expression of target genes.（*Gene. 2002 285:1*）

訳 いったん活性化されると，それらは二量体化して核に移行する

付 録

付録 1

WebLSD の使い方

はじめに

ライフサイエンス辞書プロジェクトでは1993年より有志による独自の辞書構築を行ってきた．当初からコンピュータで利用できる各種の辞書を提供することを想定して，図1のようなリレーショナルデータベースとしてデータの維持・管理を行ってきた．データベースの内容を絶えず改良・拡充することにより，英和辞書，和英辞書，かな漢字変換辞書，スペルチェック辞書などの各種の辞書を常にアップトゥデートな形で提供することができる．これらの辞書は一貫してフリーウェアとして無償で提供されてきたほか，現在は市販のツール専用の辞書として販売されているものもある．

'96年11月からはウェブ上で最新の辞書を検索できるWebLSDのサービスを開始した．当初は英和辞書，和英辞書の検索と用例の表示のみであったが，その後，共起表現の表示（'97年12月），携帯電話向けの英和・和英辞書検索（2000年6月），ネイティブスピ

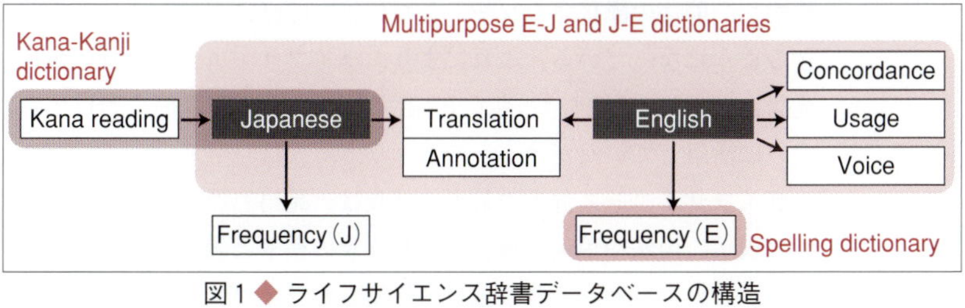

Multipurpose E-J and J-E dictionaries
Kana-Kanji dictionary
Kana reading
Japanese
Translation
Annotation
English
Concordance
Usage
Voice
Frequency (J)
Frequency (E)
Spelling dictionary

図1◆ライフサイエンス辞書データベースの構造

ーカーによる音声の提供（'02年11月），任意の語句に対するオンデマンド共起検索（'03年5月）などを順次追加し，現在に至っている．特に任意の語句に対する共起表現の検索はWebLSDならではのものであり，本書の内容を補完する意味でも活用していただきたい．スタート時には日に数百件程度であった検索件数も年を追うごとに増加し，現在では1日の検索件数が10万件を超すことも珍しくない．また，英和・和英を合わせた見出し語の数もスタート時の約6万8,000語から現在は約9万7,000語にまで増えている．

WebLSDは現在下記の3つのサーバで提供を行っている．メンテナンス等のためにやむを得ずサービスを停止することもあるので，頻繁に使用される方は複数のサーバをブックマーク登録しておくことをお勧めする．

http://lsd.pharm.kyoto-u.ac.jp/ja/
http://lsd.bioscinet.org/ja/
http://wwwsoc.nii.ac.jp/lsdproject/ja/

WebLSDの特長

WebLSDは，英和，和英，用例，共起表現，音声の5種類の辞書を統合したものであり，英和，和英の検索結果からは他の辞書をワンクリックで参照できるようになっている．また共起表現については，英和，和英検索の結果から参照できるだけでなく，ユーザが指定した任意の検索語句についてオンデマンドで高速にコーパスを検索し，前後の隣接語でソートした結果を即座に表示することができるようになっている．これにより，ライフサイエンス分野の専門英語について，「読む」「書く」「話す」のいずれの場面においても大いに活用していただけるものと思う．

ライフサイエンス辞書の最大の特長は，見出し語の選択，使用頻度の表示，用法や用例の選択，共起表現の検索と表示など，あらゆる面において，われわれが独自に構築したライフサイエンスコーパス（本書について2を参照）をベースにしている点である．ライフサイエンスコーパスは毎年更新されており，これに伴ってWebLSDの内容も年2回程度のペースで改訂されているので，常に最新の動向を反映した活きた英語を知ることができるようになっている．

もう1つの特長として，オンライン辞書ならではの双方向性がある．各検索結果の表示画面には新しい用語や対訳を受け付けるためのボタンを配し，ユーザからの意見を辞書にフィードバックできるようにした．ユーザから寄せられた新しい用語や対訳は，プロジェクトメンバーによる精査を経たのちデータベースに登録され，その後に製作される各種辞書にも反映される．ユーザからの新しい用語，対訳の提案は現在毎月約250件に達しており，われわれ自身による最新コーパスの解析と合わせて，辞書の充実に大きく貢献している．

英和・和英辞書の使い方

1）画面構成

日本語版のWebLSDのトップページのウィンドウは上下2つのフレームに分かれており，上のフレームには検索の結果が，下のフレームには検索語や検索オプションを入力するためのフォームが表示される（図2）．

2）検索語の入力

下のフレームの入力欄に検索語を入力しSearchボタンを押す．

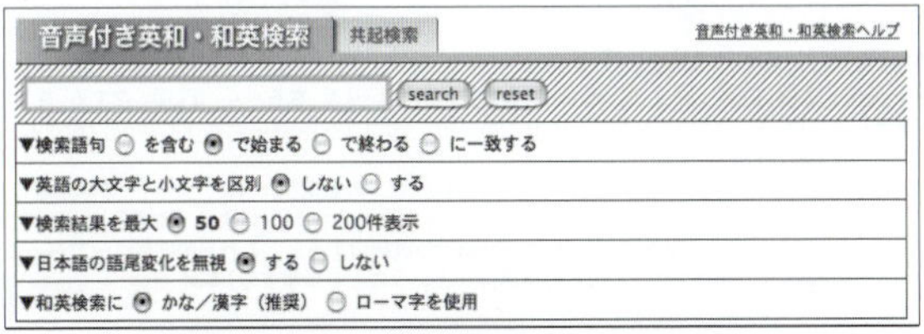
音声付き英和・和英検索
共起検索
音声付き英和・和英検索ヘルプ
search
reset
▼検索語句 を含む で始まる で終わる に一致する
▼英語の大文字と小文字を区別 しない する
▼検索結果を最大 50 100 200件表示
▼日本語の語尾変化を無視 する しない
▼和英検索に かな／漢字（推奨） ローマ字を使用

図2◆英和・和英検索の入力フォーム

検索語の文字種によって，英和検索か和英検索かが自動的に判断され実行される．英和検索を行うにはアルファベットを入力し，和英検索を行うには日本語の読みをひらがなで入力する．和英検索は漢字やカタカナでも可能であるが，「タンパク質」「たんぱく質」「蛋白質」のように表記のゆれがある場合はひらがなで入力するのがもっとも無難である．

3）オプションの選択

検索語の入力欄の下には，選択可能なオプションの一覧が表示されている（図2）．それぞれチェックのついた項目がデフォルトの設定である．例えば，標準では入力した文字で始まる語句を検索するようになっているが（前方一致），それ以外に，部分一致（検索語句を含む），後方一致（検索語句で終わる），完全一致（検索語句に一致する）を選ぶことができる．その他のオプションについてはオンラインヘルプを参照されたい．

4）英和検索結果の見方

英和辞書の検索結果の一例を図3に示す．見出し語に続く各項目について簡単に説明する．

- **頻度**：見出し語の右肩の＊印は，コーパス中での見出し語の出現頻度を5段階に分類した結果を示す．＊印の数が1つ多くなると，その語句の出現頻度はおよそ10倍になる見当である．
- **共起検索**：見出し語を検索語としてライフサイエンスコーパスを

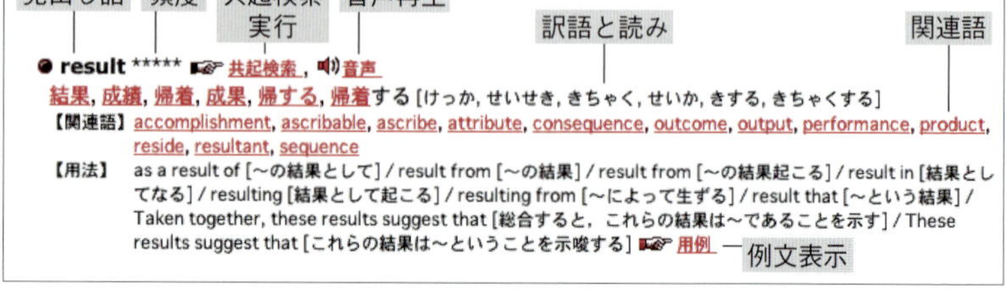

見出し語
頻度
共起検索実行
音声再生
訳語と読み
関連語
● result ＊＊＊＊＊ 共起検索，音声
結果，成績，帰着，成果，帰する，帰着する [けっか，せいせき，きちゃく，せいか，きする，きちゃくする]
【関連語】 accomplishment, ascribable, ascribe, attribute, consequence, outcome, output, performance, product, reside, resultant, sequence
【用法】 as a result of [～の結果として] / result from [～の結果] / result from [～の結果起こる] / result in [結果としてなる] / resulting [結果として起こる] / resulting from [～によって生ずる] / result that [～という結果] / Taken together, these results suggest that [総合すると，これらの結果は～であることを示す] / These results suggest that [これらの結果は～ということを示唆する] 用例
例文表示

図3◆英和検索結果の一例

検索し，左側の隣接語でソートした結果を表示する．より高度な検索が必要な時は後述のオンデマンド共起検索を利用するとよい．

- **音声**：ネイティブスピーカーによる発音例を再生する．現在はイギリス人男性によるもののみであるが，今後徐々に拡充の予定．
- **訳語と読み**：訳語をクリックすると和英の逆引きができる．
- **関連語**：データベースの中で，同じ訳語を持つ英語を関連語として表示する．品詞が異なる場合もある．
- **用法**：代表的な用法を日本語の訳とともに表示する．用例が収録されている場合は用例ページ（図 4）へのリンクが表示される．

5）和英検索結果の見方

和英辞書の検索結果の一例を図 5 に示す．見出し語に続く各項目について簡単に説明する．

- **頻度**：見出し語の右肩の / 印は，見出し語の出現頻度を 5 段階に分類した結果を示す．/ 印の数が 1 つ多くなると，その語句の出現頻度はおよそ 10 倍になる見当である．
- **訳語**：各行ごとに 1 つの英語訳が表示される．訳語をクリックすると英和の逆引きができる．それぞれの英語訳について，品詞，英語の出現頻度，共起検索へのリンクが，また該当する項目がある場合には，注釈，用例へのリンク，音声へのリンクが表示される．

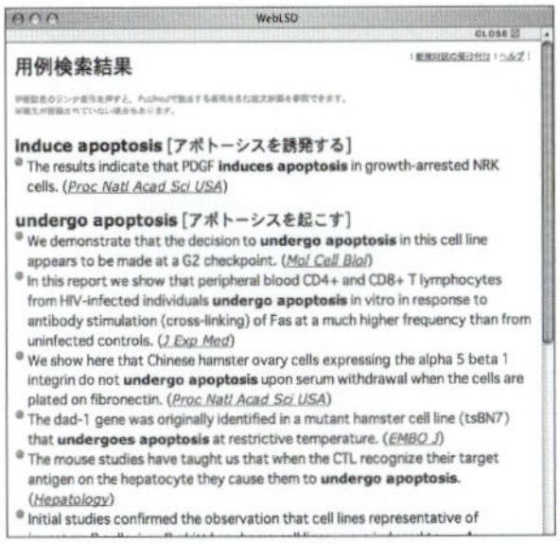
WebLSD
CLOSE
用例検索結果
ヘルプ
induce apoptosis [アポトーシスを誘発する]
The results indicate that PDGF induces apoptosis in growth-arrested NRK cells. (Proc Natl Acad Sci USA)
undergo apoptosis [アポトーシスを起こす]
We demonstrate that the decision to undergo apoptosis in this cell line appears to be made at a G2 checkpoint. (Mol Cell Biol)
In this report we show that peripheral blood CD4+ and CD8+ T lymphocytes from HIV-infected individuals undergo apoptosis in vitro in response to antibody stimulation (cross-linking) of Fas at a much higher frequency than from uninfected controls. (J Exp Med)
We show here that Chinese hamster ovary cells expressing the alpha 5 beta 1 integrin do not undergo apoptosis upon serum withdrawal when the cells are plated on fibronectin. (Proc Natl Acad Sci USA)
The dad-1 gene was originally identified in a mutant hamster cell line (tsBN7) that undergoes apoptosis at restrictive temperature. (EMBO J)
The mouse studies have taught us that when the CTL recognize their target antigen on the hepatocyte they cause them to undergo apoptosis. (Hepatology)
Initial studies confirmed the observation that cell lines representative of

図 4 ◆ 用法・用例表示の一例

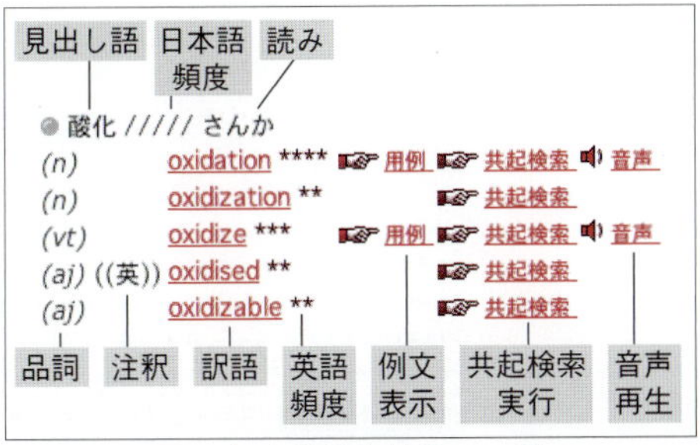
見出し語
日本語
頻度
読み
酸化 ///// さんか
(n) oxidation **** 用例 共起検索 音声
(n) oxidization ** 共起検索
(vt) oxidize *** 用例 共起検索 音声
(aj) ((英)) oxidised ** 共起検索
(aj) oxidizable ** 共起検索
品詞
注釈
訳語
英語
頻度
例文
表示
共起検索
実行
音声
再生

図 5 ◆ 和英検索結果の一例

付録1

オンデマンド共起検索の使い方

1）画面構成

WebLSDのトップページにおいて，入力フォーム（図2）の上部にある「共起検索」と書かれたタブをクリックすると，共起検索のメニューに切り替わる（図6）．すでに検索語が入力されている場合は新しいメニューにも引き継がれる．

2）検索語の入力

図6の入力欄に英語の検索語を入力しSearchボタンを押す．検索語としては単一の単語のほか，連続した語句（複合語）や簡単なフレーズを指定することもできる．デフォルトでは，各単語について語尾活用（規則活用および主要な不規則活用）を考慮した上で検索が行われる．

3）オプションの選択

検索語の入力欄の下には，選択可能なオプションの一覧が表示されている（図6）．それぞれチェックのついた項目がデフォルトの設定である．

4）共起検索結果の見方

標準ではライフサイエンスコーパスから検索語を含む文章を最大で300件抽出し，見出し語を中心にそろえて整列し，左側（直前）の隣接後でソートした形で結果を表示する．このような表示形式をKWIC（Keywords In Context）と言い，前後にどのような単語や表現が好んで使われるかを直感的かつ数量的にとらえることができ

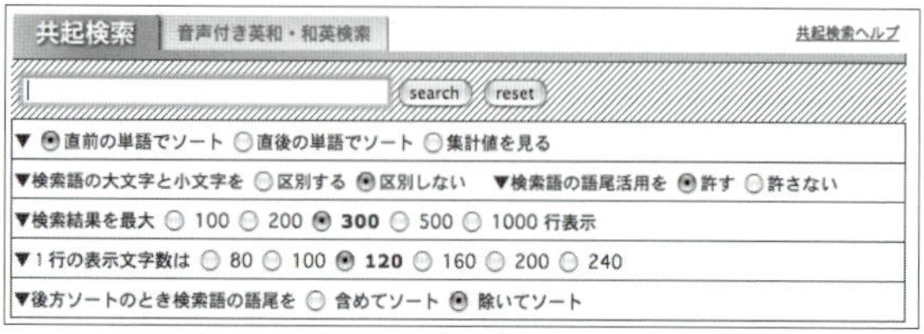

共起検索
音声付き英和・和英検索
共起検索ヘルプ
search
reset
▼ 直前の単語でソート 直後の単語でソート 集計値を見る
▼検索語の大文字と小文字を 区別する 区別しない ▼検索語の語尾活用を 許す 許さない
▼検索結果を最大 100 200 300 500 1000 行表示
▼1行の表示文字数は 80 100 120 160 200 240
▼後方ソートのとき検索語の語尾を 含めてソート 除いてソート

図6◆オンデマンド共起検索の入力フォーム

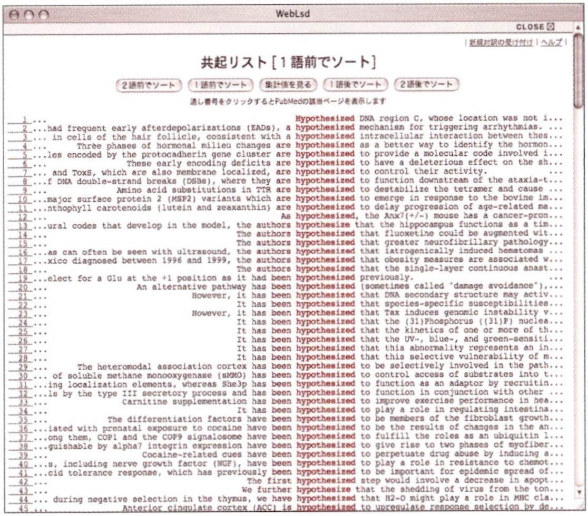
WebLsd
CLOSE
新規対訳の受け付け | ヘルプ
共起リスト [1語前でソート]
2語前でソート
1語前でソート
集計値を見る
1語後でソート
2語後でソート
通し番号をクリックするとPubMedの該当ページを表示します
1 ... Hypothesized DNA region C, whose location was not i...
2 ...had frequent early afterdepolarizations (EADs), a hypothesized mechanism for triggering arrhythmias. ...
3 ... in cells of the hair follicle, consistent with a hypothesized intracellular interaction between thes...
4 ... Three phases of hormonal milieu changes are hypothesized as a better way to identify the hormon...
5 ...les encoded by the protocadherin gene cluster are hypothesized to provide a molecular code involved i...
6 ... These early encoding deficits are hypothesized to have a deleterious effect on the sh...
7 ... and ToxS, which are also membrane localized, are hypothesized to control their activity. ...
8 ...f DNA double-strand breaks (DSBs), where they are hypothesized to function downstream of the ataxia-t...
9 ... Amino acid substitutions in TTR are hypothesized to destabilize the tetramer and cause ...
10 ...major surface protein 2 (MSP2) variants which are hypothesized to emerge in response to the bovine im...
11 ...nthophyll carotenoids (lutein and zeaxanthin) are hypothesized to delay progression of age-related ma...
12 ... As hypothesized, the Anx7(+/-) mouse has a cancer-pron...
13 ...ural codes that develop in the model, the authors hypothesize that the hippocampus functions as a tim...
14 ... The authors hypothesized that fluoxetine could be augmented wit...
15 ... The authors hypothesized that greater neurofibrillary pathology...
16 ...as can often be seen with ultrasound, the authors hypothesized that iatrogenically induced hematomas ...
17 ...xico diagnosed between 1996 and 1999, the authors hypothesized that obesity measures are associated w...
18 ... The authors hypothesized that the single-layer continuous anast...
19 ...elect for a Glu at the +1 position as it had been hypothesized previously. ...
20 ... An alternative pathway has been hypothesized (sometimes called "damage avoidance"),...
21 ... However, it has been hypothesized that DNA secondary structure may activ...
22 ... It has been hypothesized that species-specific susceptibilities...
23 ... However, it has been hypothesized that Tax induces genomic instability v...
24 ... It has been hypothesized that the (31)Phosphorus ((31)P) nuclea...
25 ... It has been hypothesized that the kinetics of one or more of th...
26 ... It has been hypothesized that the UV-, blue-, and green-sensiti...
27 ... It has been hypothesized that this abnormality represents an in...
28 ... It has been hypothesized that this selective vulnerability of m...
29 ... The heteromodal association cortex has been hypothesized to be selectively involved in the path...
30 ... of soluble methane monooxygenase (sMMO) has been hypothesized to control access of substrates into t...
31 ...ing localization elements, whereas She3p has been hypothesized to function as an adaptor by recruitin...
32 ...ls by the type III secretory process and has been hypothesized to function in conjunction with other ...
33 ... Carnitine supplementation has been hypothesized to improve exercise performance in hea...
34 ... It has been hypothesized to play a role in regulating intestina...
35 ... The differentiation factors have been hypothesized to be members of the fibroblast growth...
36 ...iated with prenatal exposure to cocaine have been hypothesized to be the results of changes in the an...
37 ...ong them, COP1 and the COP9 signalosome have been hypothesized to fulfill the roles as an ubiquitin l...
38 ...guishable by alpha7 integrin expression have been hypothesized to give rise to two phases of myofiber...
39 ... Cocaine-related cues have been hypothesized to perpetuate drug abuse by inducing a...
40 ...s, including nerve growth factor (NGF), have been hypothesized to play a role in resistance to chemot...
41 ...cid tolerance response, which has previously been hypothesized to be important for epidemic spread of...
42 ... The first hypothesized step would involve a decrease in apopt...
43 ... We further hypothesize that the shedding of virus from the ton...
44 ... during negative selection in the thymus, we have hypothesized that H2-O might play a role in MHC cla...
45 ... Anterior cingulate cortex (ACC) is hypothesized to upregulate response selection by de...

図7◆共起検索結果の一例
（1語前でのソート）

る．例としてhypothesizeを検索した結果の一部を図7に示す．「1語後でソート」と書かれたボタンをクリックすることにより，右側（直後）の隣接後でソートし直すことができる．「2語前」「2語後」についても同様である．また，「集計値を見る」のボタンをクリックすると，各位置での単語の出現数が頻度の高いものから順に一覧で表示される．

本書とともにWebLSDも活用し，活きた英語を論文に使っていただけたら幸いである．

（藤田信之）

付録1

付録 2

コーパス解析から見た日本語と英語表現の違い

ライフサイエンス辞書の作成を通して

本書やオンライン辞書サービスWebLSDの元となっているのは，独自に収集した専門文書を英語および日本語テキストとして蓄積して（これをコーパスという），それらを自作ツールなどで解析した数値データである．しかし，そのようにして収集した語句を関連づけて対訳テーブルを作成する作業は，今のところすべて人間が行っている．その際には対訳の「属性」や「意味」も考えながら対応づけをしてゆくので，作業中に日本語と英語のズレや差異に気づくことが多い．ここではそんな事例を数値データも交えながら紹介してみたい．

英語と日本語は1対1ではない ⇒ 表1

一般的には日本語のほうが英語よりも語彙数は少なく，1つの日本語がいくつかの意味を持っていることが多い．表1に示す「刺激」や「基質」の例では，文脈によっては異なる事物を意味している．

これに対して英語の場合には，語源によって意味が決まるためか，「discharge」といったよく使われて出現頻度の高い単語ほど，学問分野ごとにまったく異なる意味を有している場合が多い．ことに気をつけなくてはいけない領域は，数学や物理学，心理学や精神医学，そして植物学や農学である．

類義語は翻訳時に混同されることがある ⇒ 表2

典型的なのは，薬物を表す用語である英語の「agent」と「drug」の関係が日本語における「剤」と「薬」に対応していない例である．英語において，agentが“作用する機能を有する物質の総称”

表1◆意味が1対1にならない対訳の例

英語	日本語	区別
stimulation	刺激	行為を指す
stimulus (stimuli)		刺激そのもの
substrate	基質	酵素の
matrix		細胞の
discharge	発火	生理学
	分泌	内分泌学
	退院	臨床医学
vector	ベクター	分子生物学
	ベクトル	数学
precipitation	沈殿	生化学
	降水(量)	気象

表2◆類似している複合語の日本語訳で見られる偏り

英語	agent	drug	日本語	剤	薬
immunosuppressive	263*	229	免疫抑制	102	6
antihypertensive	107*	82*	降圧	2	71
antibacterial	**131***	55	抗菌	3	313
anti-inflammatory	197*	**411**	抗炎症	6	102
anticancer	125	121	抗癌	633	6
英語	dysfunction	injury	日本語	障害	傷害
hepatic (liver)	107*	**540**	肝	114	29
cellular (cell)	188	**491**	細胞	180	201
endothelial	**280**	76	内皮	13	14
balloon	0	107	バルーン	14	**57**
cardiovascular※1	13	2	脳血管	82	0

数字は論文中での連接用例数
色文字は80%以上の偏り，**太字**は60～80%の偏りを表す
＊はMeSH用語として規定された統制語を表す
※1：cardiovascular event = 305

として，drugは“より狭く人体に対して主として治療効果をもたらす物質”として定義されるのに対して，日本語での「剤」と「薬」の関係は，もともと「接着剤」や「殺虫剤」などの名称から機能単位としての「剤＝agent」，また，人間に対する「薬＝drug」とそれぞれ対応していたと考えられる．しかし実際には「免疫抑

表 3 ◆ 日本語で好まれる表記と好まれない表記

英語	論文・総説	教科書	日本語	基礎系	臨床系
cancer	28,302 (43%)	5,158 (50%)	癌	13,630 (72%)	1,154 (52%)
carcinoma	5,846 (9%)	1,368 (13%)	癌腫	**29 (0%)**	**5 (0%)**
sarcoma	1,235 (2%)	340 (3%)	肉腫	289 (2%)	57 (3%)
tumor	31,207 (47%)	3,242 (31%)	腫瘍	**4,192 (22%)**	989 (44%)
neoplasm	504 (＜1%)	207 (2%)	新生物	**4 (0%)**	**6 (0%)**
oncogenesis	373 (＜1%)	6 (0%)	発癌	763 (4%)	39 (2%)
計	66,357 (100%)	10,321 (100%)		18,916 (100%)	2,250 (100%)

数字は論文中での連接用例数

色文字 =英語に比べて割合が高い，**太字** =英語に比べて割合が低い

制剤」「抗菌薬」「抗癌剤」「抗炎症薬」など，執筆者によってどちらか一方が好んで（しかし混同して）用いられる．同様な混同は「dysfunction = 障害」と「injury = 傷害」のようにたまたま訳語が同音異義語であった場合に，「injury」であっても日本語で「（機能の）障害」が用いられる例が多く見られる．

日本語が適切でない場合は淘汰される ⇒ 表 3

ほとんどの学術用語集や辞書で「cancer = 癌」「tumor = 腫瘍」「sarcoma = 肉腫」といった厳密な対応関係が整理されているが，「carcinoma」や「neoplasm」に関しては学術用語としての「癌腫」や「新生物」は廃れており，ほとんど「癌」や「腫瘍」にとって代わられている．これは日本語訳の音や表記が直ちに認識されづらく嫌われたケースのように思われる．また，「腫瘍プロモータ」が一般的に「発癌プロモータ」と表記されるような例も，よりわかりやすい言葉へと変遷していった例であろう．これらは学術的な定義よりも，表記や音の響きによって日本人が好む用語があることを意味しているように思われる．

日本語と英語は直訳関係にならない ⇒ 表 4

一般的に英語は唯物論的に事物を命名するのに対して，日本語で

表 4 ◆ 直訳的な対訳関係にならない例

英語（PubMed）	digestive/alimentary	gastrointestinal	日本語（臨床）	消化性	消化器	消化管	胃腸
(total)	440	**3,372**	（総数）	32	152	279	72
ulcer ulceration	0	24	～潰瘍	32	0	2	1
symptom	3	112	～症状	0	68	1	7
bleeding hemorrhage	0	283	～出血	0	1	41	0
disorder disease	23	**116**	～疾患	0	**10**	7	0
organ system	35	24	～系	0	9	0	0
tract	170	**1,077**	～管（路）	0	0	－	0

数字は論文中での連接用例数
色文字 ＝大多数を占める表記，**太字** ＝好まれる表記

は機能単位として事物を捉えて名付ける場合が多い．一例として「消化に関係する語句」を考えた場合，英語においては「消化」に相当する形容詞「digestive」や「alimentary」はあまり用いられず，形態学的に「gastrointestinal ＝胃腸の」を用いる例が圧倒的に多い．しかしながら日本語においては「胃腸症状」は用いられるが，「消化器系」「消化性潰瘍」「消化器症状」「消化管出血」といった機能的記述が圧倒的に多い．同様な例は「cardiovascular ＝心血管」が日本語訳では「循環」となるように数多くあげられる．

おわりに

以上，例示してきたような差異は従来の辞書ではあまり問題にされていないようであるが，論文執筆の際にこういった言語間の表記や概念の相違に配慮すると，より英語らしい英語になる．今後もこういった実例を収集することによって，日本人が書く（あるいは機械翻訳で訳される）生命科学の英語がよりネイティブに近いものとなるよう，少しでも寄与できれば幸いである．

（金子周司）

Column
コラム 1

まともな英語論文を書くための5つの鉄則

鉄則
その１：日本語で下書きをするな
その２：図を描いてイメージを膨らませよ
その３：過去の論文をまねよ
その４：文法を守れ！ 単語の文法（語法）に習熟せよ
その５：辞書を当てにするな！ 参考にする論文は慎重に選べ

★ その１：日本語で下書きをするな

もし，あなたが日本語で論文の下書きを書いて，それを英訳しようと考えているとしたら，それは大きな間違いではないだろうか？日本語で書いた原稿の言葉１つ１つを和英辞典で調べて英文を組み立てていったなら，とんでもなく滑稽な論文ができあがること請け合いだ．そしてそれを英米人に見せて英語の文法だけを直してもらったら，もう完璧なジャパニーズ・イングリッシュ・ペーパーの完成である．

なぜダメなのか．第１の理由は，日本語と英語の単語のニュアンスの違いにある．一部の専門用語を除いて，ほとんどの日本語と英語の単語の関係は１対１ではない．いくら辞書で調べて，同じか非常に近い意味の言葉を見つけたとしても，それは意味の一部が同じであるに過ぎず，たいていは逐語訳で通用するほど似ているものではない．それぞれの単語には，それぞれ固有の語法があり，使い方には一定のパターンがある．それに何を言いたいかは，文あるいはパラグラフ全体で表現するものだ．それらを無視して逐語訳をやったとしたら，とんでもない直訳英語になることに間違いない．

第２の理由は，英語で論文を書くことよりも，日本語の下書きを英訳することの方が高度の技術を要することだ．まともに日本語を英語に翻訳するには，まず日本語の文章の内容を理解し，次にその文章を全部チャラにしてから再び英文を組み立てなければならない．つまり，日本語の下書きを英語に翻訳するためには，英語を書く能力はもちろんのこと，日本語を適切な英語に置き換える能力までが必要になる．しかも，いったん日本語で原稿を書いてしまうと，日本語の呪縛を解いてから英文を書き始めねばならず，頭の切り換えが大変である．学会によっては，日本語と英語の２つの抄録を要求するところもある．たとえ日本語の抄録を先に書いてしまった場合でも，それを英訳しようとせず，全く新たに書くことをお勧めしたい．日本語と英語の抄録の内容は，必ずしも同一でなくてもいい．大筋で同じであれば，日本語の抄録や下書きと同じことを英語で書く必要はないだろう．

★ その２：図を描いてイメージを膨らませよ

では，なぜ最初から英語で論文を書かずに，日本語で下書きするのか？ それは，おそらく英語で論文を書く前に，何を書くべきかをハッキリさせるためではないだろうか．確かに何を書いたらいいかわからなければ，論文などさっぱり書けないに違いない．苦労して考えた構想は，忘れないようにメモ程度は日本語で書いてもいいだろう．しかし，それ以上はやめた方がいい．日本語の下書きは，思いのほか大きな影響を与えることがあるもので，簡単なメモですら日本語の呪縛になりかねない．「日本語で要点を箇条書きにしてから，英語で論文を書いてみなさい」と言ったら，英語論文まで箇条書きになってしまった人がいた．確かに，いきなり英語で書き始めることは難しいことかもしれないが，つたない文章でもいいからともかく書いてみて，それからブラッシュアップするようにすればどうだろうか．書き始めれば，いろいろとアイデアも湧いてくるものだ．

また，書く内容をハッキリさせたいだけなら，日本語で下書きすることよりも図や表をつくりながら論文の構想を膨らませていくことの方がお勧めだ．最近では，それほど重要でないデータでも，supplemental figuresとしてジャーナルのWebサイトに置ける場合もあるので，図はたくさんつくってみるべきであろう．そうすれば，書こうとする論文で何を提示できるのか，何を主張するのかが明確になる．また，不足しているデータにも気づくであろう．しかも，図には日本語と英語の違いがないから，共通のイメージで論文の内容を捉えることができて好都合である．描けば描くほど論文のイメージができあがっていく．図ができあがれば，おのずと論文の構成が決まってくるので，それから書き始めても遅くはない．図もつくらずに安易に書き始めるより，この段階では論文をどのような構成で組み立てていくのかをじっくり見極めることが肝心だ．

★ その３：過去の論文をまねよ

しかし，いくら論文の内容を図のイメージで捉えたからといっても，自然に英語が出てくるわけはない．きちんとした論文を書くためには，まず過去の類似の論文の研究を十分に行うことが必要だ．特に序論（Introduction）や図を含めた結果（Result），方法（Method）の部分については，過去の論文の内容とスタイルをじっくり研究したうえで，どのようなパターンの図や文章で構成できるかを考えなければならない．類似の論文でよく使われるパターンというものがあり，それをまねることによって，書こうとする論文のスタイルも見えてくるはずだ．ただし，いやしくも原著論文である以上，過去の論文の表現の丸写しは厳禁である．過去の論文と比べて，自分の論文のどこに新しさがあるのかをきちんと主張できることが重要だ．過去の類似の論文のパターンをまねつつ，書きたいことをうまく表現するように心掛けよう．

★ その４：文法を守れ！ 単語の文法（語法）に習熟せよ

大学院生に論文や学会の抄録などを英語で書いてもらうと，英語の文法がめちゃくちゃであることが少なくない．１つの文章に動詞が連続してみたり，逆に動詞がなかったり，他動詞なのに目的語がなかったり複数あったり，自動詞であるはずなのに目的語があったり，主語と動詞の人称が一致しなかったり…と，高校・大学でいったい何を勉強したのかと目を疑いたくなる．1つの文に2つも3つも文法的な誤りがあったら，とても何が書いてあるかなどわかったものではない．少なくとも，文法的な誤りがたくさんある状態で他人に原稿を見せてはいけない．理解不能な英文は内容以前の問題であり，ゴミ箱直行間違いなしだ．

さて，昔習った英語の5文型を思い出し，辞書で動詞の語形変化を確認し，他動詞と自動詞の区別を調べ，副詞の位置にも気を付けてばっちり文法どおり書いたとしても，それだけでは十分ではない．英文法の授業で習った文型は一般論であり，すべての単語に適用できるわけではない．動詞も他動詞と自動詞の区別だけでは不十分である．実際には，他動詞にも自動詞にもなる動詞が少なくなく，また，可算名詞にも不可算名詞にもなる名詞も多い．つまり，それぞれの単語ごとに語法があり，使い方のパターンがある．一般的な文法規則のどの部分を適用できるのかが異なっているので，それぞれの単語ごとに語法を調べて身につける必要がある．単語の語法は，書くためにはきわめて重要だ．

★その5：辞書を当てにするな！ 参考にする論文は慎重に選べ

単語の使い方を調べるためには，文法的な解説の多い学習辞典がいいが，用法が気になるような単語には，いろいろな意味や使い方がある場合が多い．他動詞・自動詞，可算名詞・不可算名詞の区別だけを調べてみても，どちらの表示もある単語が多い．多様な情報が記載されているわりには，結局，調べたけれどよくわからなかったで終わってしまいがちだ．また，英語の表現がわからなかったり，同じ表現の繰り返しを避けたりする目的で，和英辞典や類語辞典を使って単語を探してみても，先に述べたような理由であまり役に立たないことが多いのではないだろうか．たとえ，類語辞典でよさそうな単語を見つけたとしても，それが論文にふさわしいのかどうかを知ることは難しい．いろいろなことを調べるために辞書は不可欠だが，論文で使う単語の用法を学習するためには，実際の論文での用例を参考にすることが必要であろう．ただし，科学論文の英語のレベルは雑多であり，特に英語を母国語としない国からの論文には，間違いや非標準的な用例が少なくないので，参考にする論文を選ぶときには細心の注意が必要だ．

本書は，編者が「こんな本があったら便利だろうな」と長年考えていたことを具体化したものである．実際の論文を調査して，よく使われる表現を良質の例文とともにまとめてあるので，論文執筆の際には必ず役に立つであろう．ぜひ，有効に活用していただきたい．

★

さて，ここまで書きあげたところで重大な問題に行き当たった．この鉄則を守っても，すぐに論文は書けないかもしれないということだ．たとえ本書を参照しながらでもたやすくはないだろう．まさに，論文執筆にも王道なしだ．ふだん論文を読む際にも本書で確認して，論文執筆に使えそうな表現を頭の引き出しに蓄えていこう．そうした日々の積み重ねが大切なのだ．

（河本　健）

Column コラム2 日本語訳から見えてこない類語の使い分け

★ possible/probableはどれほど違う？

possible/probable（possibility/probability，possibly/probably）にみられるように，日本語の訳語が互いに意味が近接する場合，それらの適切な使い分けは，案外難しいものである．学習者用英語辞典として評判の高い『ジーニアス英和辞典 第3版』（大修館書店）では，possible/probableの訳語として以下のように示されている．

> **possible**：（ひょっとすると）起りうる，ありうる，可能性のある
>
> **probable**：（十中八九）ありそうな，起りそうな，十分に可能な，ほとんど確実な，たぶん…だろう

さらに，possibleの項では，「起る公算が50％より小さいと話し手が考えている場合」という解説が付記されている．くれぐれも用法を間違わないようにという配慮が感じられる．しかしながら，学習者は必ずしも訳語にこめられた細かい差異や，解説に述べられた貴重な情報に細心の注意を払ってこれらの語彙を学習するわけではない．多くの場合「可能な」という共通語を核として，さして訳語の差異を気にとめることなく，せいぜい「probableを使った場合は，possibleを使った場合よりも可能性が高い」という程度の認識でしかこれらの語を理解していないのではないかと懸念される．

★ 日本語訳に隠されたワナ

SARS（重症急性呼吸器症候群）の蔓延が心配された時期に，WHO（世界保健機構）が提示した診断確定の基準は以下のようなものであった．

Case definitions (revised 1 May 2003)

Suspect case

1. A person presenting after 1 November 2002 with history of:
   - high fever (>38 ℃)

   AND
   - cough or breathing difficulty

Probable case

1. A suspect case with radiographic evidence of infiltrates consistent with pneumonia or respiratory distress syndrome (RDS) on chest X-ray (CXR).

これに対して，国立感染症研究所のホームページには以下の日本語訳が掲示されている．

症例定義（平成15年5月1日改定）

Suspect Case（疑い例）

1. 平成14年11月1日1以降に発症して受診し，以下の項目を満たす者：

   ・高熱（>38℃）

   且つ

   ・咳嗽，呼吸困難

Probable Case（可能性例）

1.「疑い例」で，胸部レントゲン写真において肺炎の所見又は呼吸窮迫症候群（RDS）の所見を示す者

注目すべきは，Probable Caseが「可能性例」と訳出されていることである．ここで使われた「疑い例」「可能性例」という用語は専門用語に近い性格を持っていると推察するが，一般に臨床に携わる医師たちに，これら２つの日本語表現のうちからどちらがよりSARSに近い症例を連想させるか尋ねてみると，十中八九「疑い例」の方が深刻度が高い印象を直感的に受けるという反応が返ってくる．「可能性例」という日本語からは，原文に使われているProbable Caseという表現ではなく，Possible Caseという英語が生成される可能性が高い．そして，Possible Caseであれば，おそらくさまざ

まな可能性の考えられる症例で，症状をいろいろ勘案しながら診断を確定する必要があるという具合に解釈されかねないであろう．原文で使われているprobableの語感を伝えるためには，「ほぼ間違いない例」とでも訳出するのがむしろ原義に忠実で，誤解を生じる可能性が減少したのではないかと思われる．

★ ネイティブと同じような語感を身に付けよう

ことほどさように，probableとpossibleを適切な日本語に置き換えるのは難しく，これらの語の語感を習得するのは難しいと想像される．昨今，SARSの危機よりも鳥インフルエンザの危機が取り沙汰されているが，以下の3つのニュースのヘッドラインに接して，危険度を肌で感じながら読むことができる語感が，英語を外国語として学ぶ日本人には求められている．

① First probable case of bird flu transmission between humans（*2004-09-28 Asia News*）
② Woman dies of suspected case of bird flu, fourth case in two weeks（*2005-01-11 USA TODAY*）
③ Possible Case of Bird Flu First in Finland（*2005-09-26 ABC News*）

①の英文に接して，「今そこにある危機」を実感すれば，ほぼ正しい語感が身に付いていると言えるかもしれない．すでに述べたとおり，probableは可能性が非常に高く，「十中八九ありそうな（ジーニアス英和辞典）」場合に用いるのであり，③のpossibleは可能性が半分以下程度，あるいは単に「可能性がある」という意味で使われていると解釈できる．

★ ネイティブと日本人の英語論文を比べてみると

さて，possibility/probabilityの差異については，どのように学習されているであろうか．どちらも「可能性」という意味を核に持つ

表 1 ● possibilityを修飾する形容詞の頻度

miniLSDコーパス（1,188）		日本人英語論文コーパス（2,225）	
intriguing	13	high	31
new	9	strong	10
interesting	7	little	9
distinct	4	great	7
therapeutic	3	less	5

カッコ内の数字はpossibilityのそれぞれのコーパスにおける頻度数

表 2 ● probabilityを修飾する形容詞の頻度

miniLSDコーパス（643）		日本人英語論文コーパス（454）	
high	17	high	20
higher	9	higher	13
low	8	cumulative	10
cumulative	6	normal	7
equal	5	low	7

カッコ内の数字はpossibilityのそれぞれのコーパスにおける頻度数

単語という認識だけでは，両者の適切な使い分けはできない．**表1**は，われわれが独自に構築した，日本人によって書かれたライフサイエンス分野に特化した英語の論文要旨からなるコーパス（日本人英語論文コーパス）と英語母国語話者の論文コーパス（miniLSD）（両者とも約 1,000 万語からなる）を利用して，possibility の直前に出現する形容詞をそれぞれ頻度順に示したものである．同様に，**表 2**は，probability の直前に出現する形容詞をそれぞれ頻度順に示したものである．

表1から，日本人は「高い可能性」「強い可能性」という自然な日本語表現をそのまま英語に転用し，high possibility/strong possibility という表現を好んで使っていることがうかがえる．その他，little/great/less がリストにあがっていることからも，日本人は概して「可能性」の「高い・低い」「強い・弱い」という表現へのこだわりが強く，あくまでも possiblity の「程度」を表現しようとする傾向

が強い．実際には，『ジーニアス英和辞典』にも解説があるとおり，high possibility とは英語では通常言わない．

一方，miniLSD コーパスの形容詞リストを眺めてみると，「高い・強い」という「程度」を表現する形容詞は見当たらず，intriguing/new/interesting といった，「興味をそそる，新しい」という，どちらかというと possibility の「性質」にかかわる形容詞が好んで使われているのがわかる．**表2**が示しているように，「可能性」の「高い・低い」について言及するためには，probability（見込み・公算・蓋然性・確率）という語を使う必要がある．probability を修飾する形容詞については２つのコーパスの間に大きな差異は認められないが，probability の頻度は日本人英語論文コーパスでは miniLSD よりも低く，possibility については逆に miniLSD よりも非常に高い．日本人は，本来なら probability を使うべきところを，possibility で代用している可能性があるのではないかと推測される．

★ 日本語訳の呪縛から抜け出そう

最後に，possibly/probably の用法の違いに注目してみよう．**図1・図2**は２つのコーパスから，may possibly/may probably のコンコーダンス（用語索引）を抽出したものである．さて，どちらの図が日本人英語論文コーパスからのものであるか判定できるであろうか．

```
... of amyotrophic lateral sclerosis (ALS) may probably be related to the impairment...
... platelets by suction or centrifugation may probably play the most important role...
...n of RV afterload produced with an IABP may probably be due to degree of recovery...
...bed reticular formation in the midbrain may probably account for the remaining of...
```

図1 ● may probably のコンコーダンス

```
...rogesterone may stimulate breathing and may possibly improve symptoms of hypovent...
...ectin further suggests how some domains may possibly be important for protein int...
...            While various unknown factors may possibly give rise to selective activ...
...an unitary displacements, this mutation may possibly perturb the mechanical coord...
```

図2 ● may possibly のコンコーダンス

may probablyという表現は，「おそらく～かもしれない」という，自然な日本語を連想させる．しかしながら，すでにpossibleとprobableの対比で述べたように，probablyはpossiblyと違い，確率的に起こることがきわめて高いと思われるときに使う語である．「ひょっとしたら～かもしれない」という比較的確信度の低い推量を表す助動詞may〔「約5割の確率で起ると考えていることを表す」(ジーニアス英和辞典)〕とは相容れない副詞である．つまり，**図1**が日本人英語論文コーパスからのものである．**図1**の英文は微妙に奇異に映る英文として評価されるのではないかと思われる．ちなみに，ネイティブの英文からなるminiLSDコーパスでは，当然のことながら，may probablyという不自然な表現は検出されなかった．**図2**は，may possiblyを検索語としてminiLSDコーパスから抽出したものである．possiblyはmayという助動詞と親和性が高いが，probablyは通常mayとは共起しないことを確認しておく必要がある．

ここに示した事例は，文法の約束を守るだけでは，自然な英文生成が必ずしも約束されるわけではないことと，日本語の訳語のみに依存しすぎると，原義の持つ本来の意味を見失うことになりかねない場合があることを示している．論文を執筆する際にはこの点にも十分配慮することが肝要である．

(大武　博)

Column
コラム3

類語＝同義語？

★ 文法的には正しいけれど…

「お急ぎでしょうから，私のチャリンコをお使いください」

これは，確かに文法的に正しい日本語であるが，このような発話をする日本人を想像するのは難しい．「お急ぎでしょうから」「お使いください」という，いかにも目上の人に対する丁寧な表現の中に，「チャリンコ」という俗語は不釣り合いである．日本語を母国語にしていれば，上記の表現が通常では使われないことは直感的に判断できる．

同じことが英語についても言えるが，実際には次のような英文に接して即座に妙な英文を言い当てることができる英語学習者は少ない．

① Why don't you drop in at my residence on your way home?
② Why don't you drop in at my place on your way home?
③ Why don't you drop in at my abode on your way home?

上記３つの内容的に類似する英文に接して，それぞれどのような印象を抱くであろうか．residence/abode はいずれも house を意味する語である．①の英文では，why don't you や drop in といった表現が使われている非常にくだけた口語的文体の中で，residence という単語は不釣り合いなほど公式的（formal）であり異質である．日本語で，さしずめ「おいらの邸宅にこないかい」と訳出すると，この英文の奇異さが伝わるだろうか．③の英文で使われている abode は英和辞典では，「住居，住まい，居所」などの訳語を確認できる．しかし，英語を母国語としていれば，abode は通常はあまり使わない単語であるが，humble abodeという表現でなら耳にすることがあるので，「粗末な家」あるいは「拙宅」に類する意味を込めた，へりくだった表現か，ユーモアを狙った表現であろうかと推測するかもしれない．残念ながら，これらの情報を持ち合わせない日本人学習者には，②がごく自然な英語表現であって，①と③の英文は，奇異な印象を与えかねないということを，直感的に判断することはできないであろう．ことほどに，英語を学習する際には，単語の文体上の位置づけに関する情報も含めて語彙の習得を心がける必要があるが，実際にはこの側面は忘れられがちである．

★「a lot of」は「many」の代わりに使えるか

われわれが独自に構築したライフサイエンス分野の日本人の英語論文要旨からなるコーパスの中に，さすがにkidの使用例は検出されない．ライフサイエンス関連の英語論文の中で，「子供」に言及するとき，kidは学術論文にはふさわしくないという賢明な判断がおそらく日本人研究者に働くからであると思われる．kidは「日常会話ではchildより多く使われる」，と辞書（ランダムハウス英語辞典，小学館）に解説があるが，これはあくまでも「日常会話」では好んで使われるということである．

では，「多くの」という意味で通常学習されるa lot ofという英語表現はどうだろうか．日本人の英語論文コーパスを検索してみると，少なからずこの表現を使った英文が抽出される（図1）．これは，「多くの＝many＝a lot of＝a (large) number of」のような等式が

...ammatory features which are composed of **a lot of** lymphocytes and proliferated cap...
...emic autoimmune disease and consists of **a lot of** diseases with each clinical enti...
...itro was followed by the development of **a lot of** cell engineering techniques whic...
... also permits retrospective analysis on **a lot of** cases and studying different spe...
...P and it is further expected to propose **a lot of** potential applications as a new ...
... That is, in old rats **a lot of** deposits were chiefly distribute...

図1 ●日本人英語論文コーパスにおけるa lot ofのコンコーダンス一部

頭の中にあって，それらの英語表現は等価であり，置換が自在にできるという誤解に基づいているのではないかと思われる．a lot ofは，きわめて口語的な英語表現であり，会話などでは好んで使われることはあっても，英語論文などでは出る幕があまりないはずである．

類語学習のイロハのイとして，「類語必ずしも等価ならず，置換不可の場合もあり」，と肝に銘じておく必要がある．ついでながら，日本人が好んで使うa number ofは，ここに示した等式とは微妙に意味の違いがあることを確認しておかねばならない．学校英語などでは，a number ofを「多くの」と教えることが多いが，手元にある数冊の英英辞典を調べてみても，manyとして定義を与えているものはない．むしろsome/severalの定義を与えている．a number ofは，「いくつかの」という意味で再確認されることが望ましい．英語論文で発信するときに，あるいは情報を入手するときに，数の程度の認識において誤解が生じないためにも重要である．

★「～にもかかわらず」は「in spite of」？

同じように，日本語訳からほぼ同一の意味・用法の語として学習されているのではないかと懸念される別の表現を見てみよう．「～にもかかわらず＝in spite of＝despite」という等式は，英語表現を効率よく学習するのに役立つかもしれないが，実際には学術論文の中では，出現頻度に歴然とした違いが認められる．**表1**は，日本人の英語論文要旨からなる英文コーパスとネイティブスピーカーの英語を中心に構築されたminiLSDコーパス（両者とも総語数約1,000万語）を利用して，in spite ofとdespiteの出現頻度を調査した結果を示したものである．

表1 ● in spite of / despite の出現頻度

	日本人英語論文コーパス	miniLSDコーパス
in spite of	641	72
despite	736	2,101

この表から，日本人研究者は，「〜にもかかわらず」という意味を表す英語表現については，ここで取り上げた2つの表現の使用頻度に大差がなく，両者の用法の差異を特に意識していないことがうかがえる．一方，miniLSDの方では，in spite ofに比べてdespiteの使用頻度が圧倒的に高い（約30倍）．in spite ofは『ジーニアス英和辞典』でも「略式」と分類されているように，フォーマルな学術論文では文体上の制約から，あまり好んで使われないということを確認しておく必要がある．

★ oldとelderlyの使い分けにも要注意！

最後に，やはり類語として分類される英語old/elderly/seniorの使い分けを確認しておこう．**図2**は，日本人英語論文コーパスで抽出されたoldのコンコーダンスの一部を示したものであるが，いずれもelderlyによる代用が望ましいのではないかと思われる．通常「年老いた（患者）」という意味では，old（patients）よりも婉曲的で丁寧な表現であるelderly（patients）が好んで使われる．LSDコーパスでoldの用法を調べると，oldが使われる場合は，原則的に次

...	**Old patients** had the same tendency as pat...
...toposide daily showed a good outcome in	**old patients** with malignant lymphoma. ...
...VLCFAs unveiled the diagnosis of AMN in	**old patients** with spinal spondylosis or w...
...al (113 males and 330 females) and some	**old patients** without having any diseases ...
...ant therapy should be considered in the	**old patients**, but not always be indicatio...
...hmias were serious problems in managing	**old patients**. ...
...in damage and mortality, especially for	**old patients**. ...
...tivities, however, sexual activities in	**old people** might decrease because of medi...
... the aged people because there were few	**old people** more than 70 years of age and ...
...n addition, clinical characteristics of	**old people** with irritable bowel syndrome,...
...tible with the previous studies on very	**old people**, they are different from those...
...or symptoms was higher in Koto-ku among	**old people**. ...
... more effective for cancer screening in	**old persons** (65 years or older), where ca...

図2 ● 日本人英語論文コーパスにおけるoldのコンコーダンスの一部

の3つの場合である．

① 年齢に言及する場合
② young に対比して言及する場合
③「古い」ものに言及する場合

「年長者・年長の」の意味で使われる senior の用法は，日本人学習者に時折混乱が見られる．senior は通常「上級の・上位の」という意味で使われ，resident/doctor/citizen などを修飾することが多く，階級（レベル）を意識した表現として使われる．日本人英語論文コーパスから検出された以下の表現は，いずれも「年長の」という意味で使われたものと思われるが，階級（レベル）にかかわる名詞を修飾していないこれらの用法は，英語表現としては不自然である．

blood pressure in **senior infants**
senior patients with neurological diseases

以上，例を示しながら「類語必ずしも等値ならず」を解説してきたが，類語を賢く使い分けることに習熟することは重要である．

（大武　博）

索引

欧文

【A】

【B】

品詞が2つに分かれる場合は以下の略号で示した
動（動詞） 名（名詞） 形（形容詞） 副（副詞） 接（接続詞/語）

INDEX

[E]

[F]

【Q】

【R】

【V】

【W】【Y】

和 文

【あ】

【か】

【さ】

【た】

■ 編者略歴

河本　健（かわもと・たけし）

広島大学大学院医歯薬学総合研究科講師．広島大学歯学部卒業，大阪大学大学院医学研究科博士課程修了，医学博士．高知医科大学助手，広島大学助手を経て現職．専門は，生化学・口腔生化学．時計遺伝子群による概日リズム調節，間葉系幹細胞の再生医療への応用などを研究している．

ライフサイエンス英語類語使い分け辞典

2006年 6月20日　第1刷発行
2018年 3月30日　第9刷発行

編　集　河本　健
監　修　ライフサイエンス辞書プロジェクト
発行人　一戸裕子
発行所　株式会社　羊　土　社
〒101-0052
東京都千代田区神田小川町2-5-1
TEL　03(5282)1211
FAX　03(5282)1212
E-mail　eigyo@yodosha.co.jp
URL　www.yodosha.co.jp/
印刷所　株式会社 平河工業社

Printed in Japan
ISBN978-4-7581-0801-0